"十二五"国家重点出版规划图书项目

中国当代建筑地图

ATLAS OF CONTEMPORARY CHINESE ARCHITECTURE

支文军 / 戴春　编

常文心 / 贺丽 / 张晨　译

辽宁科学技术出版社

CONTENTS 目录

008 EDITOR'S WORDS
编者的话

NORTH CHINA 华北

BEIJING 北京

010 National Stadium of 2008 Olympic Games, Beijing
2008 年北京奥运会国家体育场 (鸟巢)

012 National Swimming Centre of 2008 Olympic Games, Beijing
2008 年北京奥运会国家游泳中心 (水立方)

014 National Centre for the Performing Arts
国家大剧院

016 National Library of China
中国国家图书馆

018 Digital Beijing
数字北京

020 China National Convention Centre (Beijing 2008 Olympic Games Multi-purpose Arena)
国家会议中心 (北京 2008 奥运会多功能场馆)

022 Public Service Platform at Beijing Contemporary Art and Cartoon Industry Centralised Area
北京当代艺术与卡通产业集聚区公共服务平台

024 Raffles City Beijing
北京来福士广场

026 Huamao Centre
华贸中心

028 Linked Hybrid
万国城

030 Beijing Publishing House
北京出版创意中心

032 China Youth Travel Service Plaza
中青旅大厦

034 Beijing Nexus Centre
北京嘉盛中心

036 GreenPix – Zero Energy Media Wall
零耗能多媒体幕墙

038 Family Box
家·盒子文化有限公司

040 Shangri-La Hotel of Supershine Upper East
阳光上东香格里拉酒店

042 LATTICE in Beijing
北京格子

044 No.23 Qianmen East Street
北京市前门东大街 23 号

046 XIU
"秀"吧

048 National Museum of China
中国国家博物馆

050 Beijing International Automotive Expo Centre Automotive Museum
北京国际汽车展览中心汽车博物馆

052 China Science and Technology Museum
中国科学技术馆

054 National Zoological Museum of China and Institute of Zoology, Chinese Academy of Sciences
中国科学院动物所及国家动物博物馆

056 Beijing Hutong Bubble 32
北京胡同泡泡 32 号

058 Placid Rivers Club
泰禾俱乐部

060 Hongluo Clubhouse
红螺会所

062 Qingcuiyaun Clubhouse
晴翠园会所

064 No.18 Guanshuyuan Hutong
官书院胡同 18 号

066 Living in Seclusion – "Xun" Private Lounge
"蛰居"之处——"旬"会所

068 G-Dot Art Space
贵点艺术空间

070 Inside-out Art Centre
中间艺术馆

072 Songzhuang Artist Residence
宋庄艺术公社

074 Songzhuang Art Museum
宋庄美术馆

076 Qi Zhilong Residence and Gallery
祁志龙画室

078 Hejingyuan Contemporary Art Gallery
北京和静园现代美术馆

080 Westhill Art Workshop, Beijing
北京西山艺术工坊

082 MOSAIC in Beijing
北京马赛克

084 Bamboo Wall
竹屋

086 BUMPS in Beijing
北京冲击

088 The Villas A+B of the Lushi Mountain Villas
庐师山庄别墅 A+B

090 Beijing Citic Guo'an Conference Centre Courtyard-style Guestroom Buildings
北京中信国安会议中心庭院式客房

092 SIEEB
清华大学环境楼

094 Students' Comprehensive Service Building at the New Campus of Beijing Institute of Civil Engineering and Architecture
北京建筑工程学院新校区学生综合服务楼

096 Beijing South Station
北京南站

TIANJIN 天津

098 Bridge Culture Museum
天津桥园桥文化博物馆

100 Sports Arena at Tianjin University
天津大学体育馆

102 Fusion Square (Business District and Five-Star Hotel in Binhai New Area)
融和广场（滨海新区商务街区及五星级酒店）

104 Meijangnan Lake Residential Development
美江南湖住宅区

106 West Railway Station
天津西站

108 Tianma International Club
天津天马国际俱乐部销售中心及会所工程

110 ICTC
国际经济贸易中心

112 Costal Building
滨海浙商大厦

114 Idea Valley Building
慧谷大厦

116 City Centre
赛顿中心

118 YJP Administrative Centre
于家堡工程指挥中心

120 Gui Yuan Atelier
圭园工作室

122 Zhangjiawo Elementary School, Xiqing District
天津西青区张家窝镇小学

124 Feng Jicai Literature and Art Academy at Tianjin University
天津大学冯骥才文学艺术研究院

HEBEI 河北

126 Tangshan City Hall
唐山市城市展览馆

128 Qinhuangdao Bird Museum
秦皇岛鸟类博物馆

130 Tangshan Museum
唐山博物馆

132 Wonder Mall
万象城

134 Complex of Hebei Education Press
河北教育出版社综合楼

136 Tangshan Exhibition Centre
唐山会展中心

138 Ming-Tang Hot Spring Resort
河北茗汤温泉度假村

SHANXI 山西

140 Shanxi Grand Theatre
山西大剧院

142 Yungang Grottoes Museum
云冈石窟博物馆

INNER MONGOLIA 内蒙古

144 Dalai Nur Nature Reserve Education Centre
达里诺尔自然保护区宣传教育中心

146 Site Entrance of XANADU
元上都遗址大门

148 Ordos Art Museum
鄂尔多斯美术馆

150 Ordos Museum
鄂尔多斯博物馆

152 Daihai Hotel
岱海宾馆

154 Baotou Children's Palace and Library
包头市少年宫、图书馆

NORTHEAST CHINA 东北

LIAONING 辽宁

156 Number Two Middle School at North District, Shenyang
沈阳市第二中学北校区

158 Northeast Yucai Bilingual School
东北育才双语学校

160 Luxun Academy of Fine Arts
鲁迅美术学院

162 Kindergarten for the Dalian Software Park
大连软件园幼儿园

164 Howard Johnson Parkland Hotel
百年汇豪生酒店

166 Maritime Museum of Arts
海中国．美术馆

JILIN 吉林

168 Museum of Culture, Fine Arts and Science, Changchun
长春美术馆、科技馆及文化馆

EAST CHINA 华东

SHANGHAI 上海

170 Shanghai World Financial Centre
上海环球金融中心

172 You You Grand Sheraton International Plaza
由由喜来登国际广场

CONTENTS 目录

174 Wanxiang Plaza
万向大厦

176 Riviera TwinStar Square
浦江双辉大厦

178 BEA Financial Tower
东亚银行金融大厦

180 City Hall of the Shanghai Nanhui District
上海南汇区政府办公楼

182 Liantang Town Hall, Qingpu
青浦练塘镇政府

184 Jinqiao Office Park
金桥开发区研发楼

186 Cobest Design & Manufacturing
科倍集团

188 Henkel Asia-Pacific and China Headquarters
汉高亚太中国总部

190 Shanghai International Cruise Terminal
上海国际港客运中心

192 Maritime Museum
中国航海博物馆

194 Museum Shanghai-Pudong
上海浦东博物馆

196 Shanghai Museum of Glass
上海玻璃博物馆

198 Zhujiajiao Museum of Humanities and Arts
朱家角人文艺术馆

200 SPSI Art Museum
上海油雕院美术馆

202 Shanghai Museum of Contemporary Art
上海当代艺术馆

204 Zhou Chunya Art Studio
周春芽艺术工作室

206 Beicai Culture Centre
北蔡市民文化中心

208 Jintao Village Community Pavilion
金陶村村民活动室

210 Culture and Sports Centre, Anting
安亭镇文体活动中心

212 Hanbiwan Holiday Villa
涵璧湾花园

214 World Expo Village
世博村

216 Shanghai Oriental Sports Centre
上海东方体育中心

218 Renovation of Qingpu Stadium and Training Hall
上海青浦体育馆、训练馆改造

220 Kindergarten in Jiading New Town
嘉定新城幼儿园

222 Xiayu Kindergarten, Qingpu
青浦夏雨幼儿园

224 Botanical Gardens "Chenshan"
辰山植物园

226 Sino-French Centre, Tongji University
同济大学中法中心

228 West Shanghai Holiday Inn
竞衡西郊假日酒店

230 Plaza 353
353 广场

JIANGSU 江苏

232 Chervon International Trading Company
泉峰国际贸易公司

234 Huawei Research and Development Park
华为软件研发中心

236 POD – Jiangsu Software Park
建筑豆——江苏软件园南京徐庄基地

238 CIPEA No.4 House
CIPEA 四号住宅

240 Concrete Slit House
混凝土缝之宅

242 Poets Residences
诗人住宅

244 Three-Courtyard Community Centre
三间院

246 Dashawan Beach Facility
大沙湾海滨浴场

248 Pavilions at Lake Yangcheng Park, Kunshan
昆山阳澄湖公园景观建筑

250 Kangju Community Centre
昆山康居社区活动中心

252 Sichang-Road Teahouse, Kunshan
昆山思常路茶室

254 Chinese Fourth Army Jiangnan Headquarters Exhibition Hall
新四军江南指挥部展览馆

256 Suzhou Museum
苏州博物馆

258 China Museum of Sea Salts
中国海盐博物馆

260 Jiangsu Provincial Art Museum
江苏省美术馆新馆

262 Nanjing Sifang Art Museum
南京四方美术馆

264 Fei Xiaotong Memorial Museum
费孝通江村纪念馆

266 Nanjing Massacre Memorial Hall Extension Project
侵华日军南京大屠杀遇难同胞纪念馆扩建工程

268 Nanjing Changfa Centre
南京长发中心

270 Nanjing International Conference and Exhibition Centre
南京国际会展中心

272 Suquan Yuan, Suzhou
苏泉苑

274 Nanjing University Performing Arts Centre
南京大学表演艺术中心

276 Nam Gallery
南画廊

278 Spgland Xi Shui Dong Sales House
无锡西水东售楼处

ZHEJIANG 浙江

280 Ningbo Southern Business District
宁波南部商务区

282 Congress Centre Hangzhou
杭州会议中心

284 China Textile City International Convention and Exhibition Centre
中国轻纺城国际会展中心

286 Liangzhu Museum
良渚文化博物馆

288 Ningbo Fellowship Museum
宁波帮博物馆

290 Yu'niaoliusu Street in Liangzhu Culture Village
杭州万科良渚文化村商业街区 "玉鸟流苏"

292 Ninetree Village
九树公寓

294 Tianhe Residence
天河家园

296 Banyan Tree Hangzhou
西溪悦榕庄

298 Ningbo Wulongtan Reception Centre
宁波五龙潭接待中心

300 Xixi Artists' Clubhouse
西溪湿地三期工程艺术集合村 J 地块会所

302 N Plot of Hangzhou Xixi Wetland Art Village
杭州西溪湿地艺术村 N 地块

304 Xiangshan Campus, China Academy of Art, Hangzhou
中国美术学院象山校区

306 Alibaba Headquarters
阿里巴巴总部

308 Park Block Renovation, Luqiao Old Town
路桥旧城小公园改造项目

ANHUI 安徽

310 Momentary City
瞬间城市

312 Hefei Art Gallery
合肥美术馆

314 Crossing Battle Memorial Hall
渡江战役纪念馆

316 Anhui Museum
安徽省博物馆新馆

SHANDONG 山东

318 Jinan Sports Centre
济南奥林匹克体育中心

320 Qingdao Grand Theatre
青岛大剧院

322 Sailboat Hotel
帆船酒店

324 Mount Taishan Peach Blossom Valley Visitor Centre
泰山桃花峪游人中心

FUJIAN 福建

326 International Tourist Passenger Terminal, Xiamen Port
厦门港国际旅游客运码头

328 Fujian Provincial Electric and Power Company
福建省电力公司大楼

330 Bridge School
桥上书屋

TAIWAN 台湾

332 Lanyang Museum
兰阳博物馆

334 Hunya Chocolate Museum
宏亚巧克力博物馆

336 Bellavita
宝丽广场

338 Glassware on Water
亲家爱敦阁接待中心

340 Star Place
台湾大力精品购物中心

342 Kelti International Group Corporate Headquarters
克缇企业总部大楼

344 Hsinchu Xiang Duan Village Community Centre
新竹香村段闲谷小区中心

346 Huga Fab III and Headquarters Building
广镓光电中科三厂暨总部大楼

348 W Hotel Taipei
台北 W 酒店

CENTRAL CHINA 华中

HUNAN 湖南

CONTENTS 目录

350 Northstar Delta Project Exhibition Centre
北辰长沙三角洲项目展示中心

352 Hallelujah Concert Hall
哈利路亚音乐厅

354 Maoping Village School
毛坪村浙商希望小学

356 Jishou University Reserch and Education Building and Huang Yongyu Museum
吉首大学综合科研教学楼及黄永玉博物馆

HUBEI 湖北

358 Wuhan CRland French-Chinese Art Centre
武汉华润中法艺术中心

360 Emperor of Ming Dynasty Cultural Museum
明代帝王文化博物馆

362 Wuhan New Railway Station
武汉火车站

364 Zhongxiang Culture and Sports Centre
钟祥市文化体育中心

HENAN 河南

366 Zhengdong District Urban Planning Exhibition Hall
郑州郑东新区城市规划展览馆

368 Luoyang Museum
洛阳博物馆新馆

370 Culture Museum of Zhengzhou
郑州中原文化博物馆

SOUTH CHINA 华南

GUANGDONG 广东

372 Futian Sports and Entertainment Complex
福田文体中心

374 Bao'an Stadium
宝安体育场

376 Universiade 2011 Sports Centre
深圳 2011 年世界大学生运动会体育中心

378 Century Lotus Sports Park
世纪莲花体育中心

380 Guangdong Xinghai Performance Group Office
广东星海演艺集团办公楼

382 R&F Centre
富力中心

384 Vanke Centre, Shenzhen
水平摩天楼 / 深圳万科中心

386 Gemdale Meilong Town, Shenzhen
深圳金地·梅陇镇

388 Shangri-La Hotel, Guangzhou
广州香格里拉酒店

390 Driving Range – Mayland International Golf Resort
美林国际高尔夫度假村

392 Sheraton Dameisha Resort Hotel
大梅沙喜来登度假酒店

394 Dafen Art Museum
大芬美术馆

396 OCT Art and Design Gallery
华·美术馆

398 Guangdong Museum
广东省博物馆

400 OCT Design Museum
OCT 设計博物館

402 University Town Library
大学城图书馆

404 Liberal Art Department, Dongguan Institute of Technology
东莞理工学院文科系馆

406 The New Conghua Library
从化市图书馆新馆

408 Guangzhou Opera House
广州歌剧院

410 Canton Tower
广州塔

412 Guangzhou South Railway Station
广州南站

414 Dongguan Toy Factory
东莞玩具工厂

HAINAN 海南

416 InterContinental Sanya Resort
三亚洲际度假酒店

HONG KONG 香港

418 Creative Media Centre, City University of Hong Kong
香港城市大学创意传媒中心

420 Hong Kong Polytechnic University – Hong Kong Community College
香港理工大学——香港专上学院

422 The Hong Kong Design Institute
香港知专设计学院

424 Clinical Science Building at Prince of Wales Hospital for The Chinese University Hong Kong
香港中文大学威尔斯亲王临床医学大楼

426 The Arch
凯旋门

428 12 Broadwood Road Residential Development
跑马地 12 号乐天峰住宅开发项目

430 One Island East, TaiKoo Place
港岛东中心，太古坊

432 Landmark East
城东志

434 Enterprise Square 5, MegaBox
企业广场第五期红点购物中心

MACAO 澳门

436 City of Dreams
新濠天地

SOUTHWEST CHINA 西南

CHONGQING 重庆

438 Chongqing Library
重庆图书馆

440 Huxi Campus Library of Chongqing University
重庆大学虎溪校区图书馆

442 Science and Technology Museum, Chongqing
重庆科技馆

444 Chongqing Grand Theatre
重庆大剧院

446 Beity Hot Spring Hotel
贝迪颐园温泉度假酒店

SICHUAN 四川

448 Jianchuan Mirror Museum and Earthquake Memorial
建川镜鉴博物馆暨汶川地震纪念馆

450 Xinjin Zhi Museum
成都新津县芷博物馆

452 Relics Exhibition Hall at the Jinsha Site Museum
金沙遗址博物馆文物陈列馆

454 Chengdu Fluid Core-Yard
置信综合办公楼

456 Urban Planning Exhibition Hall at Beichuan Antiseismic Memorial Park
北川羌族自治县抗震纪念园——城市规划展览馆

GUANGXI 广西

458 Nanning International Convention and Exhibition Centre
南宁国际会展中心

460 Planning and Architectural Design of the Saloon Street, Yangshuo, Guilin, Guangxi
广西桂林阳朔酒吧街规划与建筑设计

462 Riverside Restaurant
天门山"山之港"临江餐厅

464 Niyang River Visitor Centre
尼洋河游客中心

TIBET 西藏

466 Namchabawa Visitor Centre
西藏林芝南迦巴瓦接待站

468 The Apple Elementary School in Ali, Tibet
西藏阿里苹果小学

470 St. Regis Lhasa Resort
拉萨瑞吉度假酒店

YUNNAN 云南

472 Banyan Tree Lijiang
丽江悦榕庄

474 Water House
淼庐

476 Gaoligong Museum of Handcraft Paper
高黎贡山手工造纸博物馆

478 Yun Tianhua Group Headquarters Office Building
云天化集团总部办公楼

NORTHWEST CHINA 西北

NINGXIA 宁夏

480 Yinchuan Cultural and Art Centre
银川文化艺术中心

SHAANXI 陕西

482 The Silk Road Site Museum of West Market in Chang'an, Sui and Tang Dynasties
隋唐长安城西市及丝绸之路博物馆

484 Fuping Museum of Ceramic Art
富乐国际陶艺博物馆群主馆

486 Jia Pingwa Literary Art Museum
贾平凹文学艺术馆

488 Xi'an Horticultural Exposition Garden Reception Centre
西安园艺博览会精品酒店

490 Shangri-La Hotel, Xi'an
西安香格里拉酒店

492 Flowing Gardens
流动花园

494 Xi'an Television Broadcasting Centre
西安广播电视中心

496 Well Hall, Lantian
蓝田井宇

498 Mount Hua Forum and Ecological Square
华山论坛及生态广场

500 Aler Museum
三五九旅屯垦纪念馆

502 INDEX
索引

EDITOR'S WORDS 编者的话

Chinese contemporary architecture has gained rationality and confidence, after integration and foreign impact. The thinking following the designs seems especially precise in this age of vanity, under the stylist and exaggerated look of the architectural works.

Looking back at the history of Chinese architecture, as one of the three greatest architectural systems in the world, classical Chinese architecture has witnessed the vitality of traditional Chinese culture, and has kept its uniqueness during the long development process. What contemporary architects and decision makers face is the tricky game of tradition and modernity, due to the architectural vocabulary that cannot be duplicated nor exceeded and the spirit of traditional Chinese architecture it implied. The development of contemporary Chinese architecture has always been associated with the discussion on the issue of discarding and reserving,

After the reform and opening-up, Chinese architects have shown unprecedented enthusiasm and started bringing in foreign architectural theories on a large scale following a comprehensive introduction of the USSR architectural theories, under the climate of the global coexistence of various architectural theories. As the reform and opening-up continued, the architectural industry gradually expanded, and buildings rose across the country overnight. However, a series of problems were revealed during this process, a most typical one is the contradiction between the international convergence and local characteristics of the architectural form. Regardless of the size of the cities, the repetitive international pattern of architecture has limited the development of local architectural forms in the rapid construction process, lacking reflection about the forms of architecture.

When the process of urbanisation proceeded at full speed, the conflicts between original urban structure and modern lifestyle have become increasingly acute, as old streets are getting overburdened with the growing traffic, skyscrapers of power and wealth keep pushing the boundaries of the skyline, and the exploitation and conservation of historic structures conflict with the renovation of the old city. The architecture, as a form of physical image of the ideology, has influence on people's life and thinking more or less. This is when the architects started collective reflection in the design process, which is mirrored in the following designs, focusing on the complicated relationship between the buildings, the people and the cities.

This book is a comprehensive review of contemporary Chinese architecture displaying the achievement gained in the rapid process of China's urbanisation, with nearly 300 representative architectural works selected from 34 provincial-level administrative divisions across China. Among these, there are both works of local architects and cooperation projects with overseas companies. It can be seen that among the mega-cities and their surrounding areas on the east coast of China, three places get the most attention. They are the Beijing-centred capital circle and the Beijing-Tianjin city cluster facing the Bohai Bay; the Shanghai-centred Yangtze River Delta compact regions, with the Yangtze River as the backbone and Shanghai-Ningbo-Hangzhou as its main

中国当代建筑在经历了舶来与整合的过程之后，逐渐变得理性与自信，在时尚与张扬的外表下，建筑背后的思考在这个浮华的时代显得弥足珍贵。

回顾中国建筑的发展历程，曾经作为世界三大建筑体系之一的中国古典建筑有效见证了中国传统文化强大的生命力，并且在随后漫长的演变过程中始终保持着自身的独特性。这种不可复制与超越的建筑语汇以及蕴含其中的中国传统空间精神，留给当代建筑师及决策者一个传统与现代博弈的难题。有关种种舍弃与保留的讨论一直伴随着中国当代建筑发展的始终，从"民族固有式"的探讨到建筑文化的反思，再到多元化的探索与整合，直至现阶段理性的创作，中国当代建筑师们始终在一种矛盾的、思辨的状态中寻找一种值得认可的价值体系。

改革开放以后，中国的建筑师们表现出空前高涨的创作热情，在世界范围内的建筑理论多元的背景下，开始了自1950年代全面引进苏联社会主义建筑理论之后的又一次大规模引进外国建筑理论的进程。随着改革开放的不断推进，建筑业的规模逐渐扩大，全国各地的建筑如雨后春笋般拔地而起。然而，在这个蓬勃发展的过程中却暴露出了一系列的问题。其中最典型的问题便是建筑形式的国际化趋同与地域特征之间的矛盾。在高速建设的过程中，不论大中小城市，千篇一律的国际化风格建筑束缚了本土建筑形式的发展，缺乏对建筑形式的反思过程。

中国高速城市化推进过程中，原有的城市结构与现代生活模式之间的矛盾日益尖锐，比如不断增长的交通量超过了传统街道的负荷，象征着权利与财富的摩天高楼不断刷新着城市的天际线，旧城改造更新过程中历史建筑的开发与保护等。而建筑作为意识形态的物质表象，又或多或少地影响着现代人的生活和思想。建筑师在创作的过程中开始了集体的反思，这些思索回馈到下一步的创作之中，最终的落脚点应该是建筑、人、城市之间复杂有机的关系。

本书作为一次中国当代建筑创作的整体性回顾，从全国34个省级行政区当中精选出近300件最具代表性的建筑作品，以反映中国在现代化的快速发展进程中所取得的成绩。在这当中有本土建筑师的项目也有与境外事务所合作的项目。通过众多建筑作品的展示我们可以看出，在中国东部沿海特大城市地区当中，最为引人注目的地区有三个：以北京为中心，面向渤海湾的大首都圈和京津城镇群；以上海为中心，长江黄金水道为主干，沪宁杭三足鼎立的长江三角洲城镇密集地区；以广州为中心，

forces; last but not least the Guangzhou-centred Pearl River Delta export-oriented urbanised region, with axis development facing Shenzhen and Zhuhai, integrating resources in Hong Kong and Macao. The economic, geographic and cultural advantages of these three regions are obvious, and to a certain extent in architecture, they represent the height of contemporary Chinese architectural development. This is also reflected in the fact that a considerable number of design companies from home and abroad are located in the eastern coast of China, expanding their business to the west.

Although there are still significant deficiencies in contemporary Chinese architecture due to relatively short period of development, it can be observed through the successful projects that there has been an increasing local influence. Architects focus more on local culture in their designs and research deeper into the natural and cultural elements of the architecture. Some try to find breakthroughs in traditional local materials, construction modes, and new construction forms through a large number of experiments, while the scholars work on the constitutional creativity balancing form and function as well as establish a building evaluation system with more Chinese features.

From the prospective of macro-development of architecture, the cultural symbols of historical characteristics have clearly been repeated, yet their content has changed in an unnoticeable way. During this process, contemporary Chinese architecture has turned from modern architecture symbolising the achievement of the 20th century industrial revolution to intelligent buildings of the 21st century of information. We look forward to finding constant resonance of memories and perception in the future development of contemporary Chinese architecture.

Zhi Wenjun

Professor and PhD supervisor at College of Architecture and Urban Planning, Tongji University
Editor in Chief of Time Architecture
16/11/2012

向深圳、珠海方面轴向发展，汇集港澳一体化的珠江三角洲外向型都市化地区。这三个区域的经济、地理、文化区位优势都非常明显，它们的建筑从某种意义上来说，可以代表中国当代建筑发展的高度，近几年国内外设计团队立足于东部沿海，向西部拓展业务范围的市场表现也反映了这一发展现状。

尽管中国当代建筑由于发展的时间较短，仍然存在一些明显的问题，但是通过一些相对成功的案例我们可以看出，中国当代建筑的地域性特征正在逐渐增强。一些建筑师在创作的过程中更加注重从地域文化出发，并对建筑特定的自然因素以及人文因素进行着更加深入的研究。一些实践建筑师通过大量实验，在当地传统材料、建造模式和新的建筑形式之间寻找切入点，学界探索建立更加具有中国特征的建筑评价体系，在形式与功能之间寻求原发性创造力。

从建筑宏观发展的角度出发可以清楚看到，象征历史特征的文化符号似乎一直在重复，但其本质的内容已经发生了悄然的改变，在这种进化的过程当中，中国当代建筑正在从象征20世纪工业革命成果的现代建筑转向21世纪信息化时代下的智能建筑。我们期待在中国当代建筑不断前行的过程中，能够不断的找到某些记忆和知觉的共鸣。

支文军

同济大学建筑与城市规划学院教授、博导
时代建筑杂志主编
2012-11-16

National Stadium of 2008 Olympic Games, Beijing

2008年北京奥运会国家体育场(鸟巢)

Location: Beijing Olympic Park, Beijing
Area: 98,000 m²
Architect: Herzog & de Meuron, Li Xinggang, China Architeture Design & Reserch Group
Completion Date: 2007
Client: National Stadium Corporation Ltd.

地点:北京,奥林匹克公园
建筑面积:98,000 平方米
建筑师:赫尔佐格和德梅隆,李兴钢,中国建筑设计研究院
建成时间:2007 年
客户:国家体育场有限责任公司

Beijing, China
中国,北京

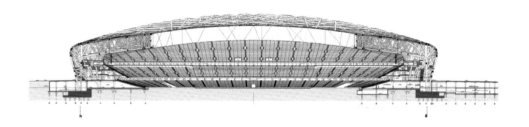

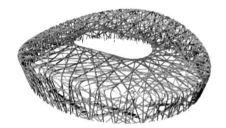

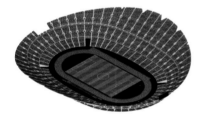

The National Stadium of China is located at the south of the Olympic Park central area. It is to the east of the 200-metre wide green pedestrian on the central axis, to the west of the Dragon. Shaped Water System, to the south of the Zhongyi Road and to the north of the Nanyi Road. The Stadium is built on a sloping plinth, via which spectators can enter the Stadium from the Olympic Green. To the south of the platform is the sinking warming-up area, which is connected with the main field in the stadium by the players' tunnel. The complex annex facilities and commercial outlets are arranged underground, which provides more reasonable access for different crowds and keeps the stadium clean and integrated. The main body of the National Stadium is an elliptic structure, 333 metres long and 69 metres tall from north to south, 294 metres wide and 40 metres tall from east to west. Its central opening is 182 metres long from north to south and 124 metres wide from east to west. The main steel structure, a colossal large-span grid formation just like a bird's nest, and the stand, a concrete bowl structure, are separated from each other. The roof is covered by a double-layer membrane structure, with a transparent ETFE membrane fixed on the upper part of the roofing structure and a translucent PTFE membrane fixed on its lower part. A PTFE acoustic ceiling is attached to the side walls of the inner ring. The spacious grid encircles the capacious concourse, which runs full circle around the stand and functions as an open urban space with restaurants, VIP reception and lounge. The seating bowl, like a classical arena, is divided into three tiers: upper, middle and lower. Between the upper and the middle tier of the stand, there are the VIP box area and VIP seats. All the seating area can be easily accessed from the corresponding concourse.

中国国家体育场位于北京奥林匹克公园中心区南部。西侧为 200 米宽的中轴线步行绿化广场，东侧为湖边西路龙形水系及湖边东路，北侧为中一路，南侧为南一路。国家体育场坐落在由地面缓坡缓起的基座平台上，观众可由奥林匹克公园沿基座平台到达体育场，从基座平台上可俯视奥林匹克公园。基座北侧为下沉式的热身场地，通过运动员通道与主场内的比赛场地连通。国家体育场复杂的附属部分和赛后商业等安排在升起的地面之下，使不同人群进入的方式更加合理，同时保持了主体建筑外观的清晰、纯粹和完整。体育场主体建筑为南北长 333m、东西宽 298m 的椭圆形，最高处高 69m、最低处高 40m；中间开口南北长 182m、东西宽 124m。主体钢结构形成整体的巨型大跨度钢桁架编织式"鸟巢"结构，体育场看台为混凝土碗形结构，两部分在结构体系上脱开。屋顶维护结构为钢结构上覆盖双层膜结构，即固定于钢结构上弦之间的透明的上层 ETFE 膜，和固定于钢结构下弦之下及内环侧壁的半透明的下层 PTFE 声学吊顶。开敞的钢结构网格包围着宽阔的集散大厅，它环绕着碗形看台，是一个开放的城市空间，设有快餐、商店等各类观众服务设施，并设有餐厅层和为主席台观众服务的贵宾接待区和休息区。犹如古典竞技场般效果的碗形看台分为上、中、下三层坐席，并在中、下层看台之间设置包厢层及其坐席区，分别由各自对应的集散大厅进入。

National Swimming Centre of 2008 Olympic Games, Beijing

2008年北京奥运会国家游泳中心（水立方）

Location: Beijing Olympic Park, Beijing
Area: 87, 000m²
Architect: PTW Architects + CSCEC+ CCDI Design + Arup — John Bilmon, Mark Butler, Chris Bosse, Zhao Xiaojun, Zheng fang, Wang Min, Shang Hong, Tristram Carfrae, Peter Macdonald, Kenneth Ma, Haico Schepers
Completion Date: 2008
Client: Beijing State-owned Assets Management Co., Ltd.

地点：北京，奥林匹克公园
建筑面积：87, 000 平方米
建筑师：PTW 建筑师事务所 + 中建国家游泳中心设计联合体 + CCDI 中建国际设计 + 奥雅纳工程公司 + 中建总公司 –John Bilmon、Mark Butler、Chris Bosse、赵晓钧、郑方、王敏、尚宏、Tristram Carfrae、Peter Macdonald、Kenneth Ma、Haico Schepers
建成时间：2008 年
客户：北京市国有资产管理有限公司

Beijing, China
中国，北京

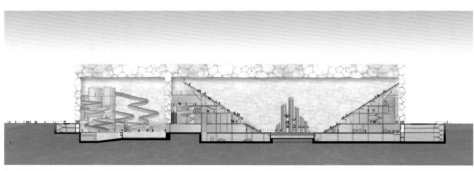
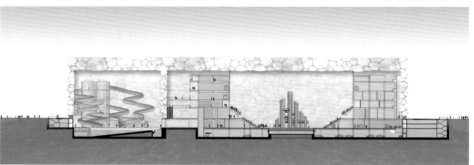

The National Swimming Centre (NSC) will be central to the success of the 2008 Olympic Games, with all pools and functions that FINA needs to conduct every session of Swimming, Diving, Water Polo, and even Synchronised Swimming. NSC is also a truly multi-function leisure centre for Beijing City in the 21st century. It encapsulates every aspect of water – hot and cold, shallow and deep, lazy rivers and pounding surf beaches, competition water, and even has ice.

The design process of National Swimming Centre exhibited a concerted collaboration of international teams. Architects from PTW helped a lot on the skin material ETFE; engineers from ARUP bring out a perfect structural model based on foam theory; and the design group from CCDI gave the idea of "Aquatic Cube" and endowed it with much meaning of Chinese traditional philosophy. The design-union collaboration itself is a joyful journey, which inosculated the different ways of art, culture, technique, as well as the 2008 Olympic Games in Beijing.

国家游泳中心是2008北京奥运会重要场馆之一。它将作为北京2008奥运会游泳、跳水、花样游泳和水球的比赛场馆，并将在赛后被改造成为多功能的大型水上运动中心。中心内包括游泳池、跳水池、热身池、嬉水乐园，以及关于"水"这一母题的方方面面。于是，建筑师以"水立方"为设计理念——这是一个简洁明快又富有神秘感的蓝色方盒。

"水立方"是国内外一流建筑设计公司充分合作的结晶。正因如此，它才得以融会中国传统文化和现代科技，并显深邃的文化内涵和技术高度。来自中建国际（CCDI）、PTW、ARUP三方面的团队构成了一个充满创造力的合体。PTW的建筑师引入ETFE膜作为建筑表皮，使其具有独特的视觉效果，并使建筑变得柔和；ARUP的工程师基于Kelvin"泡沫"理论，为该建筑设定了一种完美的结构形式。而"水立方"的基本哲学理念则来自CCDI对中国文化的真切把握。CCDI作为设计总负责单位，不仅在建筑学意义上为水立方赋予许多中国元素的含义；也在实践层面完成了深化、详实的工程图设计；更借该项目机缘，为国家完成了多项创新和研究任务。

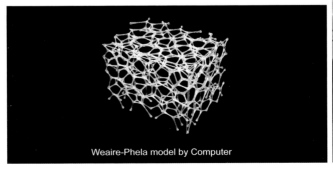

Weaire-Phela model by Computer

National Centre for the Performing Arts

国家大剧院

Location: Beijing
Area: 172,800m²
Architect: Paul Andreu Architect Paris
Completion Date: 2007
Client: Client's Committee of National Centre for the Performing Arts

地点：北京
建筑面积：172,800 平方米
建筑师：巴黎保罗·安德鲁建筑师事务所
建成时间：2007 年
客户：国家大剧院工程业主委员会

Beijing, China
中国，北京

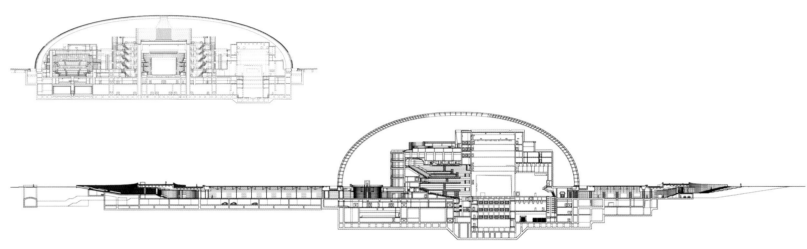

The National Centre for the Performing Arts (NCPA) complex houses a 2,500-seat opera house, a 2,000-seat concert hall, a 1,200-seat drama hall and a 300- to 500-seat seats multi-functional theatre. With the improvement of living conditions in China, people are in pursuit of better performing effect. They are no longer satisfied with the effect in one multi-functional theatre which can provide all kinds of performances. The NCPA, as a complex for performing arts which comprises different special theatres that can provide various performances and bring best effect, intensively embodies the cultural life of the country or the region.

The proscenium of the opera house is 18.6 metres wide. As a national opera house, the size of its proscenium shall be higher than regional ones, which is convenient for troupes to set up sceneries. The opera house has a seating capacity of 2,500, which includes the seats in the pit and the last three rows of standing seats behind the seats on the first floor. Such an arrangement can provide not only better audio and visual effects, but also enough seats.

国家大剧院包括一个 2500 座的歌剧院，一个 2000 座的音乐厅，一个 1200 座的戏剧院和一个 300~500 座的多功能小剧场，通常称为"三大一小"。这种综合趋势的出现并非偶然，一个主要的原因是功能的要求，文化生活水平的提高使人们对演出效果的追求亦越来越高，已经不满足那种可供多种演出的多功能剧场演出的效果，那种作为一个国家或一个地区的文化生活集中体现，建造一个由专业剧场组成的可满足各种形式演出的并达到最佳效果的演出综合体建筑应运而生是很自然的事。

国家大剧院台口宽度为 18.6m。作为一个国家级的剧院台口尺寸不应低于地方剧院台口 18m 的尺寸，这也便于全国各地剧团的舞台布景通用。大剧院座位是 2500 座，这里包括乐池部位的座席和二层楼座后三排的站席的数量，这样视听效果上容易处理一些，又能保证一定数量的观众。

National Library of China
中国国家图书馆

Location: Beijing, China
Area: 77,000m²
Architect: KSP Jürgen Engel Architekten
Photographer: Hans Schlupp, National Library of China
Completion Date: 2008

地点：北京
建筑面积：77,000 平方米
建筑师：KSP 尤根·恩格尔建筑事务所
摄影：汉斯·斯库普、中国国家图书馆
建成时间：2008 年

Beijing, China
中国，北京

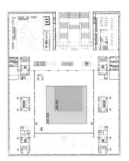
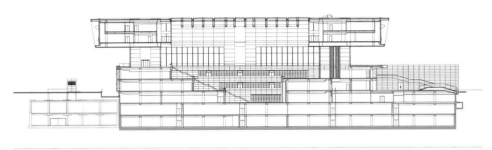
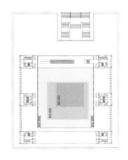
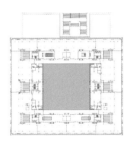
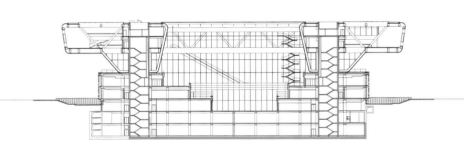

The task was to extend the existing library by 77,000m² in order to accommodate approximately 12 million books, some 12,000 people per day visiting the National Library. An important future-oriented part of the extension of the existing library is the "Digital Library". Unparallelled in its historical and cultural significance, the "Si Ku Quan Shu" collection is also housed in the new building. The design is based on this collection, which is housed in the building's base.

The enormous rectangular building ensemble, measuring 90 by 119 metres, is horizontally organised, and developed. The library is surrounded by an austerely laid-out garden, which serves to further underline the building's well-defined appearance. It is through the garden and a geometrically arranged grove at the entrance to the library that the visitor approaches the building. The entry to the library is via broad steps set between two lower building sections, which lead into the second storey. From here the visitor has a view out over the entire library and into the extensive steel structure spanning the roof. This structure rests on the base at six points, and these support the roof, while leaving the remainder of the large space open.

As the centre and core of the library, the glazed vault for the "Si Ku Quan Shu" collection of China's written cultural heritage, bound in wooden panels, is always visible, and transparent. Looking at the library from outside, the upper two storeys have the appearance of a large book with gently rounded contours, held by slanting supports, which towers over the storeys below.

这一设计任务为将原有图书馆扩建77,000平方米,以存放大约1200万册图书并满足每日约12000人来访的需求。扩建部分最为重要的当属"数字图书馆"。此外,《四库全书》作为历史及文化宝典也将被存放在新建筑内,而建筑的设计也以其为基本理念。

巨大的长方形建筑体四周被造型规则的小花园包围,进一步突出图书馆本身的造型特色。穿过花园以及大门处形状规整的小树林便可达到图书馆门前,入口处设有宽大的台阶,可一直通向三层。从这里,便可以看到图书馆的全景以及钢结构屋顶。上面两层采用柱结构支撑,似乎与下面结构分离开来。

图书馆内部中央区玻璃拱顶上收藏着《四库全书》,清楚可见。从外面望去,图书馆上面两层犹如一本巨大的书籍,四周曲线柔和。

Digital Beijing
数字北京

Location: Beijing
Area: 98,000m^2
Architect: Zhu Pei, Wu Tong, Wang Hui / Studio Pei-Zhu, Urbanus Architecture & Design
Completion Date: 2007

地点：北京
建筑面积：98,000 平方米
建筑师：朱锫、吴桐、王辉 / 朱锫建筑设计事务所、都市实践
建成时间：2007 年

Beijing, China
中国，北京

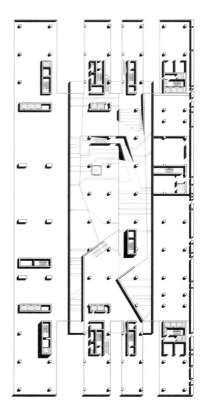

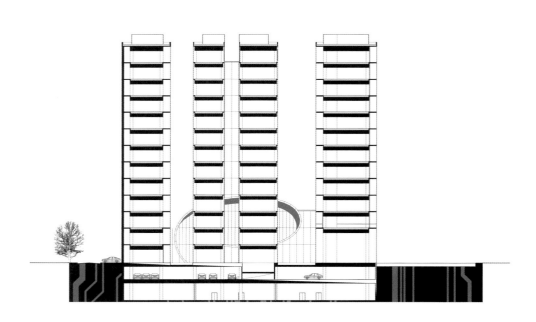

Modern urban space shares common characteristics with microchips from a specific aspect. Fluidity and diversity are the two main conceptual ideas for this project.
Digital Beijing is far beyond a distinctive architecture. It is a building to store data and manage information during the Olympic Games as well as an imformation centre for the city. Based on these functions, the entire volume is divided into four blocks, the east block with good natural lighliting and perfect view as office while the middle and west blocks as digital infirmation space. The "digital carpet" expands from the ground floor and unfolds and rolls into walls and finally forms a virtual museum in the air. Visitors walk across the floating bridge and then go to the inside to feel the changes visully, the contrast between scenes and the dialogue between nature and technology.

从某种意义而言，现代城市空间与微芯片结构所反映出来的共同特征，流动性、层次多样性是该设计的基本概念。
"数字北京"已远远不只是一栋有特色的建筑物。一方面它作为奥运信息的储存器，另一方面它更应是城市的一个信息中心场所。沿着这个思路，建筑师成功、逻辑地将"数字北京"的形体切割为四个板块，东侧的一块为办公区，具有良好的采光和视野，中间和西侧的板块为数字机房。四个信息块通过入口首层的网络桥塔进而被激活，承担着展示功能的"数字地毯"从地下一层渐渐升起变成为墙面，再不断延伸和卷起，构成了空中的奥运数字虚拟博物馆。水平流动的"数字地毯"，快捷有效的网络桥，悬浮在空中的博物馆，它们之间的透明介质形成了层次多样的平面叠加关系。当人们行走在自然水面上的浮桥，进入建筑内部，在移动中会获得丰富的视觉变化感受，强烈的场景对比，自然与科技之间的对话。

China National Convention Centre (Beijing 2008 Olympic Games Multi-purpose Arena)

国家会议中心（北京 2008 奥运会多功能场馆）

Beijing, China
中国，北京

Location: Beijing
Area: 270,000 m²
Architect: RMJM
Photographer: Ben Mcmillan
Completion Date: 2007
Awards: 8th Tien Yow Jeme Civil Engineering Prize 2008

地点：北京
面积：270,000 平方米
建筑师：RMJM 建筑师事务所
摄影：本·麦克米兰
建成时间：2007 年
获奖：2008 年第八届詹天佑土木工程大奖

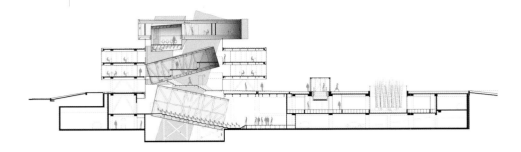

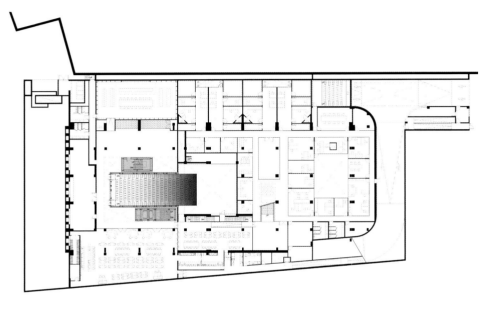

The China National Convention Centre (Beijing 2008 Olympic Games Multi-purpose Arena) in Beijing played a key role during the 2008 Beijing Olympics. Located at the top of Olympic Boulevard, it housed the Media Centre and the International Broadcasting Centre, and hosted fencing and air-pistol events.

The development covers a site area of 81,000m². The highest point of the building is 43m. The Centre boasts an iconic roof that was inspired by the flying roof structure of Chinese architecture, and is a metaphor of a "bridge" – symbolising the unification and communication of China. The floor-to-roof glass façade both represents the minimalist architectural style, and allows natural daylight to penetrate the depths of the interior.

The development is linked together by internal promenades and interface zones – also known as "The Street" which runs from one side of the Centre to the other. Sub-buildings include an eight-storey retail centre and main exhibition centre.

The Centre is a feat of civil engineering as the space is characterised by long-span, column-free spaces which are necessary for a successful exhibition space. However, given Beijing's seismicity, innovative design solutions were required to satisfy the seismic building codes. The final design considered and effectively addressed the vibration performance of long-span suspended slabs.

With its prominent place in the Olympic Park, the Centre is a green oasis. Its roof is also a sustainable feature; the swooping structure acts as a giant funnel for rain water, which is collected and stored in a treatment facility for irrigation.

国家会议中心位于奥林匹克大街起点，在2008年奥运会期间发挥了重要作用，内部设有媒体中心、国际广播中心、击剑及气手枪比赛场馆。

建筑占地面积达81000平方米，高达43米。其屋顶设计源于中式建筑中的"飞檐"结构，并运用了"桥"的比喻，代表着国家的团结与交流。全玻璃外观突出了极简建筑风格，同时将光线引入到内部空间。

建筑空间通过室内"回廊"（街道）连通，贯穿整个结构。辅助楼包括8层的零售中心以及主展区。开敞而无梁柱支撑的空间设计更能体现土木工程的特质，同时实现展览的功能。但是由于考虑到北京地区地震活动较为强烈，因此设计中要求满足抗震需求。最终采用悬浮板支撑结构，有效地实现了抗震功能。

此外，建筑位于奥林匹克公园内，因此其本身更像是一处绿洲。屋顶设计突出环保特色——弯曲的造型结构犹如一个巨大的雨水收集池，经储存之后用于灌溉。

Public Service Platform at Beijing Contemporary Art and Cartoon Industry Centralised Area

北京当代艺术与卡通产业集聚区公共服务平台

Beijing, China
中国，北京

Location: Beijing
Area: 12,770m²
Architect: Kang Kai, Jin Dayong, Li Yu
Photographer: Zhang Guangyuan
Completion Date: 2011

地点：北京
建筑面积：12,770平方米
建筑师：康慨、金大勇、李宇
摄影：张广源
建成时间：2011年

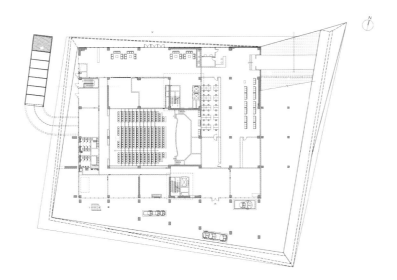

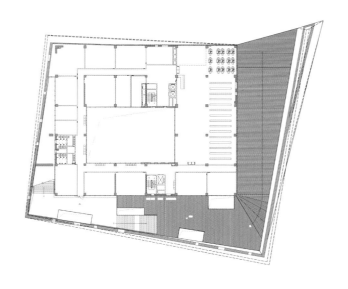

022 / NORTH CHINA

Beijing Songzhuang Service Platform is located at Songzhuang Creative Industries Area. The design overlays functional rooms including the exhibition hall, library, media centre, IDC in vertical direction. It also attempts to make effective use of the land to provide an open urban public space. The service platform is clipped among the various public functions in vertical direction, and allows sight and special infiltration among the various functions, which makes many activities including exhibition, communication, media and working take place at the same time in different parts of the building. The concept of the architectural design is to create a space, a place, which means to set up or cancel a boundary, while the boundary is the join of two worlds, two spaces and two materials. The concept design reshapes diversified spaces creating inorganic changes of the exterior interfaces by overlaying.

The architectural design of the service platform is actually an interactive dynamic design, for the sight focus is not fixed on one point. This indicates that the visitors and beholders are not in abstract space, but in a space of natural light and scenery. Thus, many courtyards, open yards, outside terraces and other outside spaces "tile" into an intact sight and logic process. During this process, the integration with nature and the materials with time dimension play a very important role, including concrete, black brick, log, bluestone batten, glass and metal. These materials integrate together in a harmonious contrast, which makes the visitors and beholders always find something interesting when walking and finally seek the unity with the cultural connotation, form and functions of Songzhuang.

北京宋庄服务平台，位于宋庄创意产业区中。方案将展示交流、图书馆、媒体中心、IDC等功能用房在垂直方向上叠加，并试图最有效地使用地块提供开放的城市公共空间。服务平台在垂直方向上被夹在各种公共功能之间，并且允许在不同的使用功能之间有视觉和空间上的渗透。使展览、交流、媒体和办公等多种活动可以同时在建筑的不同部位发生。建筑设计理念旨在创造一个空间、一个场所，意味着设定或取消界限，而界限是两个世界、两个空间、两种物质的交汇。概念设计中通过折叠从外部无机的创造出界面变化，重新塑造了丰富多样性空间。

服务平台建筑设计其实也是一个互动的流动设计，视线目标不是固化在某一点上。由此引申行者和观者不是在抽象的空间之中，而是在有自然光与景的空间之中。因此，许许多多院落、天井、室外露台、悬挂在外的空间拼贴成一个完整的视线和逻辑的流程，在这种拼贴过程中，与自然相融并具有时间维度的材料起着重要的作用，如混凝土、青砖、原木、青石板条、玻璃、金属板等，它们在和谐的对比中能相互融合，使行者和观者在三点透视的游弋中，眼帘中总有有趣的东西，最终与宋庄文化内涵、形式和功能上得到统一。

Raffles City Beijing
北京来福士广场

Location: Beijing
Area: 145,928m²
Architect: Yang Ke / Sparch architecture
Completion Date: 2008

地点：北京
建筑面积：145,928 平方米
建筑师：杨克 / 思邦建筑设计咨询有限公司
建成时间：2008 年

Beijing, China
中国，北京

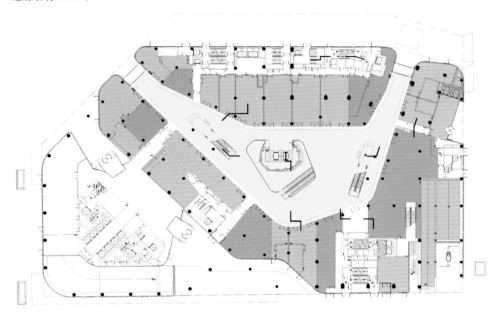
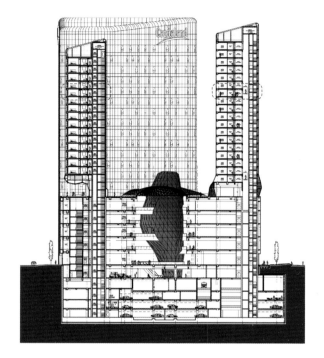

Located at Dongzhimen Transportation Hub of Beijing with convenient and easy access to the Subway, Raffles City Beijing is surrounded by high-class communities and office buildings and considered as an iconic integrated project of the East 2nd Ring Road. In order to eliminate the sense of depression caused by the large high-rise building (especially for the Dongzhimen overpass), the whole building is divided into several parts according to the functional requirements of the proposal. These parts include a series of 90-metre-high office block with triangular plans, two slab-type serviced apartments in different scales with an equal height of 90 metres, and a 30.3-metre-high business podium building that is composed by connecting three blocks. Echoing with each other, they completely embody the characteristic of the whole project and provide the viewers with more changeable perspectives and perspective depths. Additionally, the designers deliberately evaded the southwest extension of the "Dongzhimen Transportation Hub" and designed an imaginary axis to echo with the "Hub", so as to integrate the two buildings placed at the diagonal of the "Dongzhimen Transportation Hub" organically.

This project is composed of business and office buildings, apartments, clubs and underground garage as well as the ancillary space. The proposal aims to create an efficient, functional, technological, open, flexible and flowing construction. The façade design strategy highlights the sense of simplicity and elegance, integrating it with the numerous surrounding buildings seamlessly. This project has been awarded as the "Best Green Architecture" by the National Federation of Real Estate Chamber of Commerce.

北京来福士广场位于北京市"东直门交通枢纽",地理位置优越,交通便捷,是市中心区重要的商业地块,也是东二环沿线开发建设的高端项目之一。为了避免过大的高层建筑体量对城市产生压抑感(特别是对东直门立交桥),因此根据项目设计任务书的功能要求,对整体建筑进行了分割,使其成为一组90米高的平面呈直角三角状的办公楼体块,90米高的大小两个薄板式酒店式公寓体块和将三个建筑体块连为一体的30.3米高商业裙楼体块,建筑整体形象通过相呼应的群体建筑体现,而且为人们观赏建筑提供了更为变幻的视角和透视深度。为了与东直门东北角的"东直门交通枢纽"相呼应,设计特意将三个高大的建筑体块避开"东直门交通枢纽"的西南延长线,而是以一个虚轴与之呼应,从而使东直门对角线两个建筑成为有机动态的连接。

该项目功能由商业、办公楼、酒店式公寓、会所、地下车库及配套用房几大部分组成。设计以高效的建筑性能、更多的功能选择、多种科技的普遍利用、空间的开放性及灵活性、交通可达性为设计目标。立面处理简洁大方,使之能够与毗邻的"城市文明"和众多在建筑形态上难以协调且鳞次栉比的城市建筑群落巧妙而完美地融合。该项目获全国工商联房地产商会"最佳绿色环保奖"。

Huamao Centre
华贸中心

Beijing, China
中国，北京

Location: Beijing
Area: 619,400m²
Architect: Kohn Pedersen Fox Associates PC (Design Principal: James von Klemperer)
Photographer: Fu Xing, Kohn Pedersen Fox Associates PC
Completion Date: 2007
Client: Guohua Real Estate Company
Awards: 2007 MIPIM Asia Awards, Mixed-Use Category
Shortlisted for Asia Pacific Commercial Real Estate Award

地点：北京
建筑面积：619,400 平方米
建筑师：美国 KPF 建筑设计事务所（设计总监—詹姆斯·凡·克莱普尔）
摄影：傅兴、美国 KPF 建筑设计事务所
建成时间：2007 年
客户：北京国华置业有限公司
获奖：2007 亚洲国际房地产大奖"综合建筑"提名亚太商业物业大奖

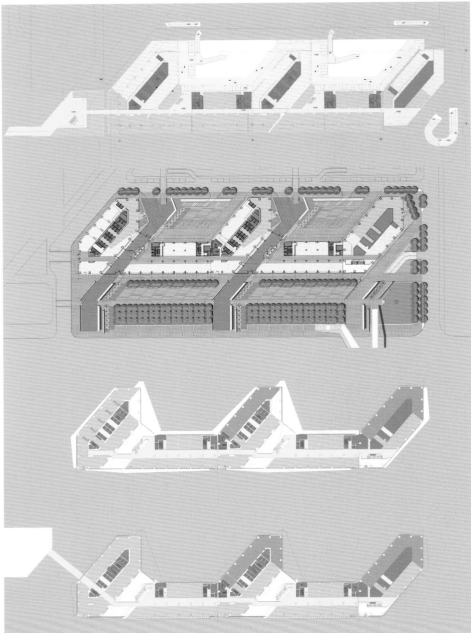

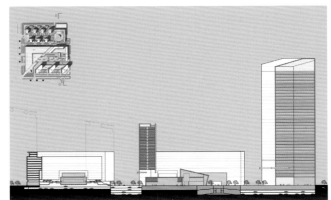

026 / NORTH CHINA

The scheme's organising principle creates a series of urban spaces, defined by mixed-use buildings that respond to the surrounding context, including Beijing's evolving Central Business District (CBD) to the west, an industrial site to the north and east, and Jianguo Road and a planned park to the south. The programmatic composition is arranged in three zones layered across the site: office programme to the south, residential to the north, and retail/hotel between the two.

The office programme is divided into three glass-skinned towers, each at a different height. Tower 1 (591 feet/180 metres), the tallest and the farthest to the east, is the dominant form viewed on the eastern approach to Beijing; when combined with a recently completed footbridge across Jianguo Road, it creates a gateway into the city. Tower 3 (394 feet/120 metres) marks the prominent southwest corner of the site with a sculptural form; its shape deflects the visual focus into the centre of the site and the retail plaza. The bases of the buildings are connected by a five-storey retail structure that allows users to reach all three towers from the subway without exiting the complex.

The retail programme is marked by a galleria that defines the circular plaza. This plaza is reinforced by the hotel, a circular reflecting pool, and a landscaped amphitheatre. These elements are designed to sponsor an active urban space used by the public, office workers, and residents.

The residential programme occupies a series of linear bar buildings and vertical towers that front onto the major east-west tree-lined boulevard.

这一设计规划原则为在高楼耸立的背景中打造一系列的城市空间，西侧毗邻中心商务区，东、北两侧为工业园，南侧为建国路及公园。这是一个综合性项目，包括由南向北依次为办公区、商业店铺／酒店、住宅区。

办公区由三幢全玻璃外观的建筑构成，其高度各不相同。最高的一幢高达180米，与横跨建国路的步行桥一起构成了北京的"东大门"。最低的一幢为120米，雕塑般的建筑形式使其成为西南角的坐标。独特的造型将视线引入到大厦内部以及附近的商业广场内。三幢大厦通过底部5层高的商业店铺连接，可从地铁通道直接走入进来。

商业区因商业街廊的设计而格外引人注目，并在中央形成了环形广场。酒店、环形空中泳池、景观剧场则为公众、员工以及附近的住户带来了便利。

住宅区由一系列的线性建筑及塔楼组成，面朝东西向的林荫大道。

Linked Hybrid
万国城

Location: Beijing
Architect: Steven Holl Architects
Photographer: Iwan Baan, Steven Holl Architects, Shu He
Completion Date: 2009

地点：北京
建筑师：美国斯蒂文·霍尔建筑事务所
摄影：伊万·班、美国斯蒂文·霍尔建筑事务所、舒赫
建成时间：2009年

Beijing, China
中国，北京

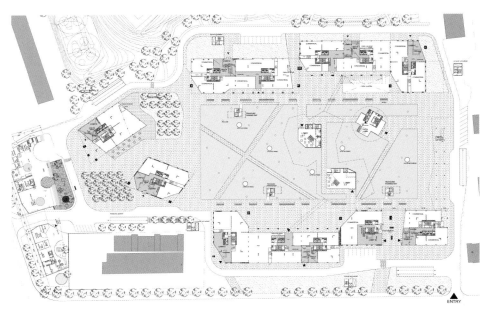

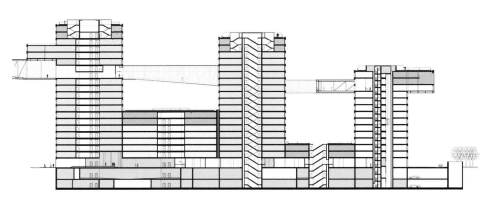

Focused on the experience of passage of the body through space, the towers are organised to take movement, timing and sequence into consideration. The point of view changes with a slight ramp up, a slow right turn. The encircled towers express a collective aspiration; rather than towers as isolated objects or private islands in an increasingly privatised city, architects' hope is for new "Z" dimension urban sectors that aspire to individuation in urban living while shaping public space. Geo-thermal wells (660 at 100 metres depth) provide Linked Hybrid with cooling in summer and heating in winter, and make Linked Hybrid one of the largest green residential projects. The large urban space in the centre of the project is activated by a greywater recycling pond with water lilies and grasses. In the winter the pool freezes to become an ice-skating rink. The cinematheque is not only a gathering venue but also a visual focus to the area. The cinematheque architecture floats on its reflection in the shallow pond, and projections on its façades indicate films playing within. The ground floor of the building, with views over the landscape, is left open to the community. The polychrome of Chinese Buddhist architecture inspires a chromatic dimension. The undersides of the bridges and cantilevered portions are coloured membranes that glow with projected nightlight and the window jambs have been coloured by chance operations based on the "Book of Changes" with colours found in ancient temples.

万国城的设计以"人体通过空间的体验"为理念，将"运动、时序、次序"引入其中。"Z"形结构展现出"集合的氛围"，与城市中不断发展的独栋建筑不同，既能突出私密性，又可营造公共生活空间。视野随着略微向上倾斜的坡道而不断变化，极具特色。

地热井夏天供应冷气，冬季提供热水，使该建筑成为城市中大规模绿色项目之一。建筑中央公共区域设有生活废水回收池，里面栽种着睡莲和水草。冬季，池面结冰便形成了滑冰场。

电影馆不仅仅是集会的场所，更成为了这一区域的视觉中心。影院似乎悬浮在其投射在池塘内的影子上，外观投射出内部放映的电影画面，影院一层面向公众开放。

中国传统寺庙建筑的多彩性在这一设计中充分展现。空中连接桥的底部以及悬臂部分采用彩色薄膜板打造，在夜晚灯光的照射下闪闪发光；窗框采用古代庙宇装饰色彩粉饰，别具一格。

Beijing Publishing House
北京出版创意中心

Location: Beijing
Area: 10,980m²
Architect: Zhu Pei
Project Leader: Li Shaohua, Jiao Chongxia
Project Team: Mark Broom, Lu Wei, Frisly Colop-Morales, He Fan, Dai Lili, Xi Weidong, Yang Chao
Completion Date: 2009

地点：北京
建筑面积：10,980平方米
建筑师：朱锫
项目负责人：李少华、焦崇霞
设计团队：马克·布鲁姆、卢蔚、弗里斯莱·克洛普－莫拉莱斯、何帆、戴丽丽、席伟东、杨超
建成时间：2009年

Beijing, China
中国，北京

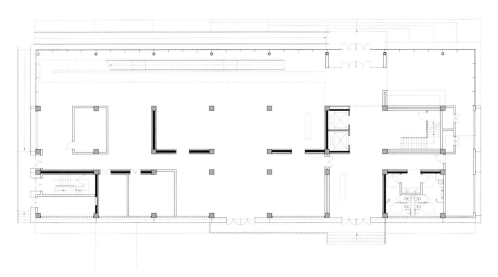

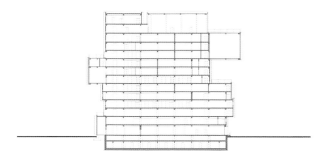

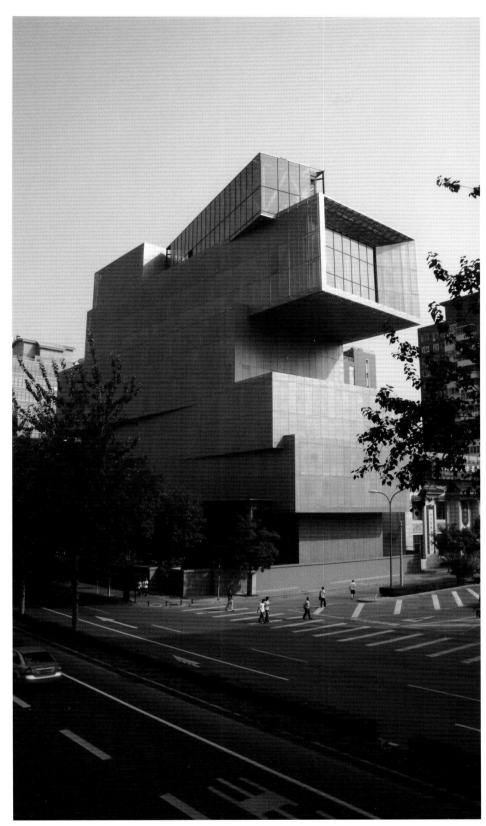

The objective is to renovate the original narrow, rigid office building into an open, flexible and creative facility. Different office spaces and the lower commerical floor as well as the top cultural floor are linked through a public space system. The limits between floors are broken through vertical streamline, combined with exterior terrace and roof gardens to form a "micro city" and to bring rich spatial experience.

The existing steel and concrete structure is retained to a large extent and the floor slab is stretched to build exterior terraces of diversified scales and on diferent levels. Public spaces and staicase (ramps) are constructed on the newly added volume. The façades are more transparent and vivid compared with the previous closed and flat appearance and look like copies of books arranged in a staggered way. Now the building has become a dynamic urban medium and stands out among the ordinarily-looking buildings located surrounding the Third Ring Road.

在设计中，通过对现有封闭狭小刻板均质的办公空间的改造，形成开放灵活的新型创意办公空间。不同办公空间以及与底层商业、顶层文化空间之间又通过不同层次公共空间系统相联系。通过灵活的竖向流线设计，打破楼层之间的阻隔，同时结合室外平台、屋顶花园等室外公共空间，形成"微观城市"的公共空间系统，带来丰富的空间体验。

为实现此公共空间系统，在原有钢筋混凝土框架结构体系的基础上，在尽可能少改变现有结构的前提下，延伸楼板设置了一系列不同尺度的悬挑，在不同高度上创造出尺度丰富的室外平台。公共空间和联系不同楼层的楼梯或坡道就集中设置在建筑的北侧和西侧这些新加建的部分。建筑外立面处理上，尽量保留原有结构的同时，结合功能组织和公共空间系统打破原有封闭平板的立面轮廓，形成通透生动而强烈的外部形态，仿佛几本交错放置的图书，连同一起成为三环周边平庸立面中与众不同的闪亮片断，此时建筑本身已经成为新的极具活力的城市媒介。

China Youth Travel Service Plaza
中青旅大厦

Location: Beijing
Area: 65,000m²
Architect: gmp Architekten – von Gerkan, Marg and Partners
Photographer: Christian Gahl, Ben McMillan
Completion Date: 2006

地点：北京
建筑面积：65,000 平方米
建筑师：gmp 建筑设计事务所
摄影：克里斯琴·加尔、本·麦克米伦
建成时间：2006 年

Beijing, China
中国，北京

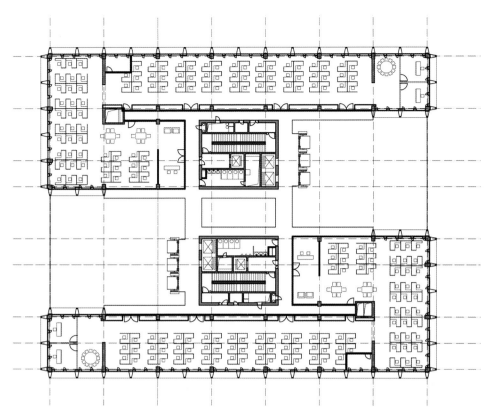

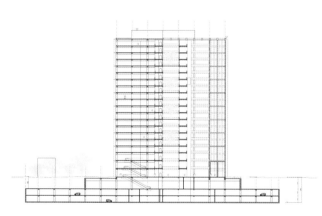

The site of CYTS Plaza is situated on the eastern highway-like Second Ring Road, which follows the course of Beijing's former city wall. As a reminiscence to the ancient wall's impression in the urban fabric, a strip parallel to the Ring Road is due to be built with approx. 80-metre-high buildings forming a clear contrast to the low buildings around the Forbidden City inside the Ring.

The design of CYTS Plaza follows this motif and – like a city gate – links the historical Emperor's City to the Central Business District being under construction at the moment. A passageway divides the 75-metre-high building, thus connecting the two city districts. Two building-high, glazed atria form the entries on both sides of this passage. They serve as foyers to the public visitors' zones and offer unique views of historic and modern Beijing on all floors. The diagonal layout of the atria has created two L-shaped structures, which accommodate optimally lit offices. Between the central cores, four-storey-high halls serve as communication zones for all people working in CYTS Plaza.

The façades of CYTS Plaza with their large-scale vertical grid clearly stand out against the surrounding buildings. The large glass surfaces and the dark aluminium structure give the building a transparent elegance, which fully expresses the company's philosophy.

中青旅大厦位于北京二环路（沿北京老城墙方向），作为对于北京老城墙的回忆，与大街平行的区域将建造一系列80米高的建筑，与紫禁城周围的低矮建筑形成鲜明对比。

中青旅大厦的设计以此为主题，犹如城门一般。一条通道犹如纽带一般将这一75米高的大厦"切开"，同时将两个城区连结，两个玻璃心房分别位于通道两侧，形成入口，同时用作通往公共区的大厅，在这里可以欣赏到北京城的景色。此外，心房对角线位置呈现两个L造型结构，用作办公区。中央区域内四层高的大厅用作员工交流区。

建筑外观呈现垂直格子结构，在周围的建筑群中脱颖而出。大幅玻璃以及铝材表皮赋予建筑透明的美感，诠释出公司的理念。

Beijing Nexus Centre
北京嘉盛中心

Location: Beijing
Area: 10,500m²
Architect: Goettsch Partners
Photographer: Doug Snower / Doug Snower Photography
Completion Date: 2008
Client: R&F Beijing AE

地点：北京
建筑面积：10,500 平方米
建筑师：美国 gp 建筑事务所
摄影：道格·斯诺尔、道格·斯诺尔摄影工作室
建成时间：2008 年
客户：富力地产开发

Beijing, China
中国，北京

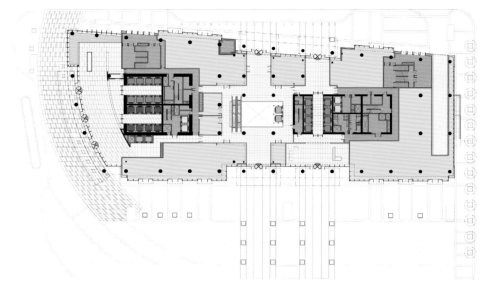

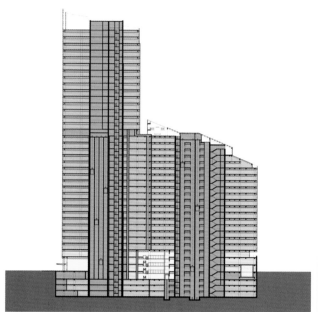

The office complex is located along the East Third Ring Road on a highly visible site between the airport and the main business district. The project is part of the northward expansion of downtown, an area that includes some of the most significant office and cultural developments in China's capital.

The structure comprises two office buildings, connected at the base by a three-storey restaurant and retail podium.

Located on a long, linear site, the dual-tower complex slopes south to north from its 551-foot apex in order to limit shadow casting on its northern residential neighbours. The eastern façade fronts the heavily trafficked ring road and gives retailers prominence along the base. The western façade, by contrast, overlooks the serene setting of an urban park. On the south, the building curves in relation to its corner site, opening up to a large landscaped plaza that segregates vehicular and pedestrian traffic while creating an inviting amenity for tenants.

The design of the complex uses a combination of stainless steel and granite to define distinct fenestration that visually breaks down the overall mass into more slender elements. Expansive floor-to-ceiling glass is applied throughout in order to bring natural light deep into the office floors. Extensive vertical shading devices on the east and west façades control heat gain, and operable windows on all floors promote the introduction of fresh air throughout the work environments.

嘉盛中心位于东三环机场及中央商业区之间，可见性极高。该项目是北京城市扩建计划的一部分，其所在区域还包括其他一些知名办公中心及文化中心。嘉盛中心共由两幢办公大厦构成，下面通过三层的餐饮及零售空间连接。

建筑建造在一处狭长的地块上，建筑顶端呈现由南向北倾斜的坡度造型，旨在减少在北侧住宅区上的阴影投射。东侧立面朝向繁忙的交通要道，因此对于底层的零售店铺来说带来了巨大效益；西侧立面则与幽静的城市公园相对；南侧立面呈现曲线造型，朝向开阔的景观广场，这一广场将车辆通道与人行道隔离开来，为业主营造了一个舒适的环境。

建筑外观采用不锈钢及大理石打造，决定了其独特的开窗方式，在视觉上将整体结构分割成细长的单体。大幅玻璃材质的运用旨在将自然光线引入到室内。东、西两侧立面上安装遮阳结构，减少阳光照射带来的热量。活动窗的安装确保室内空气流通，营造舒适的工作环境。

GreenPix – Zero Energy Media Wall

零耗能多媒体幕墙

Location: Beijing
Architect: Simone Giostra & Partners
Photographer: Simone Giostra & Partners, ARUP, Zhou Ruogu
Completion Date: 2008

地点：北京
建筑师：西蒙·季奥斯尔塔工作室
摄影：西蒙·季奥斯尔塔工作室、ARUP、周若谷
建成时间：2008年

Beijing, China
中国，北京

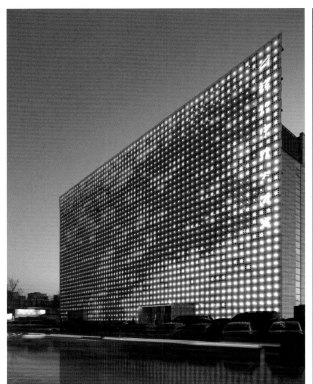

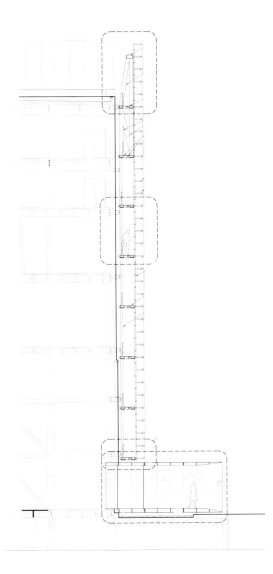

GreenPix - Zero Energy Media Wall - is a groundbreaking project applying sustainable and digital media technology to the curtain wall of Xicui entertainment complex in Beijing, near the site of the 2008 Olympics. Featuring the largest colour LED display worldwide and the first photovoltaic system integrated into a glass curtain wall in China, the building performs as a self-sufficient organic system, harvesting solar energy by day and using it to illuminate the screen after dark, mirroring a day's climatic cycle. The Media Wall will provide the city of Beijing with its first venue dedicated to digital media art, while offering the most radical example of sustainable technology applied to an entire building's envelope to date.

With the support of leading German manufacturers Schueco and SunWays, the architect Simone Giostra with Arup developed a new technology for laminating photovoltaic cells in a glass curtain wall and oversaw the production of the first glass solar panels by Chinese manufacturer SunTech. The polycrystalline photovoltaic cells are laminated within the glass of the curtain wall and placed with changing density on the entire building's skin. The density pattern increases building's performance, allowing natural light when required by interior programme, while reducing heat gain and transforming excessive solar radiation into energy for the media wall.

GreenPix——零耗能多媒体幕墙是一项可持续能源和数字媒体技术的创新性项目，是为了位于2008奥运主办场地附近的北京西翠娱乐中心而设计。这一项目拥有世界上最大的彩色液晶显示屏之一和中国第一套集成到玻璃幕墙的光电系统。整个建筑通过白天吸收太阳能、晚上用它来照亮屏幕，反映出一天的天气循环，从而以自给自足的能源组织系统来运作。 这一项目旨在推广节能技术在中国新一代建筑中的集成，这一做法势在必行，有力地回应了当前业界激进、无节制、常常以环境污染为代价的经济发展模式。

在德国顶级制造商Schueco and SunWays公司的支持下，建筑师Simone Giostra 及Arup公司开发了一项将光电单元层压到玻璃幕墙的新技术，并由中国制造商尚德太阳能电力有限公司检查了第一批玻璃太阳能面板产品。多晶光电单元层压到幕墙玻璃当中，并于整个建筑的表面上通过变化的密度进行放置。密度设计允许在内部序需要时获得自然光，从而降低热能获取和太阳能向幕墙所需电能的过多转化，增强了整个建筑的表现力。

Family Box

家·盒子文化有限公司

Location: Beijing
Area: 5,625m²
Architect: Crossboundaries Architects+BIAD International Studio
Design Team: Binke Lenhardt, Wang Fang, Zheng Feng, Giacomo Butte, Huang Shanyun
Photographer: Yang Chaoying
Completion Date: 2009

地点：北京
建筑面积：5,625 平方米
建筑师：科罗斯邦建筑师事务所 BIAD 国际工作室
设计团队：蓝冰可、王芳、郑峰、Giacomo Butte/薄昭昭、黄珊芸
摄影：杨超英
建成时间：2009 年

Beijing, China

中国，北京

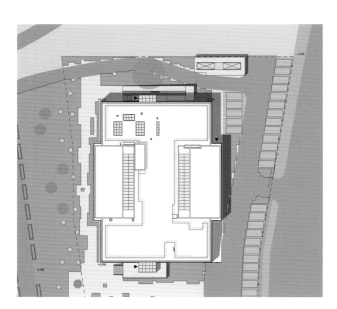

Family Box is something between an indoor playground and a kindergarten for children up to twelve years old, but it also accomodates their parents' needs. It hosts different kinds of activities – from swimming, playing games to various classes ranging from music, dancing, crafting to cooking. Furthermore, it has a big playframe, a reading area and a generous café area. Located at the outer corner of a park, it is placed in a natural environment, which enhances the visibility of the building. The use of independent rooms in the shape of freestanding boxes allows the activities to run parallel and it offers the most suitable environment for each. The rooms have their own programme or theme, all differing from the outside space in terms of colour and furniture. They have their own story and inside life and allow the children to concentrate on the programme offered. At the same time, small square window openings allow maintaining contact with the outside, and parents can have a peep inside to see what is going on. The box locations are meant to break the rigid layout of the concrete columns, which is also camouflaged with a series of arches that give a different rhythm to the environment.

Visually, the common areas are treated with low contrast finishes in order to enhance and balance the space and equipments for the children.

家盒子的定位介于游乐场和幼儿园之间，在为12岁以下儿童创造一片天地的同时，也为家长提供相配套的服务。在这里，会有不同的活动——游泳、游戏，以及涵盖音乐、手工、厨艺等内容的课程班，乃至一个巨大的攀登架、书店和宽敞的咖啡休闲区。它位于环境优美的城市公园的边角上，具有良好的景观视野。设计师把多种多样的活动装进各自独立的盒子，以使它们能够不互相干扰并拥有各自适合的空间。每一个房间都有其特定的功能和主题，从房间的外观颜色和家具能很轻易将其区分。不同的房间有不同的故事和活动，独立的空间环境使孩子们得以集中注意力于他们自己的活动。盒子上的小方窗让房间内外保持着联系，家长们可以从方窗窥见盒子内孩子们的活动。

盒子的布局打破了混凝土柱网的严整规矩，另外，装饰性的拱型型也赋予了空间多变的节奏。

在视觉上，普通区域被处理成反差较小的效果，目的是强调和平衡孩子使用的区域。

Shangri-La Hotel of Supershine Upper East

阳光上东香格里拉酒店

Location: Beijing
Area: 70,000m²
Architect: Ricardo Bofill Architecture
Completion Date: 2007
Client: Beijing Xintai Real Estate Development Co., Ltd.

地点：北京
建筑面积：70,000 平方米
建筑师：里卡多波菲（北京）建筑设计咨询有限公司
建成时间：2007 年
客户：北京新泰房地产开发有限公司

Beijing, China
中国，北京

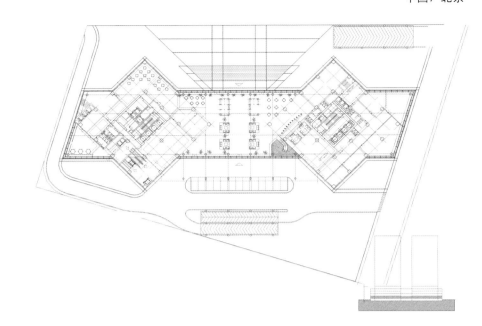

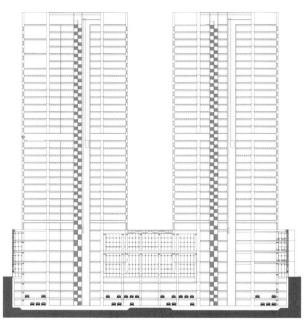

Luxury garden hotel complex. The programme forms two twin towers that act as a gateway on the northeastern artery connecting Beijing with the international airport. Five-star hotel brand services provide exclusive service to both 400 rooms and 200 apartments. A glazed podium contains all common areas such as hotel lobby, apartment lobby, business centre, ballroom (500 px), speciality restaurant, 24-hour dining rooms, retail, spa and fitness and parking.

这是一个奢华园林酒店综合建筑项目。该项目由两个双子楼构成，是连接北京与国际机场东北部主要干道的重要门户。酒店内设有400间客房和200个套间，享受全五星级的品牌服务。一个通透的墩坐墙内囊括了所有的公共区，其中包括酒店大堂、门厅、商务中心、宴会厅、特色餐厅、24小时餐厅、零售店、水疗馆与健身中心以及停车场等。

LATTICE in Beijing
北京格子

Location: Beijing
Area: 8,000m²
Architect: SAKO Architecture
Photographer: Misae Hiromatsu / Beijing NDC Studio, Inc.
Completion Date: 2007

地点：北京
建筑面积：8,000 平方米
建筑师：SAKO 建筑事务所
摄影：广松美佐江 / 北京和创图文制作有限公司
建成时间：2007 年

Beijing, China
中国，北京

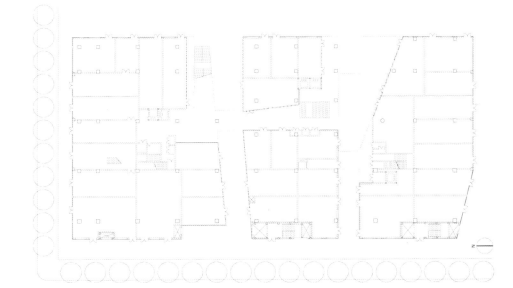

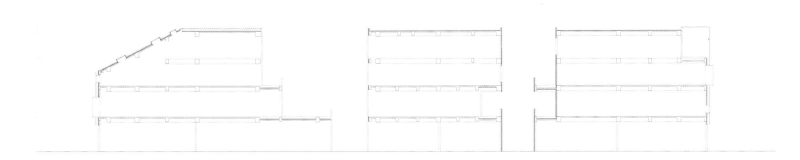

This project is a part of the redevelopment of Sanlitun, a popular entertainment area located in the embassy district of Beijing. Designers from Beijing, Tokyo and New York have come together and designed individual commercial facilities within this complex building. The dynamic mix of various styles has defined the new Sanlitun. At the beginning, the client wanted to maintain Chinese elements while incorporating bar facilities. Hence, the designers used gold-coloured mirror-finish stainless panels and cast-iron screens decorated with traditional Chinese window patterns to create a Chinese atmosphere with a sense of modernity. Some parts of the mirrors were deliberately pressed into uneven surfaces, creating a façade that shifts throughout the day with the changing light of the sun. The reflected light gives the complex completely different looks during the day and at night. At night, lights between cast iron screens and mirrors will clearly outline the cast-iron screens. Additionally, the gold and silver colour mirrors facing inside of streets were deliberately pressed. A large number of concaves create the lattice pattern like the cast-iron screens. At night, this "surface" can be sensitive to capture the lighting changes.

这是位于北京使馆区的三里屯酒吧街的重新开发项目。北京、东京、纽约等地的知名设计事务所参加了此次设计，各自的商业设计风格都极富个性。各种风格"混和"产生的活力，构成新三里屯的特征。设计伊始，甲方向设计师提出"设计时既要将中国元素运用进去，又要考虑到它将是一座集合了众多酒吧的建筑"。为此，设计师采用了金色的镜面不锈钢，以及中国传统的棂条花格式的铸铁网。其中部分镜面不锈钢的表面被轧制成凹凸状。这样不仅可以敏锐地捕捉到不同时刻光影的变化，同时还可以将其效果加强。这样的建筑立面在白昼、黑夜可以展现给我们完全不同的风情。夜晚，在铸铁网与镜面不锈钢中间的照明的作用下，铸铁网的轮廓就浮现出来。面朝街区内侧的部分，在金银镜面不锈钢的表面做轧制加工处理。通过大量的凹面，浮现出与铸铁网同样的格子图案。在夜晚，其"表层"能够敏感地捕捉到光影的变化。

Atlas of Contemporary Chinese Architecture

No.23 Qianmen East Street

北京市前门东大街 23 号

Location: Beijing
Area: 10,233m²
Architect: Cui Kai, Ye Zheng, Li Xiaomei, Tao Jingyang, Peng Bo
Completion Date: 2008

地点：北京
建筑面积：10,223 平方米
建筑师：崔愷、叶铮、李晓梅、陶景阳、彭勃
建成时间：2008 年

Beijing, China
中国，北京

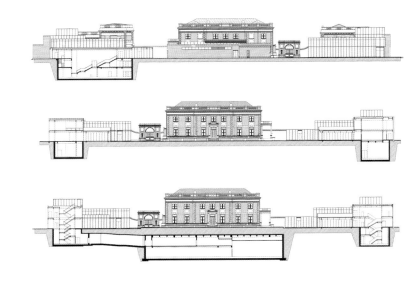

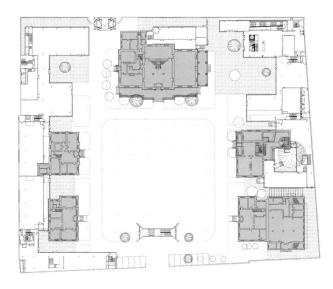

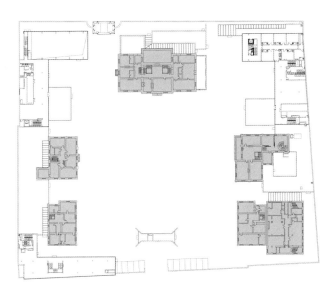

Initially built as the American Legation in the beginning of the 20th century, No.23 Qianmen East Street housed lots of institutions before the renovation, while its original layout, including five independent buildings and a quadrangle, has been damaged after several planless additions. The renovation focuses on restoring the quadrangle and making it a high-grand cultural and fashion centre in the centre of Beijing, which provides restaurants, gallery, theatre and entertainment services.

The old buildings are renewed and those low-grade constructions are cleaned up. Brick walls behind the buildings screen the disorderly parts that incorporate the exterior spaces of the quadrangle with nine aging trees. According to the preservation regulation of this historic district, the delicate north gate and north wall are restored.

The added volumes step down to make themselves subsidiaries of the old ones. Accesses for them are designed via the five old entries that provide enough chances for visitors to be inspired by the historic atmosphere. Designers juxtaposed this respect for tradition with a distinctly modern vocabulary of steel and glass structural elements, which could not breach the reserved buildings. Hiding behind the shade of trees, the crystal volumes restrain their expression and treat the historical scenes as the primary interior views. The state-of-the-art techniques and materials also give them a distinct feature from the 1920s' style, maintaining the definition of history.

项目地处北京市核心区——前门东大街23号，原址为20世纪初美国公使馆所在地，现有五栋北京市文物保护建筑、九棵挂牌古树，多家机构在此办公。改造方案通过对院落的复建和整治，形成包括餐饮、俱乐部、画廊、剧场在内的顶级文化、生活时尚中心。

首先通过恢复领事馆建筑风貌，净化周边环境，突出历史建筑的主体地位。在文保建筑后侧砌筑砖墙，遮挡零碎增建部分，重新恢复一院五楼的原貌，保留古树，拆除低质量旧建筑。恢复重建北大门和北围墙，使之符合东交民巷历史保护区整体风貌的要求。

增加的新建筑采用前低后高、缩小体量的策略，使之处于从属地位。新老建筑内部相连，主入口仍在五个老楼中，保持院落原有格局，给予人流充分体验老建筑历史氛围的机会。新建部分与老建筑相连处采用钢和玻璃的轻型构造，精心保护老建筑的完整性，并以其立面作为新建筑室内空间的主要景观。新建筑在形态上尽可能通透、轻巧，隐在大树之后，克制地表现个体。同时，简约精致的技术和材料也能明显区别建筑的年代，保持历史的清晰度和延续性。

XIU
"秀"吧

Location: Beijing
Area: 1,300m²
Architect: Zhu Xiaodi, Mi Ning, Zhong Fei, Yang Bo
Completion Date: 2009

地点：北京
建筑面积：1,300 平方米
建筑师：朱小地、宓宁、钟菲、杨波
建成时间：2009 年

Beijing, China
中国，北京

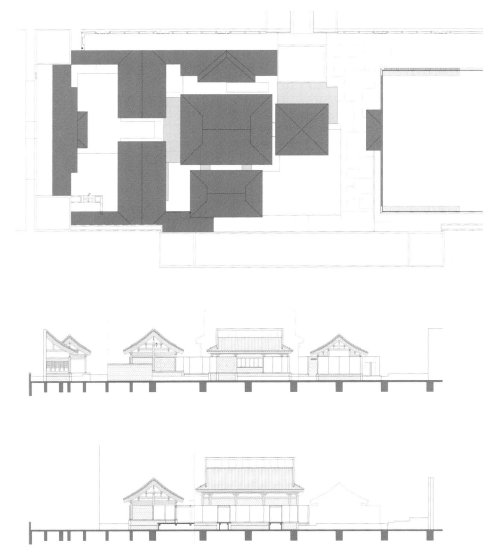

XIU is a small project with a building area of 1,300m². Beijing Yintai Centre is located in the southwest of the intersection of Chang'an Avenue and the East Third Ring road. It is a three-towered structure with the central tower rising 249.5m high and the two flanking towers about 186m. XIU rests on the roof of the annex building that is surrounded by the three high-rise buildings. Since using the traditional Chinese architectural forms to shape a bar's unique personality among highly modern buildings is regarded as a desirable way, laying the basis for a rich dialogue between the ultra-large-scale modern buildings and the intimate scale of traditional architecture becomes the key factor of design. Based on the spatial relationship controlled by the central axis of the three high-rise buildings at Yintai Centre, the designer brings this concept to the design of the bar. The design rearranges the functional areas of the bar according to the special counterpoint and transition of the axis. Thus, the dialogue relationship with environment from the aspects of plane and special layout is established. As soon as the layout is completed, this project turns to lay emphasis on the subordinate relations of the building roofs of different parts, and adopts different shapes of roofs to integrate these buildings into an entirety. This design not only gives response to the spatial relationship between the external image of the bar and the surrounding three high-rise buildings, but fully meets the needs of the bar's quality and flexible application. The bar consists of three basic functional areas, namely, reception area, bar area and assistance area. The interior roof truss woodworks adopt the rectangular cross-section, and the exposed wood components as well as the trim panels are based on the elm board and treated with antique finish to expose the wood grain and then finally covered with matte varnish.

"秀"是一个建筑面积仅为1300平方米的小工程。银泰中心位于北京市长安街与东三环路交汇处的西南方向，包括三栋超高层建筑，其中中间一栋高249.5米为酒店、公寓，东、西两栋同为高186米的办公建筑，"秀"项目就位于三栋超高层建筑围合之下的裙楼屋顶。既然在高度现代化的建筑群中以中国传统建筑形式来表现酒吧独特的个性被视为一条可取的途径，那么寻找超大规模的现代建筑群与尺度亲切的传统建筑对话的可能性就成为设计的关键。设计师将已经建成的三栋超高层塔楼强烈的轴线关系作为前提条件，进一步引申到酒吧平面中。利用居中的南北向的轴线和东西向的轴线，将酒吧各功能按照轴线空间的对位与转合进行重新布置，从而在平面和空间格局上确立与环境的对话关系。在平面布局形成之后，设计师进一步强调传统建筑群中处于不同部位的建筑屋顶的从属关系，采用不同形制的屋顶，使这组建筑形成整体。这一设计方向既回应了酒吧建筑的外部形象与周围三栋高能高层建筑的空间关系，又充分满足了酒吧建筑的品质和它灵活的使用要求。室内屋架木作均采用矩形断面，所有露明木构件、木饰面板均采用榆木板，作旧处理，见木纹，亚光清漆罩面。

National Museum of China
中国国家博物馆

Location: Beijing
Area: 191,900m²
Architect: gmp Architekten-von Gerkan, Marg and Partners
Photographer: Christian Gahl
Completion Date: 2011

地点：北京
建筑面积：191,900 平方米
建筑师：gmp 建筑设计事务所
摄影：克里斯琴·加尔
建成时间：2011 年

Beijing, China
中国，北京

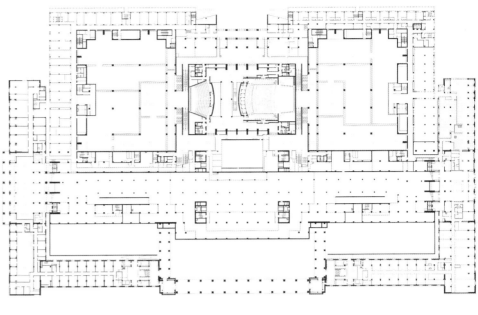

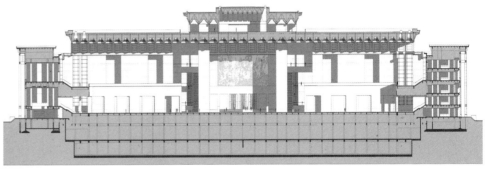

048 / NORTH CHINA

The reconstruction and extension of the National Museum of China in Beijing merges the former Chinese History Museum and Chinese Revolutionary Museum. Originally completed in 1959 as one of ten major public buildings in Tian'anmen Square in the immediate vicinity of the Forbidden City, the structure is still a milestone in modern Chinese architectural history. Elements of Chinese architectural tradition blend with a Western, neoclassical architectural idiom. The task was to combine the northern and southern wings into an integral complex of buildings by removing the central structure to make the Chinese National Museum.
The 260-metre (850ft)-long hall acts as its central access area. It widens in the centre to embrace the existing central front entrance facing Tian'anmen Square.
The materiality of the roof, which in the Forbidden City and the existing building consisted of glazed roofing shingle in imperial yellow, is re-interpreted with slightly curved, bronze-coloured metal plates. This meant the flighted roof typologies of the buildings in Tian'anmen Square and the Forbidden City were continued in the new building, yet interpreted in a contemporary fashion in the detail and materials. The harmonious use of materials in the interiors – wood, stone and glass – is found throughout the building, creating a natural feeling of identity and familiarity. Rooms of special significance are emphasised by the use of differentiated materials.

原有建筑建于1959年，前身为南翼的中国历史博物馆和北翼的中国革命博物馆。两座博物馆建筑被一分为二，泾渭分明。改建工程的目的为通过一个置于中央的空间元素将两栋建筑合二为一，促成一个宏伟庄严的综合体——中国国家博物馆。一座长达260米的艺术长廊作为建筑体的中央交通连接贯通南北。长廊正中位置扩展成为一座入口大厅，原有建筑的柱廊大门得以保留，加强了博物馆与其正对的天安门广场之间的联系。
博物馆的屋顶由略带弧度的古铜色金属铝板覆盖，是对紫禁城宫殿群和保留建筑金黄色琉璃瓦屋面的重新演绎。建筑屋顶的叠加退阶设计借鉴紫禁城殿宇重檐庑顶的同时也充分考虑到了天安门广场建筑屋面样式，但其局部细节与建筑材料均富有时代特色。
木材、石材和玻璃三种材质的搭配在博物馆内部空间的塑造上被贯彻始终，赋予了建筑独一无二的形象特征，令人印象深刻。

Beijing International Automotive Expo Centre Automotive Museum

北京国际汽车展览中心汽车博物馆

Location: Beijing
Area: 500,000m²
Architect: Henn Architekten
Photographer: Beijing Institute of Architectural Design, Henn Architekten
Completion Date: 2007
Award: Competition, First Prize

地点：北京
建筑面积：500,000 平方米
建筑师：德国海茵建筑设计公司
摄影：北京建筑设计院、德国海茵建筑设计公司
建成时间：2007 年
获奖：竞赛一等奖

Beijing, China

中国，北京

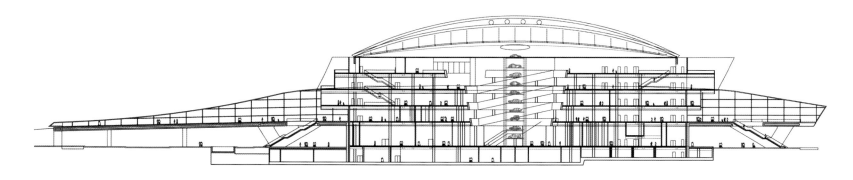

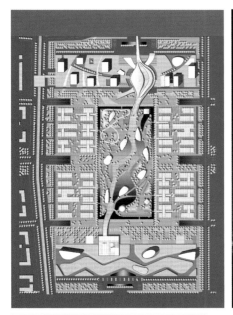

The Automotive Museum will be realised as the first building of the Beijing International Automotive Expo. The project which results from the 2004 competition award has been revised from its original planning. In view of dimension, roads, production and marketing potentials, optimised requirements have been formulated and implemented in a draft design: the ground floor is now structured into two building parts linked with one another via an open passageway. The southern and larger part is dominated by a generous entrance hall linked with the above exhibition areas by an elliptical open space widening to the top. In support of a flexible use of this space, all development elements were moved to the periphery. The immediately adjacent conference centre offers variously sized rooms and seating for approx. 650 participants. To the northern segment a 4-D cinema is planned. It can be operated under separate management.

The first floor is open to the public and in addition to ticketing and retail areas and hosts a restaurant which can be operated at two different pricing categories. The necessary supply (delivery, kitchen) as well as catering for employees is located in the same part of the building, in the basement. The museum's administration with approx. 120 employees was integrated over two storeys into the museum's building and offers direct contact with the exhibition areas. The cores are arranged to allow for an independent development. The actual exhibition areas follow on the second floor and end on the fourth floor in a stilted freely covered room.

作为北京国际汽车展览中心内第一个建成的项目，汽车博物馆的规划源于 2004 年竞赛，但其设计方案被重新修订。鉴于建筑的规模、道路的设计以及生产及营销的潜在能力，设计师试图满足及优化各种需求。一层被分割成两个结构，中间通过开放式通道连结，南侧较宽敞的部分作为入口大厅，通过椭圆形的开放式空间与上层的展区连接。为实现空间灵活性需求，所有的设施全部摆放在四周。会议中心包括不同规格的房间，可容纳 650 位与会者。北侧计划打造成独立经营的四维影院。
二层主要为公共区，包括售票处、零售店以及餐厅。博物馆办公区可容纳 120 名员工，共为两层，与展览区相邻。员工食堂等设在地下室。展览区从三层一直延伸到五层。

China Science and Technology Museum

中国科学技术馆

Location: Beijing
Area: 102,800m²
Architect: RTKL
Photographer: Fu Xing and Shu He Photography
Completion Date: 2009

地点：北京
建筑面积：102,800 平方米
建筑师：美国 RTKL 国际建筑事务所
摄影：傅兴及舒赫摄影工作室
建成时间：2009 年

Beijing, China
中国，北京

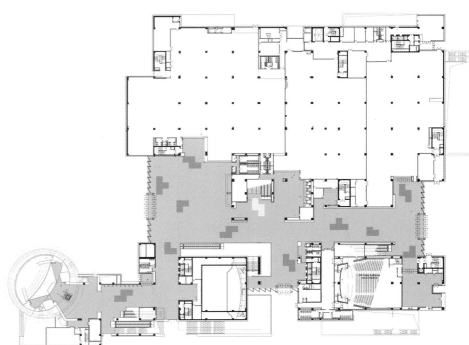

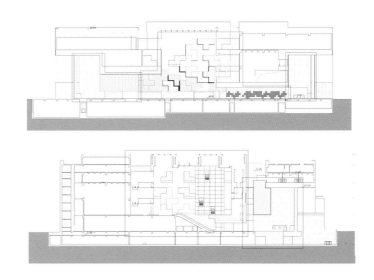

Located in the centre of Beijing's Olympic Village, the China Science and Technology Museum expresses China's accomplishments in these fields through state-of-the-art exhibits, labs and training areas, retail, restaurants and a cinema. Its main structure is in the shape of a single gigantic cube which combines the scientific thinking of the ancient Chinese with the features of a modern science and technology museum, and the entire structure is divided into a number of building blocks that occlude each other like toy bricks, making it appear like a huge cubic jigsaw puzzle. Such a structure is the embodiment of the intrinsic correlations between man and nature as well as science and technology; it is also symbolic of the fact that science has no absolute boundaries and that different disciplines intermingle and promote each other.

The conception of the new Museum is "to experience science and inspire innovations; to serve the general public and promote harmony". As the most prominent education vehicle for the new Museum, the permanent exhibitions will be organised in line with the thematic ideas of "innovation and harmony" and under six themes of display, namely "Children's Science Paradise", "the Glory of China", "Exploration and Discovery", "Science, Technology and Life", "the Challenges and the Future" and "the Beautifulness of Science", covering a total exhibition floorage of approximately 30,000 square metres.

中国科学技术馆位于北京奥运村内，主要展示国家在相关领域的成就，包括现代展示中心、实验室、培训区、零售店、餐厅及电影院。主结构是一个独立的巨大立方体，这一设计体现了中国古人思维方式与现代特色的结合。另外，整体被进一步分割成一系列的小体量结构，犹如积木一样堆砌在一起，使得整个建筑看起来更像是一个方形的拼图。这一设计体现了人与自然、科学与技术之间与生俱来的联系，更象征着"科学无界限，不同领域相辅相成"的理念。

这一建筑的设计构想即为"体验科学，激励创新，服务大众，促进和谐"。永久性展览作为博物馆内主要教育方式，以"创新及和谐"为系统理念，共分为六个主题馆，分别为"儿童科技天堂"、"中国辉煌"、"探索与发现"、"科学、技术与生活"、"挑战与未来"、"科学之美"。

National Zoological Museum of China and Institute of Zoology, Chinese Academy of Sciences
中国科学院动物所及国家动物博物馆

Location: Beijing
Area: 42,900m²
Architect: Institute of Architectural Design, Chinese Academy of Sciences
Completion Date: 2004
Award: 2009 Beijing Outstanding Engineering Construction Design First Prize, 2009 Beijing Ministry of Construction Honourable Mention

地点：北京
建筑面积：42,900 平方米
建筑师：中科院设计院
建成时间：2004 年
获奖：2009 年获北京市优秀工程建筑设计一等奖，2009 年获北京市建设部优秀奖

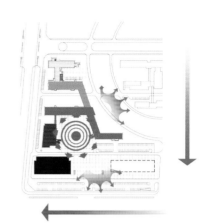

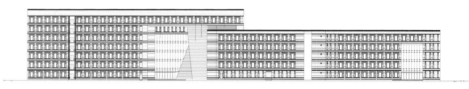
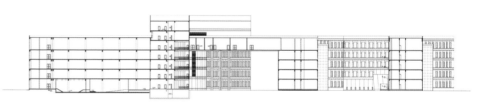

Beijing, China
中国，北京

The astronomy, geography and life park of the Chinese Academy of Sciences, located west of the Olympic Park and south of the Olympic Village, is a national-level scientific experiment park. As the largest park in the scientific community, the Institute of Zoology draws several attributes from the surrounding environment to transform it into a scientific research complex with unique scientific intentions.

Architectural form is the reflection of inherent structure and functions. Like skeleton, structure is the supporting system which pays more attention to transition and connection between the units in the architectural construction, functioning as "joints". The interface of buildings is like epidermis providing the true embodiment of skeletal system and functional logic and presenting the same skin and different skin textures in different units. The epidermis unit for standard scientific research forms a multi-level module relationship in terms of changes in the mean value, so that the classification of a stone unit can meet the coordination and uniform requirement for window opening and the completion of lighting, ventilation, and shading systems. The National Zoological Museum is harmonious but different in form, as it is a true restoration of the large space museum building. Because of its study of life sciences, the Museum applies a bent structure to construct a grand space in a gradual, regulated transformation. As a kind of frame-like system, it shows biological specimens and also reconstructs a new DNA organic building.

中国科学院天文、地理与生命公园坐落在奥运村南部的奥林匹克公园，是一座国家级的科学实验园。作为科学共同体的最大园区——动物研究所，从周边环境中吸取若干属性，并转化为具有独特科学内涵的科研群体。

建筑形式是内在结构和功能的折射。就像是骨骼一样，结构支撑着整个体系，更加注重在建筑过程中的过渡与连接，起着"关节"的作用。建筑物的表面，如同人的表皮如实的展现着骨骼系统以及逻辑功能，并且展现着在不同部分相同的与不同的肌肤肌理。表皮单位根据变化的平均值为标准的科学研究形成了一个多层次模块关系，这样的表皮能够满足开窗、照明、通风以及遮阳等要求。国家动物博物馆即是和谐的又存在着不同的形式，因为它是对于大空间博物馆建筑的准确的修复。由于它对生命科学进行研究，博物馆采用了弯曲的结构，通过平缓的规则的变化构建了宏伟的空间。作为一种框架似的系统，它展示了生物标本同时又重建了一个新的DNA有机建筑。

Beijing Hutong Bubble 32
北京胡同泡泡 32 号

Location: Beijing
Architect: MAD
Director in Charge: Ma Yansong, Dang Qun
Design Team: Dai Pu, Yu Kui, Stefanie Helga Paul, He Wei, Shen Jianghai
Completion Date: 2006

地点：北京
建筑师：MAD 建筑事务所
主持建筑师：马岩松、党群
设计团队：戴璞、于魁、Stefanie Helga Paul、何威、申江海
建成时间：2006 年

Beijing, China
中国，北京

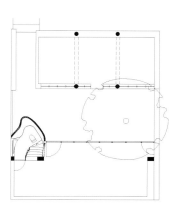
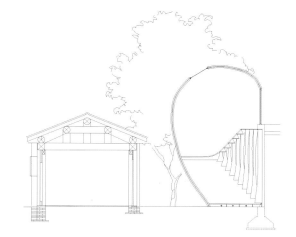

MAD's proposal for the future Beijing 2050 was first revealed at its exhibition MAD IN CHINA in Venice during the 2006 Venice Architecture Biennale. Beijing 2050 imagined three scenarios for the future of Beijing – a green public park in Tian'anmen Square, a series of floating islands above the city's CBD, and the "Future of Hutongs", which featured metallic bubbles scattered over Beijing's oldest neighbourhoods. Three years later, the first hutong bubble appeared in a small courtyard in Beijing.

China's rapid development has altered the city's landscape on a massive scale, continually eroding the delicate urban tissue of old Beijing. Such dramatic changes have forced aging architecture to rely on chaotic, spontaneous renovations to survive the ever-changing neighbourhood. In addition, poor standards of hygiene have turned unique living space and potential thriving communities into a serious urban problem. Hutongs are gradually becoming the local inhabitants' dumpster, haven for the wealthy, and a theme park for tourists. The self-perpetuating degradation of the city's urban tissue requires a change in the living conditions of local residents. Progress does not necessarily call for large-scale construction – it can occur as interventions at a small scale. The hutong bubbles, inserted into the urban fabric, function like magnets, attracting new people, activities, and resources to reactivate entire neighbourhoods. They exist in symbiosis with the old housing. Fueled by the energy they helped to renew, the bubbles multiply and morph to provide for the community's various needs, thereby allowing local residents to continue living in these old neighbourhoods. In time, these interventions will become part of Beijing's long history, newly formed membranes within the city's urban tissue.

Unexpectedly, a manifestation of this idealistic vision has sprung up in one of Beijing's hutongs, just three years after the exhibition. Hutong Bubble 32 provides a toilet and a staircase that extends onto a roof terrace for a newly renovated courtyard house. Its shiny exterior renders it as an alien creature, and yet at the same time, reflects the surrounding wood, brick, and greenery. The past and the future can thus coexist in a finite, yet dream-like world.

The real dream, however, is for the hutong bubble to link this culturally rich city to each individual's vision of a better Beijing. The bubble is not regarded as a singular object, but as a means to initiate a renewed and energetic community. Under the hatchet of fast-paced development, we must always be cognizant of Beijing's long-term goals and the direction of its creativity. Perhaps we should shift our gaze away from the attraction of new monuments and focus on the everyday lives of the city's residents.

在2006年威尼斯建筑双年展期间，MAD的城市概念作品"北京2050"首次亮相于威尼斯个展MAD IN CHINA中。其中如同水滴一样散落在北京老城区的胡同泡泡，在三年之后，出现在位于北京老城区的北兵马司胡同32号的小院里。

"北京2050"描绘了三个关于北京城市未来的梦想——一个被绿色森林覆盖的天安门广场，在北京CBD上空飘浮的空中之城，和植入到四合院的胡同泡泡。

经济发展所推动的大规模城市开发，正在逐步逼近北京传统的城市肌理。陈旧的建筑，混乱的搭建，邻里关系的变迁，必要卫生设施的缺乏，导致这种原本美好安详的生活空间变成了很大的城市问题——四合院正在逐渐成为了老百姓的地狱，有钱人的私密天堂，游客们的主题公园。

面对这种源自城市细胞的衰退与滥用，需要从生活的层面去改变现实。并不一定要采取大尺度的重建，而是可以插入一些小尺度的元素，像磁铁一样去更新生活条件、激活邻里关系；与其他的老房子相得益彰，给各自以生命。同时这些元素应该有繁殖的可能，在适应多种生活需求的基础上，通过改变局部的情况而达到整体社区的复苏。由此，世代生活在这里的居民可以继续快乐地生活在这里，这些元素也将成为历史的一部分，成为新陈代谢的城市细胞。

出人意料的是，这种微观乌托邦的理想，在展览的三年之后便开始出现在了北京的四合院中。第一个"32号泡泡"是一个加建的卫生间和通向屋顶平台的楼梯，它看上去仿佛是一个来自外太空的小生命体，光滑的金属曲面折射着院子里古老的建筑以及树木和天空；让历史、自然以及未来并存于一个梦幻的世界里。

胡同泡泡真正的城市理想是把北京的古城与每个人的梦想连接在一起，在大刀阔斧的城市巨变中，我们必须重新思考北京长期的目标和想象力在哪里。也许我们可以把目光的焦点从那些大型的纪念碑式建筑移开，而开始关注人们日常生活的改善和社区生活的重建。

Placid Rivers Club
泰禾俱乐部

Location: Beijing
Area: 12,096m²
Architect: Atelier FCJZ
Director in Charge: Yung Ho Chang
Photographer: Shu He
Completion Date: 2008

地点：北京
建筑面积：12,096 平方米
建筑师：非常建筑设计研究所
主持建筑师：张永和
摄影：舒赫
建成时间：2008 年

Beijing, China
中国，北京

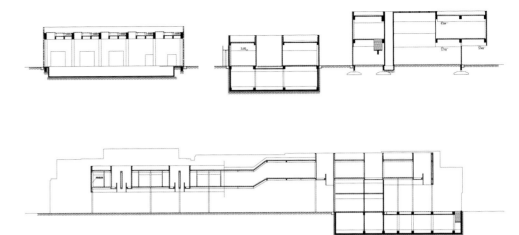

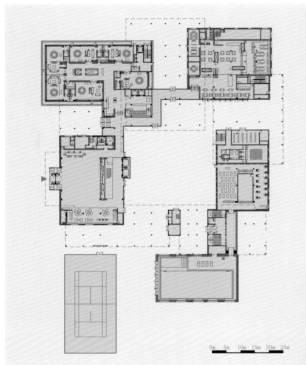

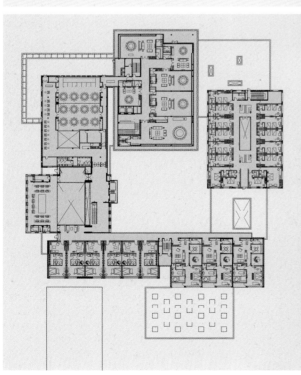

058 / NORTH CHINA

Following the architectural tradition of Beijing courtyard house, the design is based on courtyard structure. However, the present courtyard is different from the traditional one: the density of the city today does not allow for the expansion of buildings but requires them to be developed vertically. Thus, the Placid Rivers Club is divided into 12 courtyards which connect and stack with each other and meanwhile embrace a big central courtyard. The stacked structure forms a group of semi-outdoor spaces of different qualities, which can also be taken as multiple changes of traditional corridors. The guests can enjoy the fun of roaming among the courtyards, the convenience of various indoor and outdoor activities, as well as the placidity and harmony of the environment. Therefore, the architecture is also called the 12-Box Courtyard.

The 12 courtyards are endowed with four kinds of façades: concrete blocks, red cedar panel outer walls, aluminium panel walls and glass walls. All the four kinds of façades are based on the WEAVE concept, resulting in rich textural qualities in a step's distance.

The building structure consists of both concrete and steel structures, and is accompanied with shear walls. The exposed steel columns are presented in the semi-outdoor spaces in a most slender appearance. The exposed shear walls are presented with fair-faced concrete, and the hole-structure in them clearly reflects the structural rules of the principle bars inside.

延承了北京四合院的建筑传统，泰禾俱乐部的设计也以院落为空间结构。但是，今天的院落建筑与传统四合院建筑是不同的：当代城市的密度不支持建筑的水平展开，今天的院落建筑必须向垂直发展。泰禾俱乐部最终以一组十二个相互咬合叠加的院落建筑构成，它们再围合出一个中心的大庭院。院落建筑的叠摆进而构成了一系列不同质量的半室外空间，也可以看作是传统廊的多种变化。客人在泰禾俱乐部享受到的是庭院间穿梭游走的乐趣、各种室内外活动场所的便利、以及整体环境的和谐幽静。于是此项目也被称为：拾贰院。

十二个院落建筑分别赋予了四种外饰面：混凝土砌块外墙、红雪松木板外墙、铝单板幕墙、玻璃幕墙。这四种外墙的构造特点都是以编织为概念，构成近人尺度的丰富建筑质感。

建筑结构为钢筋混凝土框架与钢框架混合，并辅以剪力墙。外露钢柱以最纤细的姿态融入半室外活动空间。外露的剪力墙均以清水混凝土呈现，剪力墙的孔洞布置清晰的反映出其内受力钢筋的构造规则。

Hongluo Clubhouse
红螺会所

Beijing, China
中国，北京

Location: Beijing
Area: 189m²
Director in Charge: Ma Yansong, Yosuke Hayano
Design Team: Florian Pucher, Shen Jun, Christian Taubert, Marco Zuttioni, Yu Kui
Associate Architect: IDEA International Design Studio
Completion Date: 2006

地点：北京
建筑面积：189 平方米
主持建筑师：马岩松，早野洋介
设计团队：Florian Pucher、沈军、Christian Taubert、Marco Zuttioni、于魁
合作建筑师：主题工作室
建成时间：2006 年

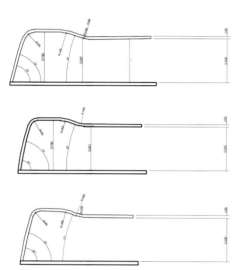
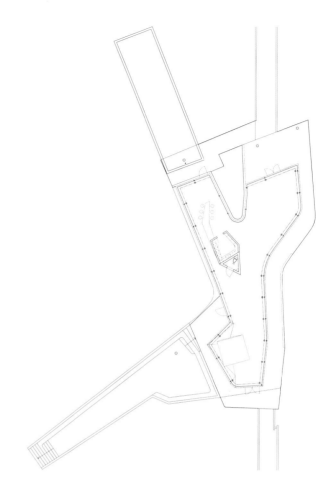

Hongluo Villa District is located to the north of Beijing, situated along the shore of Hongluo Lake with an impressive view of the mountains beyond. A wooden bridge traverses the lake, and the clubhouse is situated on an irregular platform in the middle of the bridge. The entire structure appears to float on the water, reflecting the surrounding mountains, and becoming a focal point for the whole area.

A continuous, reflective surface rises up out of the water, becoming first the roof and then the walls of the house. This surface blurs the distinction between solid and liquid states, between building and environment. An X shape is formed beneath, from which two further branches extend into the lake. A sunken garden lies 1.3 metres below the surface of the lake, giving visitors the impression of walking through water to access the building. At the other end lies the swimming pool. Inserted flush with the lake, the surfaces of both natural and artificial water are kept at the same level.

The interior of the house is equally fluid, a continuous space without internal boundaries. People are encouraged to create their own pathway through the house and experience it as they wish. Hongluo Clubhouse is an ever-changing architectural space, which reflects and responds to the beauty of its surroundings, forming a meeting point between nature and people.

距城市中心仅一小时车程的红螺湖别墅区，位于密云地区一片湖光山色的自然环境之中。别墅沿湖而建，在湖岸的一端，一座木桥通向水中一块487平方米的不规则平台，MAD第一个建成的项目——红螺会所——就坐落在这个平台之上。建筑主体是一片仿佛由流动的水面凝固而成的银色三维曲面，它在水平和竖直方向之间延伸扭转，模糊了屋顶和墙壁，室内与室外，固态和液态之间的界限。

跟随这个银色表面的运动，人们可以进入两个方向的分支空间——一个位于水平面下1.3米的下沉庭院和一个漂浮在湖水中的游泳池。下沉庭院连接着水岸的另一端，当人们从岸边拾级而下步入这里，视线与湖面在同一水平面上，身体的一半仿佛在水下。游泳池的水面和湖水也在同一平面上，在这里，人工和自然的界限再次出现模糊。这两条主要路线在主体建筑处汇合，形成一个向上生长的三维有机结构，并成为连接室内各个空间功能的线索。室内空间的使用功能由变化的屋顶高度暗示，唯一的封闭空间由屋顶下陷所形成的不规则墙体包围。

红螺会所是一个跟随人们的行走路线不停变幻的建筑空间，它摆脱了现代主义城市建筑的工业属性——被效率和利益所分割管理的"缺省秩序"。它更接近于自然的规则，一种"弱规则"。置身其中，人们被鼓励有选择地体验空间，发现新的感受。每一个人都能通过他们即时的灵感和情绪创造出新的空间路径。红螺会所以新的空间逻辑回应周围的自然文脉，成为了人与自然的交汇点。它把建筑中的生活转变为在自然中的漫步，给生活在水泥丛林里的人们以希望和启发。

Qingcuiyaun Clubhouse
晴翠园会所

Location: Beijing
Area: 2,400m²
Architect: Sunlay Architecture Design co., ltd.
Photographer: Shu He
Completion Date: 2008

地点：北京
建筑面积：2,400 平方米
建筑师：三磊建筑设计有限公司
摄影：舒赫
建成时间：2008 年

Beijing, China
中国，北京

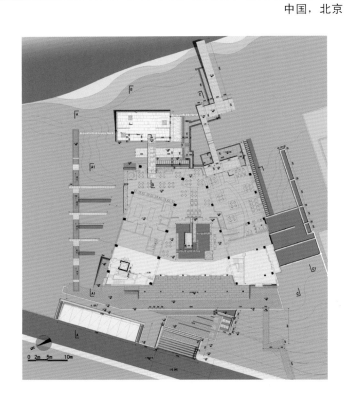

Qingcuiyaun Clubhouse is located on the west bank of Wenyu River, Chaoyang District, which has excellent condition of graceful natural landscape. The design concept is not only to create a unique building, but also to integrate it into this attractive landscape environment. The design shows two topics, Collision and Blend, the collision of different cultures and styles, the blend of time and space. It is telling a story which is ancient and modern. It is describing an aesthetics encounter between East and West.

Architectural design scheme considers "axis and symmetry" which are emphasised by classical architectural language, "fluxion and consistency" which are sought by modern architectural language. Red brick walls and transparent glazing are employed to create coagulated building mass and flowing spaces.

There is a beautiful inner courtyard which forms the centre of the whole building. It offers the scenery landscape for the entry hall and lets the light into the building as well. In this way, the project shows more meanings and it has a proper size of the centre of the building, complexing with the excellent appliance such as the inner plant garden with a sunlight dining room on the ground floor, the open café which extends itself, green landscape around the villa, water and the original natural things. All these elements make the building more green landscaped and all the elements can join together with each other very well.

晴翠园会所位于朝阳区温榆河西岸，自然景观格外优美。设计的理念即为打造一幢特色十足的建筑，同时使其完全融入到迷人的景观环境中。其设计展现两个主题"碰撞"与"融合"，突出不同文化与风格的碰撞，实现时间与空间的融合。犹如讲述着一个故事，充满着传统与前卫；更像是描述一种美观，东方特色与西方意蕴在此相遇。

古老建筑语言强调"轴线"与"对称"；现代建筑风格彰显"变化"与"连贯"。红砖墙与透明玻璃被同时运用，营造坚固建筑外观以及流畅的室内空间。室内庭院构成了整个建筑的中心，为入口大厅带来了美丽的景致，同时将自然光线引入到室内。一层，室内植物园、阳光餐厅、开敞咖啡厅、人工种植景观、自然景观交相呼应。所有这些元素相互呼应，使得建筑本身更加自然。

No.18 Guanshuyuan Hutong
官书院胡同 18 号

Beijing, China
中国，北京

Location: Beijing
Area: 160m²
Architect: Liu Yuyang, Zhao Gang / Atelier Liu Yuyang Architects
Photographer: Jeremy San
Completion Date: 2010

地点：北京
建筑面积：160 平方米
建筑师：刘宇扬、赵刚（刘宇扬建筑事务所）
摄影：杰里米·桑
建成时间：2010 年

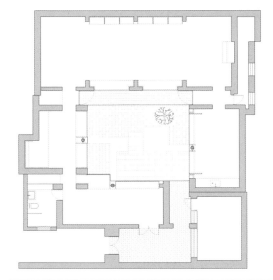

The project calls for the renovation of a small courtyard house in Guanshuyuan Hutong, Beijing, which is located between the famed Yonghe Tibetan Temple and the historical Imperial Academy. The client is an art collector from Shanghai, whose wish is to reinvent this "pocket-house" into a gallery with a "Shanghai ambiance" for his ceramic art works.

Design starts with two parallel aspects: on one hand, to infill a series of copper-plated steel display windows and veneered wood display shelves within the original wooden frames of the house, thus creating an integrated contrast of tradition and modernity; on the other hand, to recreate a courtyard reminiscent of the Jiangnan (south of Yangzi) traditional gardens. The transparent glass plays with the coppering stainless steel, echoing and contrasting with the ceramic theme display and respecting for the basic form of the quadrangle courtyard.

In view of the arrangement of the exhibition hall and lines, a French window measuring 2 metres by 2.5 metres is placed without any partition, welcoming the visitors in the reception hall from the vestibule and bringing the sense of space and scale of courtyard into the a bit crowded hall. The following series of windows change at different scale and depth, forming transparent "niches" echoing with different exhibition halls. Seen from these different perspectives, a simple yard and only one willow will present various scenic views. Changes of the paving materials for the courtyard floor have subtly deconstructed and re-interpretated the beautiful atmosphere of Jiangnan that the owner originally expects. In the wing hall at the end of the whole building, exhibition rooms have been converted into tea rooms and glazing windows have also been replaced by openable folding glass doors, favouring people wondering between the interior and the courtyard according to weather conditions.

一个建筑面积不到200平方米的小四合院，位于北京雍和宫与国子监之间的官书院胡同的近端。新业主是上海的一位文化艺术爱好者，渴望将此小小的口袋宅打造成为一座海派的陶瓷展示会所。

设计主要从两方面入手，一方面，在原四合院的建筑框架中，填充出一系列金属展示窗及木质展示柜，形成传统与现代两种风格的反差与结合；另一方面，在四合院当中营造出一种江南意境的园林景象。这些透明玻璃窗及镀铜不锈钢框的搭配，既是对陶瓷展示主题的回应和对比，也是对四合院最本质的空间形态——院子——的致敬。

在展厅和动线的安排上，当参观者穿过前庭进入接待前厅时，他首先面对的就是一扇两米乘两米五，无任何分隔的落地玻璃窗，把院子的空间感和尺度感一下子带进来原本略嫌拥挤的前厅。接下来的一系列玻璃窗，开始有尺寸和进深的变化，形成呼应不同展厅的通透"壁龛"空间。而从这些不同角度望出去，一个简单的院子和其中唯一的柳树也开始有了变化。透过院子地面铺装的材料变换，业主原先追求的江南意境也就微妙的被解构和重新诠释了。到了最后的侧厅，空间的功能已由展示转变为茶室，玻璃窗也变成可全部开启的折叠玻璃门，人们的活动也就可依天气情况自由游走于室内和院子之间了。

Living in Seclusion – "Xun" Private Lounge
"蛰居"之处——"旬"会所

Location: Beijing
Area: 1,664m²
Architect: Zhu Xiaodi, Gao Bo
Completion Date: 2010

地点：北京
建筑面积：1,664 平方米
建筑师：朱小地、高博
建成时间：2010 年

Beijing, China
中国，北京

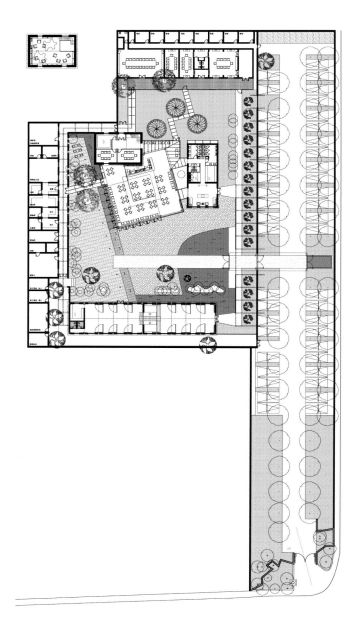

This project is located at old station maintenance buildings which are composed of four parts, namely, the south bungalow with duo-pitched roof and wooden structure, the middle bungalow and a small building with duo-pitched roof and attic, the north renovated bungalow with pent roof, and the west houses running south and north. According to the characteristics of the surrounding buildings, the duo-pitched roof house in the south is used for exhibition, for the reason that its historic gable roof can add to the space a sense of history; the renovated pent roof bungalow is used as VIP rooms for reserving the cross wall; the west houses are used for kitchens and other auxiliary function rooms. A centre of three buildings with existing and new ones is formed by setting the new building in the central area in the square site, which will be used as the reception space, main bar area, wine cellar and cigar bar, playing the role of club that goes harmoniously with the scale of the original south and north buildings. In order to highlight the main bar building which is the only new building in the courtyard and meet the requirement of viewing from the courtyard centre, the designer has deliberately rotated the main bar towards east which causes a visual conflict between the new building and the original buildings. The exterior façade of the new building adopts the cheapest material – corrugated iron sheet, which reflects the silver light and thus gives the courtyard a sense of energy with a special texture. Here, trunk, branch, leaf as well as the shadows of trees have been employed as the design elements. All of the corridors that link various functional rooms together closely connect with the landscape and create a natural atmosphere enveloping the guests. The porch of the main entrance extends from the "ceremony door" to the fence of forecourt in the east and ends with the square pool. The waving decorative fabric on the porch glows under the lights, sending its greetings to the urban guests.

这是一处旧有的车站维修用房，院子里的原有建筑大体分为四部分，南侧是一栋两坡顶木屋架平房；中部是一栋平房，另一栋是一个带阁楼的两坡顶小楼；北侧是一栋改造过的单坡顶平房；西侧则是一排南北走向的西房。根据现场建筑的特点，南部的坡屋顶房屋安排展览的功能，其颇具历史感的人字形屋架给室内空间增添了几分沧桑之感；北部的单坡屋顶的房屋由于必须保留横隔墙，不能整体使用，最适合作为VIP房间；西侧建筑可以安排厨房和其他辅助功能房间；通过一个正方的平面格局的新建筑镶嵌到院子中部、二层坡顶小楼的南侧，东面与小平顶房相连通，形成有新、旧三栋房子组合的中心，这里将是会所的接待空间、主吧空间和酒窖、雪茄吧，在平面功能上与南、北两部分原有建筑形成比例适宜的会所功能。主吧的建筑是整个院子中唯一的新建筑，为了突出这栋建筑的地位，同时也考虑院子中心观赏角度的需要，设计师将新建的主吧特意向东旋转了一个角度，造成了在外观上与旧有建筑空间格局的冲突。新建筑的外饰面采用了最低廉的瓦楞铁皮的材料，在阳关的照耀下泛着微弱的银色的光，特有的质感给院子带来了生气。设计师将树干、树枝、甚至树叶和树影都作为设计的要素，加以利用。所有联系各功能房间的通廊都与景观密切结合，尽可能穿行其间，让客人足不出门就可感受到环境的存在。主入口的门廊从"仪门"直接延伸到东侧前院的围墙处，以方形的水池结束。门廊上随风飘曳的装饰织物在夜晚灯光的照耀下，发出变幻的光芒，迎接着都市客人的到来。

G-Dot Art Space
贵点艺术空间

Location: Beijng
Area: 1,500m²
Architect: Atelier 100s 1
Design team: Peng Lele, Huang Yi, Lin Chunguang, Wang Tiantian
Photographer: Zeng Renzhen, Huang Yuan
Completion date: 2010

地点：北京
建筑面积：1,500 平方米
建筑师：北京百子甲壹建筑工作室
设计团队：彭乐乐、黄燚、林春光、王田田
摄影：曾仁臻、黄源
建成时间：2010 年

Beijing, China
中国，北京

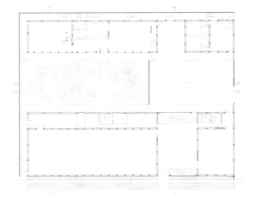
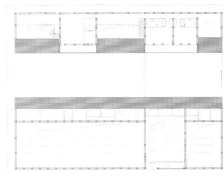

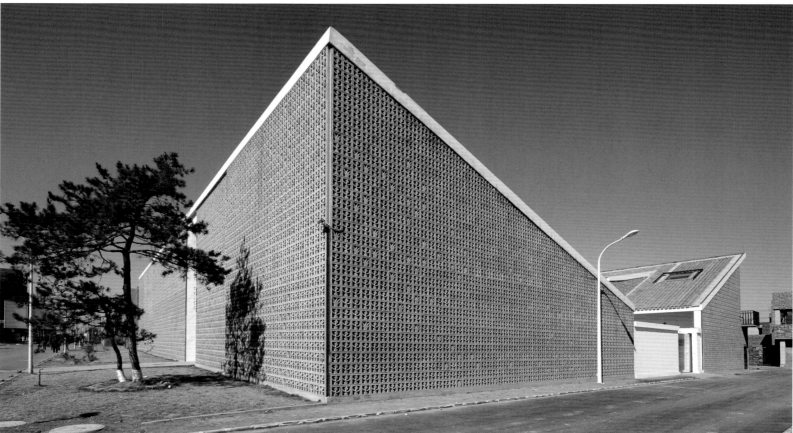

Many artists rent houses in Songzhuang for living and painting. When the architects saw the first time that artists painted in the form of grass and flowers on enclosing walls constructed by farmers with green and red colours, they found that it is very commonly seen in Songzhuang. Inspired by this, the project was finally realised with building blocks in the form of grass and flowers. The project is composed of two buildings respectively on the north and south of the site. The south building as exhibition space is a structure with the southern part higher than the northern part, accommodating exhibition in the higher space and office, storage and washroom in the lower part. The north building as residence shows a totally different form with living room and reading room in the higher part, sitting room and private gallery in the lower part. In the exhibition space, all natural light comes from skylights to leave the wall for exhibiting. In order to differentiate the spatial features among lobby, small exhibition hall and large exhibition hall, the skylights are distinctly defined and combined. Living room and private gallery are also designed with skylights to make light well distributed.

在宋庄，有许多艺术家租用农民的房子居住、画画。当发现第一块农民用于围墙的花、草图案的砌块被艺术家用红、绿颜料描绘出花、草之后，设计师发觉这在宋庄以及周边区域应用得非常普遍。"贵点艺术空间"坐落在宋庄，因此选用花、草图案的砌块。在长方形的地块上南北两边各布置一栋建筑。南侧建筑为展览空间，北侧建筑为居住功能。展览空间主体要求为高空间，为了不遮挡北侧居住部分的阳光，将展览空间设计为南高北低，高的部分为展示，低的部分为办公、储存、卫生间等辅助功能。居住部分北高南低，高的部分为小起居，主人读书室等；低的部分为单层通高的客厅，私人画室等。在长方形的地块上南北两边各布置一栋建筑。南侧建筑为展览空间，北侧建筑为居住功能。

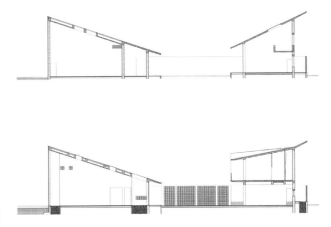

Inside-out Art Centre
中间艺术馆

Location: Beijing
Area: 2,400m²
Architect: Xu Lei, Yu Haiwei, Li Lei, Meng Haigang / Atelier 11
Photographer: Zhang Guangyuan
Completion Date: 2007

地点：北京
建筑面积：2,400 平方米
建筑师：徐磊、于海为、李磊、孟海港 / 中国建筑设计研究院拾壹建筑工作室
摄影：张广源
建成时间：2007 年

Beijing, China
中国，北京

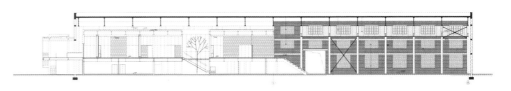

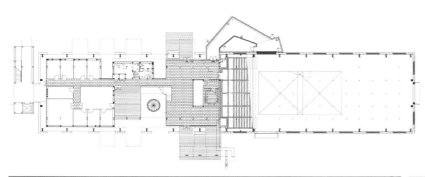

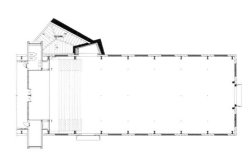

As a renovation project, the renovation started from the west side of the building and converted the space into a sales office. The design intention during this stage was to keep the original building envelope as much as possible and to use steel frame in the interior as structural elements. This allows the architects to add a box-shaped volume inside the existing building and create new rooms for sales service by extending the space outside the existing windows, as an experiment on a new approach for renovation.

In the second stage of the renovation starting one year after the renovation of the sales office, a multi-functional art centre was built on the east part of the building. The Art Centre and sales office are connected by an external passage in the shape of a zigzag tube. It is not only a functional feature for circulation, but also creates a twisted and narrow space that makes the eventual access to the grand exhibition hall an interesting experience for the visitors, with a dramatic contrast of space scale.

A simplistic aesthetic was adopted for the interiors of the Art Centre with the intention to keep some characteristics of the previous industrial building, as a reflection to the history of the site. The original windows, walls, and floors are kept with modest renovation. The combination of daylight through the high side-windows and artificial lightings on steel-framed tracks fulfills the lighting requirement for exhibitions. As a supplement to the external passage, wide steps are also built to internally connect the spaces of the exhibition hall and the first floor of the sales office.

这是一个改造项目，先期在厂房的西半部分进行改造，成为整个大项目的销售中心。这里采用的方法是保留原有的围护，在内部重新作钢结构，相当于套进去一个新的盒子，再从厂房的大窗中伸出来一些房间，在改造的模式上尝试了一种新的模式。

厂房的东侧在一年以后开始考虑，最终决定做一个多功能的美术馆。设计师通过砌筑的通宽台阶将东部大空间和西侧销售中心的二层联系起来，这不仅仅作为流线通道，同时更形成了一个通往展示大厅的蜿蜒曲折的空间，通过空间规模的对比为参观者营造了趣味十足的独特体验。

建筑内部通过简单的处理：封窗，刷白，做地面；再通过做高侧窗的光线引导，作钢结构的葡萄架承载照明等方法达到基本的物理要求。入口是从销售中心引过来的，从厂房外部折过来，既是最简单的做法，也形成了一个曲折逼仄的空间，和最终到达的展示空间形成戏剧性的对比。

Songzhuang Artist Residence
宋庄艺术公社

Location: Beijing
Area: 5,300m²
Architect: Xu Tiantian
Photographer: Zhou Ruogu, Iwan Baan
Completion Date: 2009

地点：北京
建筑面积：5,300 平方米
建筑师：徐甜甜
摄影：周若谷、伊万·班
建成时间：2009 年

Beijing, China
中国，北京

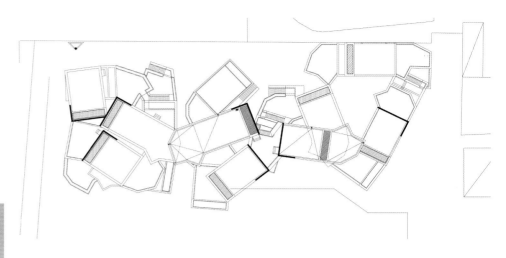

Located right next to the East Sixth Ring Road of Beijing, Songzhuang Artist Village is undergoing a dramatic expansion of artist population and increasing demand of artist's working and living space, a 20-unit artist residence facing a fishpond at a former outdoor storage lot is one of the local development targeting such demand.

The programmatic requirement of working and living defines the height and geometry of both volumes: 6m height for working and 3m for living; a simple rectangular box for studio and a complex geometry for living indicating bedroom, kitchen and toilet. Living volume is plugged into working volume either on the same level or led by stairs to upper level.

These 20 units are regarded as containers stacking up on this former industrial outdoor storage lot to express individuality of 20 units. The grey colour of painted stucco on exterior is required by planning code, but among these volumes, horizontal surfaces are coloured red to enclose a series of outdoor area; the interplay of volume and void, light and shadow allows artists and visitors to constantly explore and experiment the outdoor community space, which could be the extension of art production and presentation connecting these 20 units as showrooms on open studio days.

In other words, this complex becomes an alternative museum for living art creation and exhibitions.

随着中国当代艺术家的激增，艺术家对居住和工作空间的需求也随之扩大。坐落在北京六环以东的宋庄艺术家村就正经历着由于这种增长所带来的变化，20个面向鱼塘的艺术家工作室应运而生。

按照工作室的居住和工作两种主要功能，每一个单元都包含两种不同的尺度与组合：工作部分为6米高的简洁的方盒子，居住部分高3米，有一系列几何形空间容纳起居厨厕功能。体量高差决定了每个单元的工作和居住功能或在同层联系或通过楼梯联系。整个建筑立面采用压型钢板，水平活动平面则是当地典型的红砖。这20个工作室就像叠加在一起的集装箱，不仅和场地原有的工业气息相呼应，而且创造富有张力的外形和独特的建筑空间。体量的虚与实，光与影，创造变化的室外交流空间，为艺术展示提供了更多的可能性。当地艺术节开放工作室时，这片区域可以为参观者带来多样的空间体验，同时成为丰富的创作展览现场。这个带有艺术创作和居住功能的建筑何尝不是位于艺术村的一种另类美术馆！

Songzhuang Art Museum
宋庄美术馆

Location: Beijing
Area: 5,000m²
Architect: Xu Tiantian, Chen Yingnan, Zhu Junjie / DnA _Design and Architecture
Photographer: Zhou Ruogu
Completion Date: 2006

地点：北京
建筑面积：5,000 平米
建筑师：徐甜甜、陈英男、朱俊杰 / 徐甜甜建筑事务所
摄影：周若谷
建成时间：2006 年

Beijing, China
中国，北京

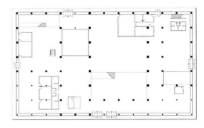
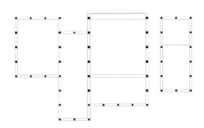
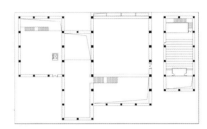
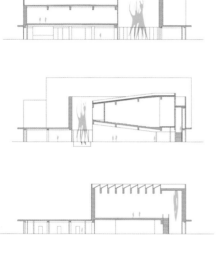

Songzhuang Art Museum is the first public art facility built in this well-recognised contemporary Chinese art village 45 minutes driving out of Beijing. With around 4,000 artists living in Songzhuang area by 2008, this Art Centre will provide a platform for local community.
The exhibition space is lifted up to create a welcoming horizontal flow with multi-purpose space on ground level, to host exhibition, events and common area. The second level is a rather introversive art space with lighting coming from the skylight and courtyard. The surrounding landscape will not intrude; here, art is the view.

在这个距离北京45分钟车程的著名中国现代艺术村内，宋庄艺术馆是首座公共艺术设施。截至2008年，约有4千名艺术家生活在宋庄地区，这一艺术中心的建成也将为当地社区提供活动平台。
展览空间设置较高，形成表现热情的水平流线。一层有多功能展厅，可举办展览、举行活动，或作为公共休闲区域使用。二层空间较为封闭，仅由天窗或庭院灯照明。在这里，周边的景观已经并不重要，因为艺术是这里最美的风景。

Qi Zhilong Residence and Gallery
祁志龙画室

Beijing, China
中国，北京

Location: Beijing
Area: 789m²
Architect: Atelier 100s 1 / Peng Lele, Huang Yi, Wan Lu
Photographer: Zeng Renzhen
Completion Date: 2009

地点：北京
建筑面积：789 平方米
建筑师：北京百子甲壹建筑工作室 / 彭乐乐、黄燚、万露
摄影：曾仁臻
建成时间：2009 年

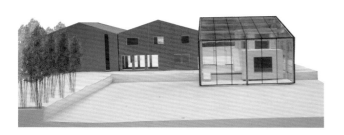
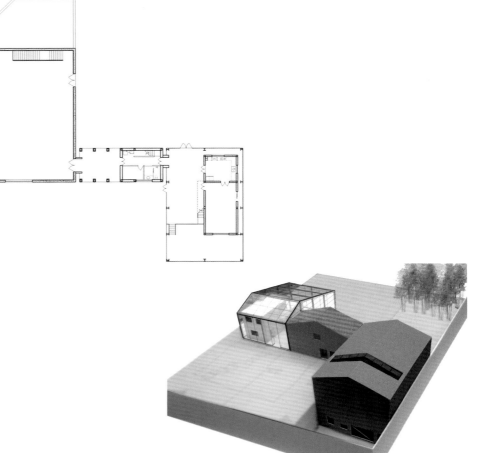

Besides referring to people inhabiting together in spite of diversified life styles and ways of living, "mix" here also means combination of buildings of different architectural styles. The project emphasises "the state of mix" by mixing itself into the existing houses with double pitch roof typically seen in Songzhuang.

The project consists of three houses: the concrete house with shearwall structure serves as gallery; the red-brick house for bedroom is a masonry structure; the glass house as living space boasts steel structure.

The three houses are combined together for different functions, of different sizes, in different structures as well as by different materials. The surfaces of the three houses are completed without any furnishes, with the construction materials exposed directly. In addition, the three houses are linked together and the reasonable organisation makes the realisation of front courtyard and back courtyard.

混杂除了指不同生活方式以及不同生存方式的人居住在一起，还指这里意识形态和物质形态同等重要的存在，也指不同的建筑形态并置在一起。"祁志龙画室"在建筑设计上强调了混杂状态，建成的建筑混入宋庄已有的典型双坡屋顶的建筑型态。

三个房子分别是：其一，混凝土房子，功能为主人画室，结构为剪力墙结构；其二，红砖房子，功能为主人居住部分，结构为砖混结构；其三，玻璃房子，功能为客人的起居居住部分，结构为钢结构。三个房子以三个不同功能要求，三个不同空间大小，三个不同结构方式及三个不同的建筑材料并置在一起。三个房子都是以构造材料直接形成外观材质。三个房子互相之间有连接体插入。三个房子在基地的合理布置而形成了前院和后院即私人院和公共院的关系，并在中间红砖部分首层前后院贯通。

Hejingyuan Contemporary Art Gallery

北京和静园现代美术馆

Location: Beijing
Area: 3,142m²
Architect: Kang Kai, Wang Jiang, Zhang Liping
Photographer: Fu Xing
Completion Date: 2008

地点：北京
建筑面积：3,142 平方米
建筑师：康慨、汪江、张丽苹
摄影：傅兴
建成时间：2008 年

Beijing, China

中国，北京

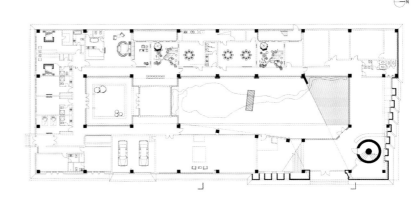

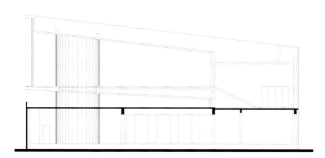

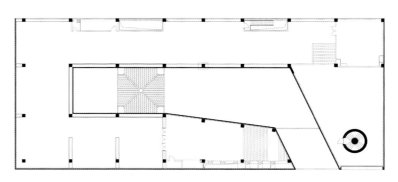

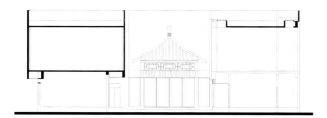

Hejingyuan Contemporary Art Gallery is located in Songzhuang Village, which is the "Artists Village" of Tongzhou Region in the east suburb of Beijing. The design traces back to the consideration of basic construction elements, namely, ratio, scale, surface, space, light, colour, etc., emphasising the transformation between design elements and objective truth.

The art gallery is composed of two floors, and the lower floor is designed to be a quadrangle courtyard, a traditional architectural form in Beijing. The entrance of the residence area is located on the southeast. Both the front courtyard and the backyard, as well as the rooms and corridors, all emphasise the integral relationship among the blue sky, lotus pond, green trees and the artificial mountain, ignoring the construction forms, leaving only one middle hall of the ancient style as the narrative theme. Ascending along the eastern public entrance and stairs to the upper floor exhibition hall, the visitors will enter a loop-shape exhibition corridor, whose wash wall daylight provides the best lighting condition for the artworks on display. The soft daylight floats naturally along the large-scale rectangular white wall that is used for exhibiting the paintings, the sculpture exhibition hall at the end rising up, and the pool that faces east stretching in a horizontal direction. By means of vertically defining, the upper conservatively traditional residence space and the lower open exhibition space are integrated seamlessly. While from the aspect of space conception, they take the opposite connotation as individual spaces. The façade of this building takes the fair-faced concrete as the material and adopts the cast-in technique, giving a sense of solidity and elegance, brilliant but not boastful, simple but not rough.

和静园当代艺术馆位于北京东郊通州区"画家村"——宋庄。设计回溯到建筑基本要素之间的关系，即比例、尺度、表皮、空间、光线、色彩等，强调设计因素与客观真实性之间的转换。艺术馆为上下两层布置，底层为北京传统的四合院格局。居住空间入口设在南面东侧，前庭后院、堂舍廊厅均强调蓝天荷塘、碧树顽石的交融关系，忽略建筑形式的存在，仅留一个仿古的中堂作为叙事性主题。由东侧公共入口沿斜坡楼梯向上进入二层展厅。封闭幽深的"回"字形展廊空间辗转闭合，仅留墙边的洗墙天光为展品提供最佳显色性。柔和的天光沿着用以展示绘画作品的大片矩形白色墙面自然流动向前，尽端的雕塑展厅向上昂起，面向东侧的池塘敞开视野。通过对空间的竖向界定，居住空间的传统内敛与展览空间的流动开放上下并置，在形态上融为一体。而它们在空间概念上却背道而驰、各自独立。建筑外立面为一次浇铸成功的清水混凝土，呈现出坚实沉稳、清雅孤傲的建筑形象，辩而不华，质而不俚。

Westhill Art Workshop, Beijing

北京西山艺术工坊

Beijing, China

中国，北京

Location: Beijing
Area: 24,225m²
Architect: China Architecture Design & Reserch Group/ Cui Shihong, Yu Guanfei, Deng Ye, Lian Li
Photographer: Zhang Guangyuan
Completion Date: 2009

地点：北京
建筑面积：24,225平方米
建筑师：中国建筑设计研究院 / 崔时红、喻关飞、邓烨、连荔
摄影：张广源
建成时间：2009年

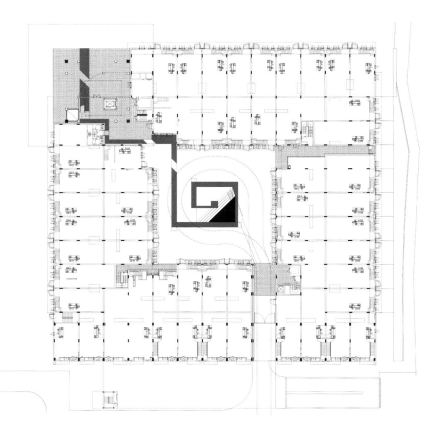

This project is designed for an art workshop that focuses on the mainstream art. It is located in the middle of Si Ji Qing Xiang, Haidian District in Beijing. This foursquare construction covers 1.29 hectares, housing 90 workshops. It is an intervenient design proposal of master planning and single planning. Above the ground are the sectional artists' workshops, which are mainly divided into lower floor workshops (ground floor and the first floor) and upper floor workshops (the second floor and the third floor); the basement part is taken up by the garage and equipment rooms as well as the storerooms for the lower workshops. Each single workshop is adjacent to the other by the way of group planning, thus forming the streets and courtyard. The ground floor is designed as an "art field" aiming at the public, and the upper floor is planned asliving spaces for the artists themselves. Meantime, these units seamlessly get together and create an identified urban architecture.

A public art exhibition hall at the northwest of the building is elevated to the height of the second floor, and then the lower part forms a dramatic entrance public space, which brings the surrounding city environment into the inner space and expresses the open attitude and the responsibility of the workshop towards the community.

这是一个定位于主流艺术的艺术工作室聚落。项目位于北京市海淀区四季青乡中部。用地约1.29公顷的正方形，需容纳90间工作室。方案介乎于规划设计与单体设计之间，地上为组合式艺术家工作室，主要分下层工作室（首层和二层）和上层工作室（三层和四层）；地下室为汽车库及设备机房，还有部分下层工作室的自用储藏室。所有的工作室单体都是通过群体规划的方式毗邻布置，形成街道与大院。底层形成面对大众"艺术圈子"，二层内街则是艺术家们自己的"生活圈子"。同时，这些单元如拼盘般聚合成一个具有标识性的城市建筑。

建筑西北角的一个公共艺术展厅被提升至三层的高度，下方形成一个戏剧化的入口公共空间，将周边的城市环境引入到整个聚落的内部，同时也表达了艺术工坊对于社区和城市应有的开放态度和责任。

MOSAIC in Beijing
北京马赛克

Location: Beijing
Area: 100,359m²
Architect: SAKO Architects
Photographer: Shu He, Misae Hiromatsu / Beijing NDC Studio, inc.
Completion Date: 2008

地点：北京
建筑面积：100,359 平方米
建筑师：SAKO 建筑事务所
摄影：舒赫、广松美佐江、北京和创图文制作有限公司
建成时间：2008 年

Beijing, China
中国，北京

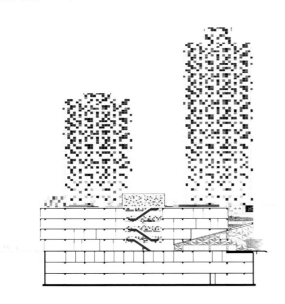
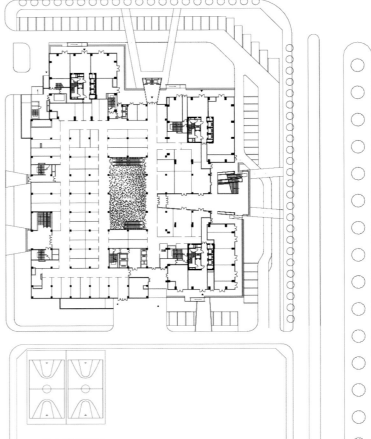

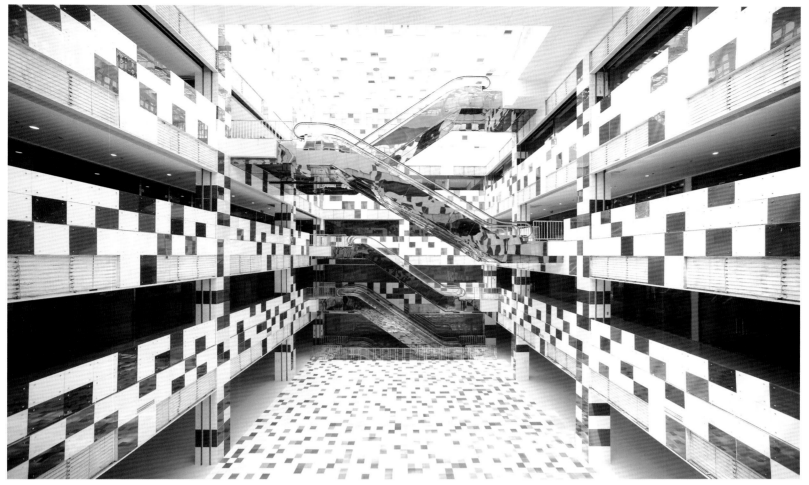

Beijing Economic & Technological Development Area has vast and flat landscape. Because of the large blocks, wide roads, offices and showrooms of major manufacturers calmly built by low density, the area feels uncomfortable. Since it is far beyond the scale with which people stand still, it cannot be said that the town is pleasant to walk at all. Since the building frame of RC construction had been built to the fourth floor, this building was left in such scenery for a long time. The emptiness worsens the impression of this town more. It is a complex facility with an office building built on a department store. Because neither of those business spheres was enough for this newborn town, the project had been stopped. It is requested to restart, renovate, and convert the unfinished building.

The motif of "Mosaic" was adopted through out the project. The big volume consisting of "mosaic" is to convey the image of the landmark filled with festivity like the spectacle of fluttering confetti. In order to enlarge the sphere of business, the purpose to demonstrate its presence in other regions has been also supposed. Two circulations were added in the department store. People can directly access from the main entrance which has the form of a huge opening mouth to the basement through the third floor. The intricate reflecting surface articulates, amplifies and emits the movement of people and the various lights. The "Mosaic" is consistently used as the motif of signs, landscape and interior designs.

北京的经济开发区，广漠而萧条。宽大的街道上，零零散散地坐落着几座大公司办公楼和样板间，寂寥而杀风景，实在是无法称它为一个能带给人们快乐和休闲娱乐的空间。这一项目正是位于这样的环境里，混凝土结构体建到四层就被停工，长期放任不管。它是由低层百货商店与高层办公楼相结合的复合设施。这个区域的商圈不完备，导致了工程中途被停工。于是，甲方委托SAKO建筑事务所重新启动、重建和修改这一工程。

充分利用这一巨大体块的特点，使它成为此地区的标志性建筑，这是否有实现的可能性呢？这不仅能扩大这一商圈，还能有效地提高它的存在感。整个工程采用了"马赛克"这一主题。彩纸屑在漫天飞舞的光景，轻飘、娇弱、空虚，但它同时也具有某种象征意义。办公楼改为一室一厅公寓，而商业部分则分割成各个店铺进行出售。小资本的划分与积累，也可称为"马赛克"式商业方式。商业增加了两条动线。其中，主要入口向外张开大口，人们可以从外部直接进入地下一层到三层的各个楼层，而入口的内装立面由很多不同角度的反射面组成，它将人们的身影、各种光线进行分割、扩大的同时，发散至四面八方。

Bamboo Wall
竹屋

Location: Beijing
Area: 32,600m²
Architect: Kengo Kuma & Associates
Photographer: Satoshi Asakawa
Completion Date: 2003
Client and Developer: SOHO China Ltd.

地点：北京
建筑面积：32,600 平米
建筑师：畏研吾建筑都市事务所
摄影：浅川敏
建成时间：2003 年
客户及开发商：SOHO 中国

Beijing, China
中国，北京

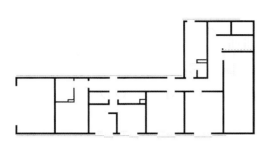

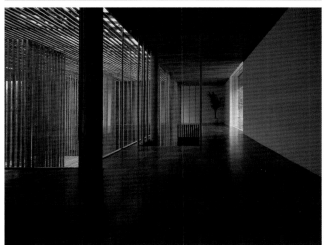

In this project, ten chosen Asian architects were commissioned to design ten residences each, creating a hundred residences all together in a forest adjacent to the Great Wall of China, an environmentally mutual commune.

The basic notion of this project was to leave the original geographical features intact and to utilise the locally produced materials as much as possible. The idea of leaving the land intact is a common consent with the planning ideology of the Great Wall.

All the 20th century houses in the suburbs had been built on flattened land. Planning and locating architectural objects like that seems to be a common way. The architects felt, however, such a method wouldn't be suitable for the beautiful land of China with an intricate undulation. In their concept, it was hence the best solution to build a wall without interfering with the original geographical feature, but instead enhancing it.

There are several reasons they chose bamboo as the principal material. First of all, they found charm in the material's weakness. The Great Wall built with solid stone and brick had been a device to sever the world of civilisation and savage, while the bamboo filter would allow light and wind to pass through. Also, the bamboo filter could work as a connection between the two worlds.

This building was intended to be a symbol of cultural interchange.

十名亚洲建筑师受邀在长城附近的森林内打造住宅，每位设计师负责十栋，总共100栋住宅将形成一个绿色集体公社。竹屋，就是在这一大背景下诞生的。

设计的基本理念即为保留场地原有的地域特色，充分运用当地石材，已与长城附近地区生态规划概念相互呼应。

20世纪所有的住宅全部在平坦的地块上建造，因此打造同类结构似乎是更为大众化的方式。然而，作为建筑师他们却不这样为。在他们眼里，这种规划方式并不适用于中国延绵起伏的地形。最后，他们决定不破坏原有的地形，而是顺势而建，整幢住宅犹如一道墙，更加突出了这里的地形特色。

竹子的选择来自于多方面的考虑。首先，材质的差异性。长城主要采用砖石打造，被用作"文明"与"野蛮"两个世界之间的防御者，而竹子则突出通透性，让光线和风自由"穿梭"。其次，竹子也可以作为连结"文明"与"野蛮"两个世界的桥梁。

这一建筑还被视为"文化交流"的象征。

BUMPS in Beijing
北京冲击

Location: Beijing
Site Area: 17,949m²
Architect: SAKO Architects
Photographer: Misa Hiromatsu / Beijing NDC Studio, Inc
Completion Date: 2008

Beijing, China
中国，北京

地点：北京
占地面积：17,949 平方米
建筑师：SAKO 建筑事务所
摄影：广松美佐江 / 北京和创图文制作有限公司
建成时间：2008 年

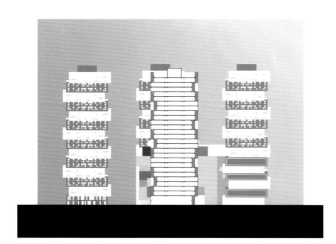

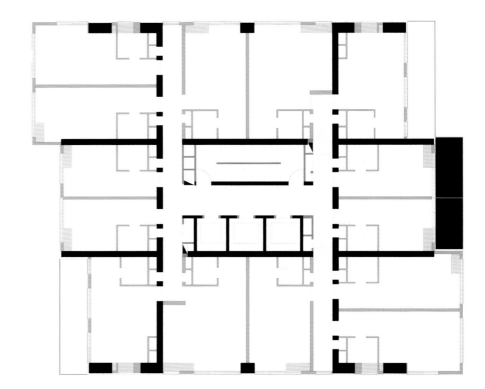

Four residential towers and shops complex 45-degree inclined from south-north axis. Roads from every corner of the project are connected to the central shop, and the design allows light from the sun to shine on every surface of the residential towers. The 80-metre-tall towers are made of cubic blocks, stacked and staggered in a repeating pattern. Each block comprises two levels and there is a 2-metre-height difference between each two blocks. The part of setback was used as terrace. Scale can be felt clearly on the tall building due to piling small blocks up. Setback of the 6-storey shop was used as terrace of the restaurant, and features a dynamic surface.

这是一个四栋公寓建筑与百货商店并存的综合项目。建筑整体沿南北方向旋转45度，这样的设计既保证了从地块四角通向中心商场的交通流线顺畅，又确保了每栋建筑各个面的日照。住宅楼楼高80米，由每2层为一个单位的体块叠加而成，每个体块上下错开2米。凹进去的部分被处理成阳台。由小体块叠加而成的高层建筑营造出有别于普通高层建筑的尺度感。最下面六层商场部分立面凹凸错落，功能上可以用做餐厅的露台，同时使得建筑物外立面充满活力。

The Villas A+B of the Lushi Mountain Villas
庐师山庄别墅 A+B

Beijing, China
中国，北京

Location: Beijing
Area: Villa A: 650m², Villa B: 640m²
Architect: Wang Yun / Atelier Fronti
Completion Date: 2006
Client: Beijing Jiangong Real Estate

地点：北京
面积：别墅 A：650 平方米；别墅 B：640 平方米
建筑师：王昀 / 方体空间工作室
建成时间：2006 年
客户：北京建工地产

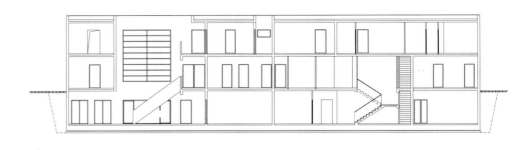

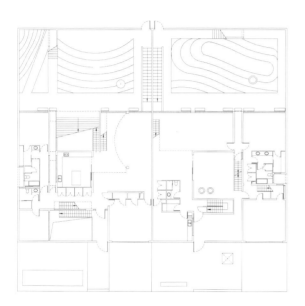

088 / NORTH CHINA

The adjacently constructed Villas A+B are located in the Lushi Mountain Villas area close to the Beijing West Mountain Badachu Park. The Lushi Mountain Villas area is composed of 52 dollhouses, among which are the Villas A+B with the characteristics of "masculine" and "gentleness" respectively, as a result of which they are named as "king" and "queen" separately. These two villas are connected together by two cubes measuring 18 metres long × 18 metres wide × 7 metres high to form a construction of two floors over ground and one floor underground. On the east side of the villas, there are two internal courtyards of 18× 21m that extend from both villas respectively, in which there are two stairways leading to the basements. In view of the whole designing, the abstract white box is adopted to construct the space. The white space could dissolve the sense of scale and distance, completely present the designer's concept, so as to provide the users with multiple spatial experiences. As for the organisation of inner space, the designers seek to convert the virtual scenery into the concrete and dynamic settings. Additionally, the indoor paths are also placed throughout the inner space, and the various sceneries that change along with moving steps bestow the abstract landscape to the users.

The courtyards formed by enclosing the eastern high walls are designed to be two "gardens in mind", rather than gardens with Chinese or Western or even Japanese styles, presenting the natural scenery by the geometric elements of cones and columns as well as walls.

住宅A+B这两个联立式的小住宅，位于北京西山八大处附近的庐师山庄别墅区中。该别墅区由52栋小住宅所构成，住宅A+B是其中的两栋，由于建筑内部空间表现出"刚"与"柔"的性格，所以这两栋住宅又被称为"王"与"后"。两栋住宅由两个长和宽均为18米和高度为7米的方盒子拼合联立在一起，并且均为地上二层，地下一层。在住宅的东侧两栋建筑分别各延伸出两个18米×21米的内院，院子中有一个楼梯可以直通住宅的地下室部分。在整体的设计上，采用抽象的白色的箱体进行空间构成。白色使空间消失确切的尺度和距离感，还原设计者纯粹的观念和对空间场景的设想，给使用者带来空间体验的多种可能性。在内部空间的组织上，建筑师力图将其意向中的风景物象化。室内穿插游走的散步路径，步移景异的空间场景，以抽象的景观作用于体验者。

在这两个住宅中东侧的两个高墙围合的院子中是一个箱庭。设计者在布置时分别设计了两个园林，这两个园林在设计时力图摒弃将其单纯地设计成中式园林，或简单的设计成西式或日式，而是力图设计一个"脑中的园林"，将自然的景致以锥和圆柱这样的几何要素来表现，加上作为背景的围合庭院的院墙，让使用者获得"无画处皆成妙境"的空间体验。

Beijing Citic Guo'an Conference Centre Courtyard-style Guestroom Buildings

北京中信国安会议中心庭院式客房

Location: Beijing
Area: 8,172m²
Architect: WSP Architects
Photographer: Shu He, Yao Li
Completion Date: 2008
Awards: 2009 Chicago Athenaeum International Architectural Award; 2009 Second Global Chinese Young Architects Award; 2008 WA Chinese Architectural Award

地点：北京
面积：8,172平方米
建筑师：维思平建筑设计
摄影：舒赫、姚力
建成时间：2008年
获奖：2009美国Chicago Athenaeum国际建筑奖；2009第二届全球华人青年建筑师奖；2008 WA中国建筑奖

Beijing, China
中国，北京

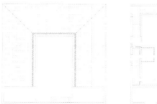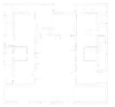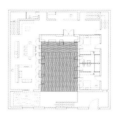
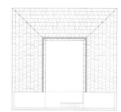

090 / NORTH CHINA

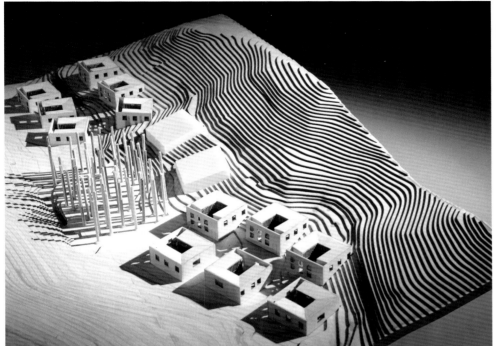

The guesrtoom buildings are located inside a valley descending from the southeast to the northwest. There is a beautiful poplar grove in the middle, which has been preserved in the design. Twelve buildings are arranged on the east and west sides of the poplar grove in two groups. Sites for two independent villas have been reserved on the north side of the poplar grove. The main road within the district is on the south side. The location and direction of each building are freely arranged according to the gradient of the hill and the direction of the road, and form rich outdoor spaces between the buildings.

There is an inner courtyard for each building. The building is arranged in U shape around the inner courtyard. Due to difference of the gradient of the hill, the entrance of the courtyard is sometimes on the underground floor, and sometimes on the ground floor. Each building is three-storey-high. On the ground floor is a lobby, a living room, a dining room and two standard guestrooms; on the first floor, there are two completely independent suites, each of which has an independent staircase; the basement is an entertainment space, including a billiards room, a bar and a cinema, in addition to equipment rooms. All main spaces like lobby, living room, dining room and guestrooms are facing the inner courtyard; and all secondary spaces like corridor, staircase, balcony, bathroom and kitchen are facing the valley.

The building is reinforced concrete shear wall construction. Outside the external walls facing the valley and the rooftop are clad with rough granite tiles, which make the buildings look like a group of "stone boxes" dispersed in the valley. The external walls facing the inner courtyard are wooden louvres and painted. GSHP (Ground Source Heat Pump System) is used to provide energy for the hot water floor radiant heating of the building. Insulation and an air cavity are set between the concrete wall (inside) and granite (outside). These four layers together constitute the outer wall. In order to improve the energy-saving effect, the architects have chosen thermally separated aluminium window frame and insulating glass in the window components.

北京中信国安会议中心庭院式客房坐落在一个从东南向西北方向跌落的谷地里，谷地的中部是一片优美的杨树林。设计保留了杨树林。十二幢点式的庭院式客房分成两组布置在杨树林的东西两侧，杨树林的北侧预留了两幢独立式别墅的基地，南侧为区内主要道路。每幢庭院式客房的位置和方向依据山地的坡度和道路的走向自由安排，形成了丰富的室外空间。

庭院式客房的内部围绕着一个内庭院成U形布置；由于地势不同，庭院的入口有时在地下层，有时在一层。建筑共分三层，其中一层为门厅、起居室、餐厅和两间标准客房；二层为两套从一层由两个不同的楼梯进入的完全独立的套房；地下室则是娱乐空间——包括桌球室、酒吧和一个家庭影院，设备用房也布置在地下室。在平面布置上，主要空间——门厅、起居室、餐厅、客房，全部面向内庭院；辅助空间——楼梯、阳台、走廊、卫生间、厨房，全部面向山谷。建筑为混凝土剪力墙结构。面向山谷的外墙面和屋顶均外挂毛面花岗岩，使得十二幢建筑像一群散落在山谷里的"石头盒子"；面向内庭院的外墙采用了木格栅和涂料。建筑使用了地源热泵低温热水地板辐射采暖；外墙在混凝土墙和外挂花岗岩之间是保温层和空气层；门窗也采用了隔热铝合金框材和中空玻璃，以保证项目的节能效果。

SIEEB
清华大学环境楼

Location: Beijing
Site Area: 17,949m²
Architect: Mario Cucinella Architects
Photographer: Daniele Domenicali, Alessandro Digaetano, MCA
Completion Date: 2006

地点：北京
占地面积：17,949 平方米
建筑师：马里奥林·库契纳拉建筑师事务所
摄影：丹尼尔·多梅尼卡利、亚历山大·迪戈塔诺、MCA 建筑事务所
建成时间：2006 年

Beijing, China
中国，北京

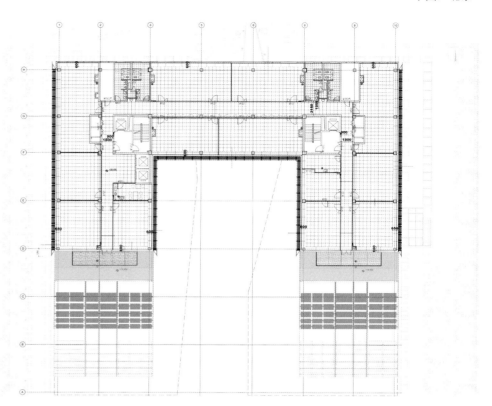

The SIEEB (Sino-Italian Ecological and Energy-Efficient Building) project is the result of a collaborative experience among consultants, researchers and architects. This integrated design process is a most distinctive part of the project and a key issue for green buildings. The building is designed to find a balance among energy efficiency targets, minimum CO_2 emissions, a functional layout and the image of a contemporary building. The SIEEB building shape derives from the analysis of the site and of the specific climatic conditions of Beijing. Located in a dense urban context, surrounded by some high-rise buildings, the building optimises the need for solar energy in winter and for solar protection in summer. Reflecting and semi-reflecting lamellas and louvres will also allow for sunshine to penetrate in the rooms in winter and to be rejected in summer, reducing the energy consumption of the building. The envelope components, as well as the control systems and other technologies are the expression of the most updated Italian production, within the framework of a design philosophy in which proven components are integrated in innovative systems.

Resources saving includes construction materials and water; Minimisation of environmental impact in both the construction and in-use stages; Intelligent control during operation and maintenance; Healthy indoor air; Environmentally sound and durable materials; Water recycling and re-use.

清华大学环境楼是专业顾问、学者及建筑师共同工作的成果，其设计过程构成整个项目中最为重要的部分，同时也是绿色建筑所需考虑的重点。这一建筑旨在寻求能源消耗的平衡，减少二氧化碳的排放，打造一个实用性格局以及现代风格建筑。建筑造型源于对其所处地形以及当地气候的仔细分析。这栋建筑坐落在拥挤的城市大环境中，四周被高楼环绕，因此设计中充分考虑冬季太阳光的获取以及夏季太阳光的阻挡。其中，反光及半反光薄板材料的运用以及天窗的设置，便可实现冬季使充足的阳光入射，夏季将强光阻隔的效果。建筑表皮元素、控制系统以及其他技术充分显示了意大利最为先进的设计理念：建筑材料及水能源的节约；建筑及使用过程中减少对环境的污染；操作及维护过程中智能系统运用；清新的室内空气；持久性材料运用；水循环及再利用。

Students' Comprehensive Service Building at the New Campus of Beijing Institute of Civil Engineering and Architecture

北京建筑工程学院新校区学生综合服务楼

Location: Beijing
Site Area: 4,443m²
Architect: Hu Yue
Completion Date: 2011

地点：北京
占地面积：4,443平方米
建筑师：胡越
建成时间：2011年

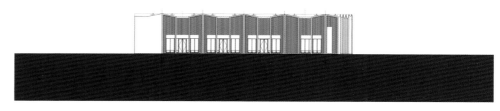

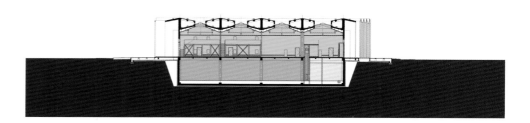

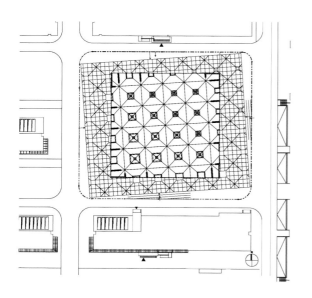

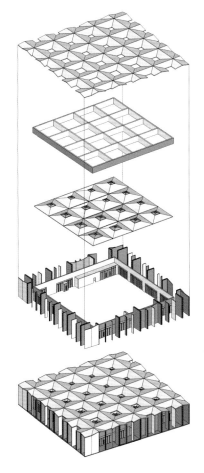

This building is settled at the new campus in Beijing Institute of Civil Engineening and Architecture, which is a small public building in the student's dormitory area. The main issue it faces is the functional transformation. In the early stages of the campus construction, the building will be utilised as a commercial facility, and then it will be converted into a multi-functional architecture as soon as the campus construction is completed. According to this situation, the designers set the free transformation of function as the starting point of the design and formulated the following key points: single storey; no column; unit module combination; the electrical equipment providing favorable conditions for function transformation; the interlayer adopting easily removable and recyclable materials; increasing devices for saving energy and protecting environment. In view of the key points of the design, the designers proposed a 60 x 60 metres square layout which is formed by 10 x 10 metres units, each of which is composed of a skylight in the middle and four sides of sloping roof. On the basis of this shape, considering the requirement of energy-saving, the designers put forward several main solutions, which helped forming the final shape.

Solution 1: Since it is a single-storey building, the thermal insulation of the roof is a tough issue. Hence, the designers added a layer of concrete roof above the four-side slope roof in the opposite direction, and hided 2-metre-high beams between two layers of the roof.

Solution 2: To place an openable skylight in the middle for ventilation.

Solution 3: A porch placed along the outside of the building could provide shade for the big windows outward and meanwhile create a semi-open space for shops and students, activities. Additionally, it would play an important role of supporting for the large-span beam. In order to enrich the façade, the porch has been processed by cutting the outside unit after turning around the axis. The whole building is based on a reinforced concrete structure; in order to express the logical relationship of the building geometry, roof and cutting surface take concrete as the main material, and the cut inside has been occupied by wooden external wall.

该建筑位于北京建筑工程学院大兴新校区内，是学生宿舍区内的一个小型公共建筑。它面临的最主要的问题是功能转变。该建筑在校园建设的初期，将做为宿舍区的配套商业设施，校园基本形成后的功能将转换成一个多功能建筑。针对这种情况，设计师将功能的自由转换做为设计构思的出发点，并制定了下列设计要点：一层；无柱；单元式模块组合；电气、设备为功能转换提供有利条件；内部夹层采用较易拆除和可回收的材料；增加节能环保设施。根据设计要点，设计师提出了一个由10米×10米单元排列形成的60米×60米的正方形平面，每个单元由中间的一个天窗和四坡屋顶组成。在这个基本形状下，结合节能要求提出几个重要措施，最后形成了现在的造型。

措施1：由于该建筑是一个单层建筑，因此屋顶的保温隔热是一个不好解决的问题。因此设计师在四面坡的屋顶上又增加了一层混凝土的反向四面坡屋顶，同时将2米高的大梁隐藏在两层屋顶之间。

措施2：中间天窗采用可开启的方式以便通风。

措施3：在建筑外侧设置了一层外廊，为外侧的大窗户提供了遮阳，同时也为店铺和学生活动提供了一个半室外空间，外廊的另一个重要的功能是为大跨度的梁提供了支撑。另外，为了丰富外立面，将外廊处理成在轴网旋转后将外侧单元进行切割的形式。该建筑为钢筋混凝土结构，为了表达建筑几何的逻辑关系，屋顶、切割面为清水混凝土，切口内为木质外墙。

Beijing South Station
北京南站

Location: Beijing
Architect: TSDI in collaboration with TFP Farrells Limited
Photographer: TFP Farrells Limited, Zhou Ruogu
Completion Date: 2008
Client: Ministry of Railways (MOR), PRC

地点：北京
建筑师：铁道第三勘察设计院集团有限公司、TFP设计有限公司
摄影：TFP设计有限公司、周若谷
建成时间：2008年
客户：中华人民共和国铁道部

Beijing, China
中国，北京

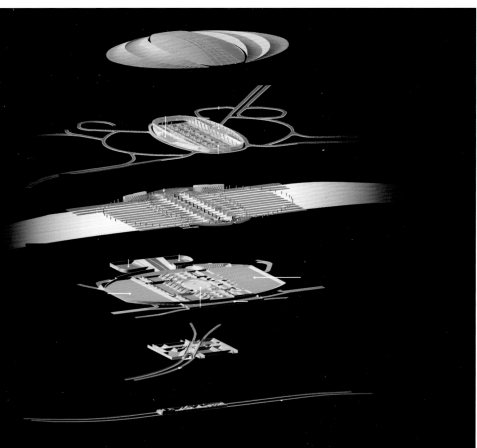

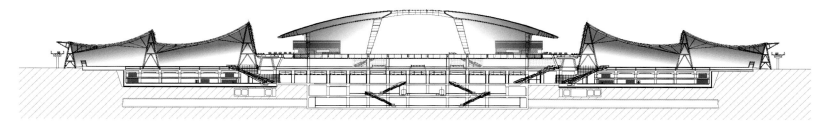

As the station is immense in scale, the architectural form and structure are clear, simple and human oriented and take into consideration the operation and management of the various rail lines, station entrances, exits, waiting areas and interchange zones taking place within – the station takes a simple ellipse form that accommodates three principal floor levels with two mezzanine floor levels for car-parking and two ancillary gateway office buildings. With such large volumes of passengers it is essential to separate the incoming and departing passengers. One of the main design objectives was to have passengers board trains with the shortest distance and time possible.

The design strategy also incorporates separate zones catering for seamless integration and transition to different types of vehicular traffic including 909 underground car-parking spaces, 28 taxi drop-off bays, 24 taxi pick-up bays with 138 queuing spaces and 38 bus spaces (12 drop-off spaces and 26 pick-up bays with 48 queuing spaces) as a comprehensive transport hub. The elliptical plan form is effective in providing an innovative solution to the station's vehicular traffic flow. The overhead road network can adjust to the traffic flows to and from the station area in all directions and assist in relieving the congestion of the surrounding urban arterial roads.

As the station is located on the existing railway land, the geometry of the site juxtaposes the diagonal fan of the railway tracks to Beijing's cardinal urban grid. The station scheme creates an urban link with the surrounding cityscape and acts as a "Gateway to the City".

火车站自身规模较大，其建筑样式及结构设计要求清晰、简约并着重强调以人为本。设计过程中还应考虑到不同线路、入口、出口、休息区、换乘区的实用性以及是否便于管理。北京南站呈现简约的椭圆造型，共包括3个主要楼层以及两个中层区（用作停车场以及辅助办公区）。鉴于车站较大的客流量，进站及出站区被隔离开来，主要目的即为让乘客能在最短时间、最短距离内乘车。

设计中还包括打造不同类型的停车区域，包括909个地下停车位、28个出租车下车区、24个出租车乘车区、38个公交车位以及1个交通枢纽区。值得一提的是，建筑的椭圆造型为车流行驶带来了极大的方便，上方的公路网伸向不同方向，在一定程度上缓解了车站附近主要道路的交通拥挤情况。

此外，由于车站选址在原有铁路旁，与北京交通网连通，更建立了北京与附近城市的连结桥梁，因此被称为"北京的门户"。

Bridge Culture Museum
天津桥园桥文化博物馆

Location: Tianjin
Area: 2,220m²
Architect: Zhang Hua, Fan Li / Sunlay Design Group Co., Ltd
Photographer: Shu He Photography
Completion Date: 2007

地点：天津
建筑面积：2,220 平方米
建筑师：张华、范黎 / 三磊建筑设计有限公司
摄影：舒赫摄影工作室
建成时间：2007 年

Tianjin, China
中国，天津

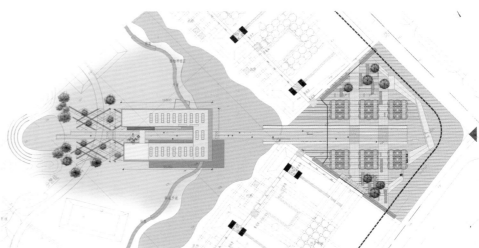

The Bridge Culture Museum is located at the centre of the Bridge Park, Tianjin. Because of its special location on the central axis facing to the main entrance, and as a landmark of the Bridge Park, this area is designed as a leisure and exhibition space in order to meet people's needs of "pleasurable show space".

The origin of the design concept is "operable bridge". Steel bridge, folded surface, variable structural columns, U-shape enclosed space and extended platform, all of these create a "Building for Strolling". This building is "accessible and passable". It brings people exciting and surprising experiences because of the conversion between interior and exterior spaces. The steel bridge between the buildings leads people to the exhibition space, which is all green; the bottom of the building is a multi-dimensional surface, which is accessible for people to go though; the sixteen structural columns have been treated in different ways to distract people's sight; the Z-shaped building, U-shaped enclosed space, the coffee break deck and the combination of the interior and exterior spaces, merged the building together with the site.

天津桥园桥文化博物馆坐落在天津桥园公园的中心。由于其坐在中轴线面向主要入口的特殊地理位置，亦由于是桥园公园的地标，博物馆设计的目标是建成一个休闲展览区域以满足人们对"赏心悦目的展示空间"的需求。

设计理念源自"可操作的桥"这一概念。钢结构桥、折叠面、多变结构柱、U形封闭空间和延伸平台等元素打造出一个"供闲庭信步的建筑"。整个建筑"易进入，可通行"，内部和外部空间的转换为参观者带来兴奋惊喜的体验。建筑中间的钢结构桥通向充满绿色的展区；建筑底部是一个多层次空间，参观者可自由通过；16个结构柱的不同使用可以转移参观者的视线；Z形建筑、U形封闭空间、茶歇休息区以及室内外设计的结合使博物馆完全融入所在场地。

Sports Arena at Tianjin University
天津大学体育馆

Tianjin, China
中国，天津

Location: Tianjin
Area: 13,835m²
Architect: KSP Jürgen Engel Architekten
Photographer: Shu He Photography & KSP Jürgen Engel Architekten
Completion Date: 2010
Awards: May 2006, Competition First Prize

地点：天津
建筑面积：13,835平方米
建筑师：KSP 尤根·恩格尔建筑事务所
摄影：舒赫摄影工作室、KSP 尤根·恩格尔建筑事务所
建成时间：2010 年
获奖：2006 年 5 月，竞赛一等奖

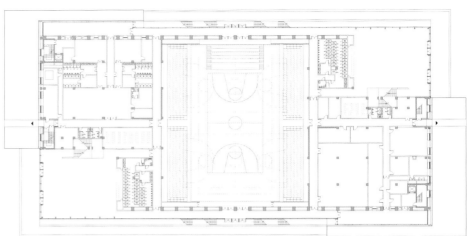

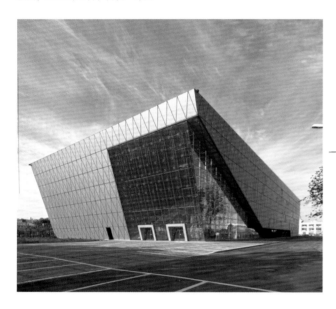

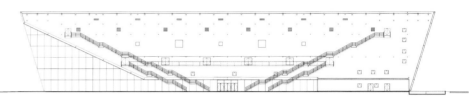

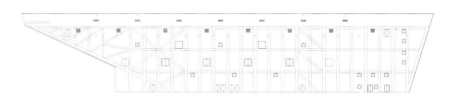

100 / NORTH CHINA

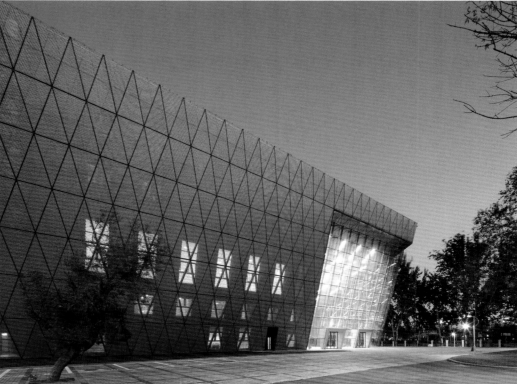

The Sports Arena has a distinctive feature in the form of a translucent sheath made of gold-coloured perforated sheet steel panels, which form a rhombic structure and emphasise the Arena's dynamic, powerful shape. Punching tools were made specially to produce the perforations, which, with their triangular shape, take up the format of the panels. A special effect adds further emphasis to the spatial depth and the lightness of the metallic façade, which is mounted some four metres in front of the actual outer wall: an enlarged likeness of the structure of the translucent perforated sheet steel façade (comparable to that of a silhouette) in the form of a black-and-white print was transferred to the outer wall of the hall. The overlapping of these two levels – the perforated sheet steel façade and the likeness of it – produces a Moiré effect, which is heightened further by the observer moving, and when it is dark, when the space between the façade and the outer wall is artificially lit.

Two spacious foyers open out the Sports Arena in the campus and a busy junction in the direction of downtown Tianjin. The two entrance halls each penetrate the corner of the outer skin and provide access to the building via the narrow sides. The large expanses of glass in the foyer extend almost the entire height of the building (24 metres) and create an entrance that is bright and flooded with light. On the longitudinal sides the glass façades mimic the inclination of the spectator stands. This way the geometry of the hall's interior and the rising spectator stands is also discernible in the design of the façade.

半透明打孔金属幕墙是该体育馆的显著特征，由此创造出来的菱形结构，使整座体育馆显得坚实而充满活力。用特殊的穿孔工具在金属板上冲制形成三角形状的孔洞。在距离体育馆外墙四米处安装空间感和明亮度均佳的金属立面，强调出一种特殊的建筑效果：半透明穿孔的金属幕墙结构是体育馆外墙放大映照在建筑外壳上的一个黑白影像（相当于外墙的剪影）。通过金属幕墙和外墙影像这两层结构的叠加，一种奇妙的云纹效果由此产生。而夜间使用的人造光源，会使这种视觉效果更加明显。

体育馆拥有两座宏伟的大厅，分别朝向天津大学校区和天津市繁华的城区。这两座大厅都建在外墙的转角处，充分利用了体育馆较为狭窄的那两侧。大厅高24米，大面积使用玻璃窗，为到访者提供了一个明亮的、日光充足的内部空间。用玻璃幕墙刻画两侧的观众看台，这样可以将看台斜坡的几何结构和幕墙设计完美结合。

Fusion Square (Business District and Five-Star Hotel in Binhai New Area)

融和广场（滨海新区商务街区及五星级酒店）

Location: Tianjin
Area: 340,000m²
Architect: Liu Shunxiao, Zhou Xiangjin / ZPLUS
Photographer: Ding Bin, Fu Jingyu, Liang Yongliang
Completion Date: 2010

地点：中国，天津
建筑面积：340,000 平方米
建筑师：刘顺校、周湘津 / 普瑞思建筑规划设计公司
摄影：于斌、付静宇、梁永亮
建成时间：2010 年

Tianjin, China
中国，天津

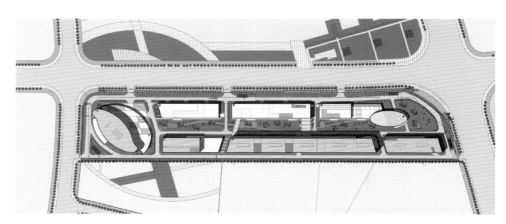

The Fusion Square is located at a prime area and has a 690-metre-long façade along the main road and facing the central lake because of the transverse site.

The planning concept of HOPSCA is adopted, integrating living, working, shopping, entertaining, socilising and playing together. The project can be conceived as a building complex of international elements and combines styles of modernity, simplicity and fashion. The design attaches great importance to the dialogue and mutuality between building and environment to create a harmonious relationship between Binhai New Area and the modern and distinct built volume. In addition, the design concept also emphasises people's direct visual feeling inside and in the public spaces outside as well as the spatial scales in order to create comfortable environment. The project has become a typical case in Binhai New Area.

The parallelled buildings respectively on the south and north boast modern European style and are constructed with modern methods of free cutting, corbelling and excavating. Metallic panels and glass are extensively used to show the solidity of the building; the rhythmic grid on the façades is either of solid wall or pure glass, adding a sculptural and modern feeling.

作为新城市结合体，融合广场享有绝佳的地理位置，而横向的基地平面，使其具有面向中心湖和主干道的 690 米长的沿街主立面及两个重要的转角位置。

运用 HOPSCA 规划理念，将居住、商务办公、购物、文化娱乐、社交、游憩等各类功能有机结合。融和广场是一个整体高度统一、国际化元素高度凝练的复合型建筑群，设计风格现代、简约、国际、时尚。项目将生态、优雅的新区环境和现代化的、特色鲜明的建筑形象巧妙结合，形成建筑与环境的互动与对话，实现区域的良性循环。设计充分考虑人在室内和室外公共活动空间的视线直观感受和空间尺度，营造宜人的环境；注重主要转角结点的塑造，形式多样而统一，成为新区高起点、高完成度的工程典范。

建筑的外部形象：南北两排板式建筑采用现代欧式风格，以自由的体量切割、悬挑、挖空等现代手法，配合金属板和玻璃等材料，突出硬朗气质，以及竖向窗格的各种韵律，局部有大的虚实对比，全玻璃或全实墙，平添雕塑感及现代感。

Meijiangnan Lake Residential Development

美江南湖住宅区

Location: Tianjin
Architect: ABSTRAKT Studio Inc.
Photographer: ABSTRAKT Studio Inc.
Completion Date: 2008

地点：中国，天津
建筑师：ABSTRAKT 工作室
摄影：ABSTRAKT 工作室
建成时间：2008 年

Tianjin, China

中国，天津

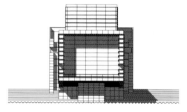

104 / NORTH CHINA

All residents have unrestricted access to water and the boardwalk running along water's edge. Meijiangnan Lake Residential Development is a part of the larger master plan by San Francisco based EDAW. It is the first of 11 satellite developments to be built around manmade lake in the suburb of Tianjin. Very stringent local building code requirements governing the orientation and distances between buildings as well as the mandatory south-facing location of living rooms influenced the master plan and its linear character.
The building types are ranging from single-family houses along waterfront to five-storey, multi-family row houses with the view of water from the top floor terraces. The multi-purpose community centre is located in the central part of the development. The average apartment area is 150 to 180m².

住宅区内的所有居民都可以自由的在湖边散步或走在沿湖铺设的木板小路上。这一项目是在城市郊区环绕人工建造11幢住宅规划的一部分。设计上严格遵循当地住宅建筑要求，住宅之间的间距以及客厅的南朝向规则影响着整体规划以及建筑的直线型布局。
住宅类型包括沿湖的单户住宅，以及面向湖水的五层多户住宅。社区中心位于场地中央，住宅面积从150平方米到180平方米不等。

West Railway Station
天津西站

Location: Tianjin
Area: 179,000 m²
Architect: Meinhard von Gerkan and Stephan Schütz with Stephan Rewolle
Chinese Partner Practice: TSDI
Photographer: Christian Gahl
Completion Date: 2011

地点：中国，天津
面积：179,000 平方米
建筑师：曼哈德·冯·格康和斯特凡·胥茨以及施蒂芬·瑞沃勒
中国合作设计单位：铁道第三勘察设计院集团有限公司
摄影：克里斯汀·格尔
建成时间：2011 年

Tianjin, China
中国，天津

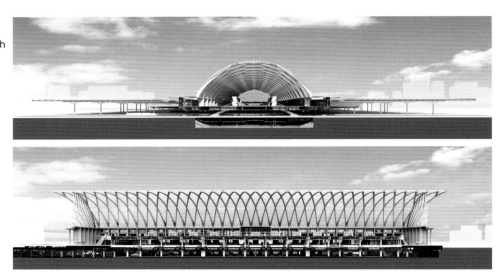

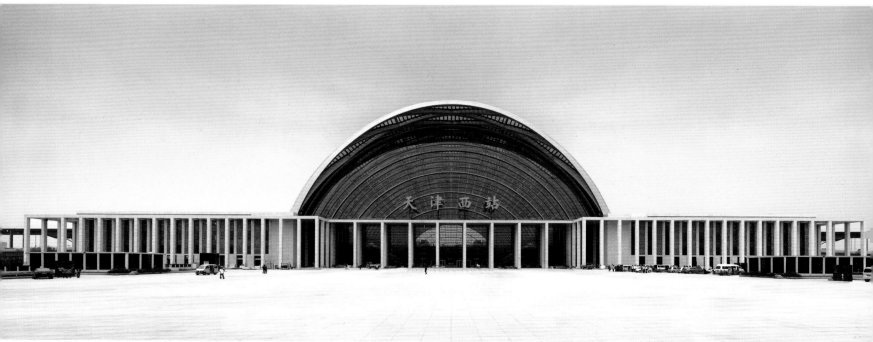

106 / NORTH CHINA

The new Tianjin West Railway Station is designed as a connecting bridge between the new city developments on both sides of the tracks. Two C-shaped, 18m high arcade buildings frame the bridge-like street fronts with waiting areas, ticketing and shops. In this way the main south entrance plaza is connected as a continuous urban space with the north plaza of the station. On top of the arcade buildings an impressive, elegant arch-shaped roof covers the bridge on a length of 360m and a span of 114m.

The filigree interwoven white steel structure of the roof will appear as one of Tianjin's new landmark structures. The classic arch shape in combination with modern engineering technique clearly indicates the function of the building as a train station and gateway to the future.

Inside the arch a 49m high column-free waiting hall provides a unique experience to the traveller with clear overview and orientation. From here escalators lead down to the tracks which are covered by a 18m high secondary steel roof.

Aboveground the station can be accessed from all four sides:

At the south and north the train station welcomes the visitor with an impressive urban arcade façade which leads into the arch.

On the west and east travellers enter the building from a drive way platform which is elevated with a long curve above the tracks.

Underground the arrival hall is linked to the Tianjin subway system.

In this way the enormous visitors flow is reduced and separated into several directions and levels.

新的天津西站将成为一座连接新老城区的桥梁。两个C形组成的、高18米的，有着街道立面一般拱廊高架桥，带有等候区、售票处和商店。南入口广场与北广场通过站房连成一个连续的城市空间。高架桥的顶部是一个令人印象深刻的，造型优雅的拱形的屋顶，南北长度为360米，跨度达114米。

金银丝交织的白色钢铁结构的屋顶将会作为天津新的地标而出现。古典的拱形与现代的工程技术相结合，明确表达了天津西站的设计理念———一座通往未来的现代化火车站。

拱桥内49米高的无柱的候车大厅为旅客提供了一个独特而又特殊的体验，有明确的总览和方向指引。从这里乘自动扶梯来到站台，站台的上方覆盖着一个18米高的二级钢屋顶。

旅客可以从四个方向进入高架层车站：

火车站的南面与北面以一个令人印象深刻的城市拱廊向游客展示着欢迎的姿态，从这两个方向可以直接进入拱廊。

从东部和西部进入的旅客可以通过车辆下客平台进入。

地下到达大厅直接连接到天津地铁系统。

通过这样的一种方式，巨大的人流量被减少和分流到了不同的方向与平台。

Tianma International Club

天津天马国际俱乐部销售中心及会所工程

Location: Tianjin
Area: 2,164.5m²
Architect: Beijing Mastubara and Architects (BMA) / Hironri Mastubara, Daijiro Nakayama, Zhang Tingting
Photographer: Shu He
Completion Date: 2006

地点：中国，天津
建筑面积：2,164.5 平方米
建筑师：松原弘典、中山大二郎、张婷婷 / BMA 北京松原弘典建筑设计咨询有限公司
摄影：舒赫
建成时间：2006 年

Tianjin, China
中国，天津

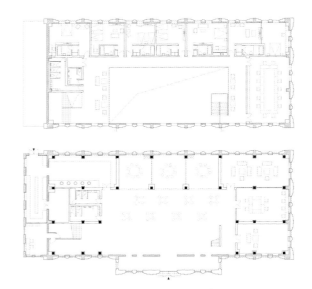

This is a condominium project invested by both Tianjin Taida Investment Company and Vantone Real Estate. The site is located in the suburb of Tianjin, where there was originally the mine factory on the large 70ha land. One of the most significant points for this renovation project is to revitalise the building by wrapping the existing building with red bricks. On the old meeting building that was renovated to accommodate the club facility, considering the clients' programmatic request, the architects added a mezzanine floor and transformed the one-storey building into a two-storey building. BMA also modified the shape of its roof. The interior of the old office building also had drastic changes to become a sales centre, although the building is still one storey. Exterior walla have been newly built for both buildings.

BMA started designing the building under the requirement of the use of traditional bricks. Consequently, they created zigzag surfaces on the exterior wall with bricks, which they called, Brick Skin. This idea of Brick Skin is applied on both club facility and sales centre. After the adjustment of the location of the original building windows, three to four layers of bricks were added to the façades. On the building elevation, zigzag patterns appear along the window, and stair-like lines for the exterior walls appear on the plan. This was a great challenge for BMA. They needed to find the answer for how they can create a modern building by using traditional building materials such as bricks.

The office building that was renovated to be a sales centre has the typical floor plan with the corridor located at the centre of its plan. Several rooms lined up along both sides of the corridor and created a closed impression for the interior space. The structural walls on both sides of the central corridor cannot be tore down easily. Therefore, if we kept this plan with the central corridors, it was obvious that the space on the building would become much oppressed. To solve the problem, the architects removed a part of interior walls, but as little as possible so that the design would not affect its original structure, and created another corridor next to the original one. By those two adjacent corridors, without any doors between them, an open impression is provided to the original closed space. The architects designed the wall for the meeting rooms along the new corridor to make the space more open. The plan for this double corridor loosens up the closed space for the building, and succeeded to create an open space centralising those two corridors.

The landscape is designed to be related to the Brick Skin façade. The most outer edges of the brick skin façade continue on to the striped floor made out of the old bricks that were left over from the construction site. The architects bedded the grey granite stone between those stripes. LED light is installed on the ground and the existing vegetation was left where it used to be.

本项目为天津泰达股份有限公司与万通地产共同投资的别墅项目。开发地位于天津郊区，用地为原天津地雷厂的开阔土地（70公顷）。本改造设计项目最大的特点是将原建筑用红砖重新包裹。因为功能的需求，会所部分需要重新加设夹层，并要调整屋顶形状，BMA的设计提案决定在原墙外侧加建新的砖墙。针对此项目中甲方对砖墙建筑的强烈喜好，BMA尊重了甲方的意见，通过改造设计去实现传统风格的红砖建筑。这对一贯承袭现代设计风格的BMA来说，是一次全新的挑战。

在利用传统建材红砖的前提下，设计工作开始进行。设计师最终找到的设计灵感就是做"砖皮"，即用红砖做有凹凸变化的外墙。销售中心、会所外立面均采用"砖皮"的设计概念。原有窗户的位置、尺寸经过调整后，在外墙部位加设3-4层新的红砖墙面。外立面产生有秩序的凹凸感。从平面图上看为台阶状的形式。这种设计构思即保证新墙面的坚固度，解决保温性能，同时实现薄砖墙层积的效果。

原销售中心的平面为典型的"布袋"式，房间规则地排列在走廊两侧，形成一个非常封闭的空间。由于走廊两侧的墙为承重墙，因此不能破坏和移动。于是设计师尝试打破这种"布袋"的平面形式，在走廊南侧设计了一个有4个"玻璃盒子"和一条可供人通行、停留的"类走廊"，借助"双重走廊"的平面，打破原有的呆板的空间，形成一个以走廊为中心的、吸引客人脚步的开放式空间。

在园林设计中引入会所的立面"砖皮"。选用与会所立面的柱子相同宽度的砖作地面，其余部分采用机刨石铺地，做成地面线性肌理。地面照明采用LED防水地灯。设计师尽量保留原有树木，树下设置照明，夜晚，自下而上将树木照亮。

ICTC
国际经济贸易中心

Location: Tianjin
Area: 90,000m²
Architect: Liu Shunxiao, Zhou Xiangjin / ZPLUS
Completion Date: 2003

地点：天津
建筑面积：90,000 平方米
建筑师：刘顺校、周湘津 / ZPLUS 普瑞思建筑规划设计公司
建成时间：2003 年

Tianjin, China
中国，天津

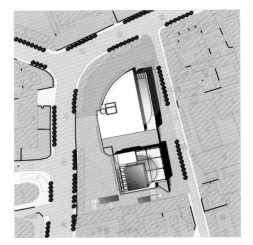

The task of the project was to refurbish the existing façades of the building based on the original graphic design and to endow it with a completely new, elegant and unique image in the CBD of Tianjin.

The building was wrapped with aluminium panels of different colours and low-energy glass windows, forming interesting patterns. The contrast of "square and circle" as well as "line and curve" was highlighted on certain places to bring warmth and intimacy to this area and to be in harmony with the British-styled architecture near by.

ICTC is located in the prime area of the city, adjacent to the Xiaobailou Downtown on the east and Fifth Avenue on the west and just one minute's walk to the bustling Nanjing Road. The ground and first floor serve as public lobby, including CEO Privilege Lounge and Café. The multi-functional conference room and fitness centre are respectively located on the second and third floor, staff restaurant on the fourth floor and luxury restaurant on the fifth floor. On the upper floors, the building is divided into two parts: tower A and tower B. On the sixth to 29th floors of tower A, there are standard separate rooms for rental companies of different scales, while on the sixth to 29th floors of tower B, there is office space specially designed for large companies.

本案在原有平面设计的基础上进行了新的外部形象的创意设计，目的在于使该办公大楼在天津CBD商务中心区内，以新颖独特的形象，高雅舒展的气质独树一帜，成为引领时代潮流的商务办公大楼。

建筑的表皮以不同颜色的铝板加上LOW—E玻璃窗，形成有趣的图案，适当的部位突出了方与圆、曲与直的对比，为CBD中的办公大楼带来了温暖的人情味，并与租界的英式风格建筑取得协调。

国际经济贸易中心东临天津市小白楼商业区，西接著名的五大道，与繁华的南京路举步之遥。这里是外汇金融街和解放路的交汇处，是海河规划的重点地区，毗邻地铁一号线，成就了大厦不可多得的黄金地段。1-2层为公共大堂，设有CEO商务酒廊、CEO咖啡厅等；3层为多功能会议厅；4层为康体健身中心；5层为大厦员工餐厅；6层为高档豪华餐厅；a座7-29层隔离式标准房间，openhouse空间结构适应不同规模公司的办公需求；b座7-29层为金融商务广场，专为大型公司设计的整体办公空间。

Idea Valley Building
慧谷大厦

Location: Tianjin
Area: 60,000m²
Architect: ZPLUS
Completion Date: 2007

地点：天津
建筑面积：60,000 平方米
建筑师：普瑞思建筑规划设计公司
建成时间：2007 年

Tianjin, China
中国，天津

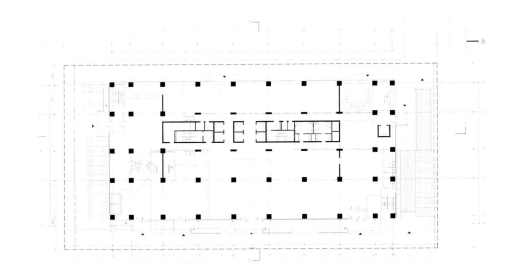

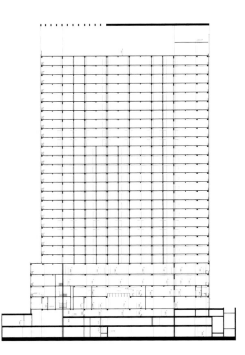

The exterior appearance of an office building always conveys the information of high-tech enterprises, and thus the design intentionally adopts the regular and rigid form with only one red-coloured structure to boast contrast and dynamic feeling. Patterns on the façades are diversified as well as delicate while the division on the façades and adornments on the detail take the effects of "landscape" and "sightseeing".

Aluminium panels and glass are extensively employed and red aluminium panels on the certain part makes the finishing point. A kind of texture is formed through the backbones, divisional joints and sunshade blinds arranged in perfect order. At the entrance, a large expanse of louvre contrasts sharply with the metallic columns, showing the elegance and liveliness of modern architectural aesthetic.

The façades present three different styles vertically: the bottom is podium and visually transparent; the middle boasts surface made of rhythmic horizontal lines; the top highlights grid pattern composed of horizontal and vertical lines to hide the facilities inside as well as to conclude the entire design. The south and west elevations are installed with aluminium curtain wall to form clear contrast with the main structure. The whole building is elegant and modern.

项目的外在形象担负着传达高科技企业信息的职责，故采用理性、严谨的板式造型，只在框架内插入一红色斜向体量，产生对比和运动感，而整体的肌理图案则细腻多变，立面的分格和细部的点缀考虑了大厦"景观"和"观景"的双重效果。建筑外表面采用铝板、玻璃等现代材质，并大胆使用了局部红色铝板，达到了画龙点睛的效果，通过排列疏密有致的龙骨和分格缝或遮阳百叶形成表面肌理，尤其入口处大面积细腻的玻璃百叶的处理，与金属圆柱形成清新明快的对比效果，表达了建筑外观优雅而不沉闷，耐人回味的现代建筑美学审美趣味。

立面的竖向划分沿用古老的"三段式"格式，下部为裙房，处理成较为通透的视觉效果，中部标准层为富于韵律的水平线条所构成的表面；顶部为加密的水平线条与竖向线条的透空格架，既遮住内部设备用房，又成为建筑顶部完美的收束。建筑西南侧结合功能做大面积的铝板幕墙，与主体形成强烈的虚实对比，使建筑呈现雅致、现代的造型特点。

City Centre
赛顿中心

Location: Tianjin
Area: 160,000m²
Architect: Liu Shunxiao / ZPLUS
Completion Date: 2006

地点：天津
建筑面积：160,000 平方米
建筑师：刘顺校 / 普瑞思建筑规划设计公司
建成时间：2006 年

Tianjin, China
中国，天津

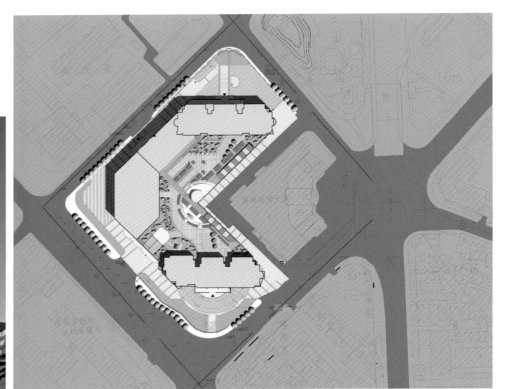

The City Centre is composed of one office tower and two apartment towers, respectively located on the three corners of the site to optimize the inner space. Each tower is designed with a unique frontage facing streets in order to create new starting and receiving point in the city layout from the visual aspect as well as bring a new axis and landmark to the city.

Aesthetic principles of contrast and compromise, scale and sequence are employed in modelling the building in order to construct three-dimensional space boasting spirit of new time. The glass curtain wall facing the street can reflect the surroundings and as a consequence weakens the feeling of being pressed caused by the large volume of the building. Diversified visual effects are formed when moving along the buildings.

Appropriate adjustments of simplicity and sobriety constitute the main style of the project, highlighting the harmonious and elegant features an urban building complex should possess. Aluminium panels of golden and champagne colours are arranged on the surface to form delicate patterns; structures resembling lighthouse and hollowed wings on the top enrich the unique and distinct image of the whole project.

该群体由三个高层建筑组成，两幢国际公寓和一幢办公大楼，分别把住基地的三个街角，最大限度地营造建筑群体内部空间，且使每个街道转角都有突出的造型，在城市格局中营造新的视觉出发点和接收点，为城市创造新的轴线，新的地标。
建筑的造型充分运用对比、调和、等级、秩序等美学法则，力求创造富有时代精神的三维空间。建筑转角处的玻璃幕墙通过对周围环境的反射达到了虚化建筑体量的作用，有助于减弱高层建筑的压迫感，随着人们视点的移动，造就出动态多变的视觉效果，并强化了挺拔感。
建筑整体的风格是在简洁庄重之中做出适度的变化，以表达大体量的都市建筑群体所应有的平和典雅而又内具不凡的气质。表皮采用深浅有别的咖啡金色和香槟色铝板的拼贴图案，在建筑顶部以高科技杆件、灯塔、镂空的机翼造型加强外观的独特性与地标性，并通过与立面造型紧密关联的屋顶构架形式，使建筑整体达到和谐统一的效果。

YJP Administrative Centre
于家堡工程指挥中心

Location: Tianjin
Area: 18,000m²
Architect: HHD_FUN
Photographer: Wei Gang, Wang Zhenfei
Completion Date: 2009

地点：天津
建筑面积：18,000 平方米
建筑师：华汇设计
摄影：魏刚、王振飞
建成时间：2009 年

Tianjin, China
中国，天津

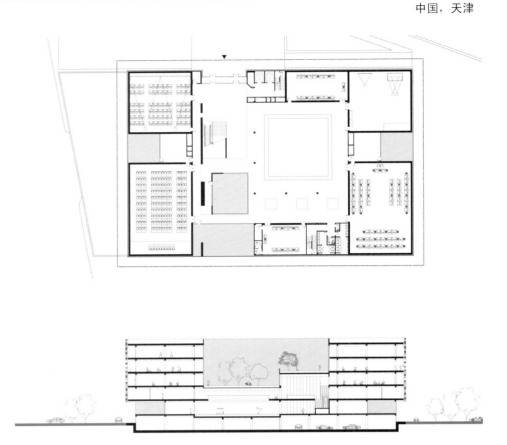

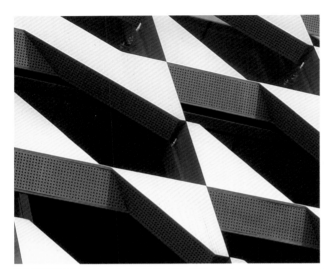

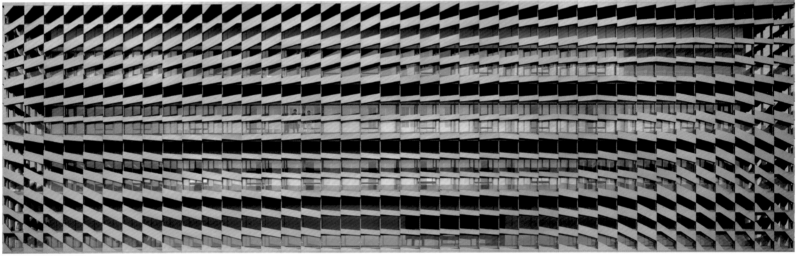

YJP Administrative Centre is a temporary building in Tianjin Binhai CBD. Enclosed verandahs are arranged around the building to afford visual access to the CBD area from within the building. This allows the occupants to survey the surrounding construction site. The size of openings within the façade relates directly to the lighting requirements for particular activities within different areas of the building. The porosity of the façade is designed to produce the required conditions for these activities. The integration of the density of the patterned façade with the various inner functions forms a key focus of the project.

The façade apertures serve as view frames. Aperture size and orientation is varied in a continuous manner introducing topological difference across the façade. The whole façade is constructed from six forms, reflected to give twelve types of identical components, making the building process highly efficient. This meant that the building to be constructed in less than seven months.

于家堡工程指挥中心是位于天津滨海商务区内的一幢临时建筑，封闭的阳台设计使得从建筑内部便可看到商务区的景象，同时便于工作人员研究周围的建筑条件。

在设计之初，就将采光、观景及私密的因素一同考虑，使立面设计能一次性满足所有要求。根据内部功能不同，立面按相应采光量需求分类区域。将立面样式与内部功能相结合的设计方式构成了整个项目的核心。

外廊外立面处理以此为基础形成不同的开孔率，形成一层开口的孔洞设计，以适应内部功能的不同要求，同时还为外廊生成了不同的景窗，使得建筑内部、游廊和周边大区域间建立起了视觉上的联系，观赏过程变得更加有趣。立面方案限制在6种不同的模块中，这6种连续变化的模块代表了6种不同的采光率。

Gui Yuan Atelier
圭园工作室

Location: Tianjin
Area: 2,100m²
Architect: Zhou Kai
Photographer: Wei Gang
Completion Date: 2008

地点：天津
建筑面积：2,100 平方米
建筑师：周恺
摄影：魏刚
建成时间：2008 年

Tianjin, China
中国，天津

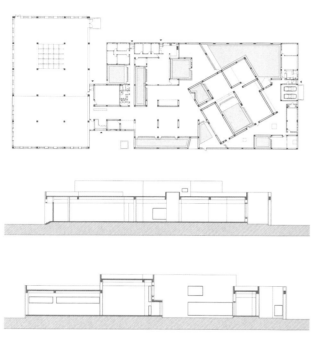

Gui Yuan Atelier is located in the Balitai Industrial Park of Jinnan District in Tianjin City, which is adjacent to a factory. The atelier's higher single-layer space is quite similar to many industrial buildings in the park in terms of scale. Its pure and simple external image goes harmoniously with the surrounding buildings, only with the deep warm brick colour to present its distinct personality among the grey environment.

Gui Yuan Atelier presents an integral shape of rectangle of 78 metres x 40 metres. The strategy of digging a yard inside the space is adopted in order to organise the space's functions and resolve the problems of daylighting and ventilation. The yard dug out allocates the space and becomes a scenic of the whole space, creating a calm and idyllic interior environment in the external environment of closure. The cubic of north-south direction inserting in the main body of the building in a 30 degree takes a higher position than the surroundings, creating some interesting space transformation while highlighting the major space.

The whole building takes bricks as the main building material. The crozzle is used inside and outside of the frame construction, where is filled with building blocks and heat insulation material, in order to create a natural and pure feeling and a kind of clear and genuine construction expression.

This project is originally built for an intimate friend who engages in interior designing. However, with the completion of this building, both designer and client unite thoroughly, forming the present space of architectural landscaping interior designing. Hence, there is one more paradise for spending free time with friends.

圭园工作室位于天津津南区八里台工业园区，与相邻厂房彼邻而建，其较高的单层空间与园区内大量的工业建筑尺度相近，单纯朴素的外部形象也与周边建筑一脉相承，只有其深沉的暖砖色在灰灰的邻里中透露出一些不同的个性。

圭园外形完整，是一个约78米X40米的长方形，以在内部挖院子的方法组织功能，解决采光通风。挖出的院子组织空间并形成了空间中的景观，在较封闭的外形里形成一个安静而富有意境的内部环境。平面中间与外部呈30度插入的南北向正方体，体形略高于周边，在强化主要空间的同时亦形成一些有趣的空间变化。

圭园以砖为主要建筑材料，在框架结构的内外均以过火砖砌筑（中间填充砌块及保温材料），力图营造出自然质朴的感受，清晰真实的建筑表达。

圭园本是为好友室内设计师所建，伴随着房子的建成，甲方乙方合二为一，形成了现在的建筑景观室内设计的空间所在。从此工作之余朋友相聚便又多了一个好去处。

Zhangjiawo Elementary School, Xiqing District

天津西青区张家窝镇小学

Tianjin, China
中国，天津

Location: Tianjin
Area: 18,000m²
Architect: Dong Gong, Xu Qianhe / Vector Architects + Lü Qiang / CCDI
Photographer: Shu He
Completion Date: 2010

地点：天津
建筑面积：18,000 平方米
建筑师：董功、徐千禾 / 直向建筑 & 吕强 / 中建国际
摄影：舒赫
建成时间：2010 年

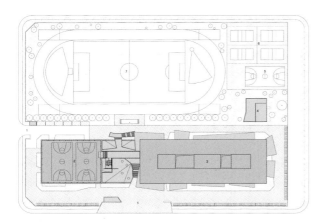
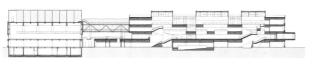

The goal is to establish a unique place within the school that encourages interaction between the students and teachers through their daily learning and teaching life.

The design process starts with an analytical research of the spatial pattern of interactive activities, both in plan and in section. A series of physical study model were built along the process, in order to seek the most reasonable spatial and programmatic layout. Eventually the best location of the primary interactive space is discovered to be on the second floor, sandwiched by regular classroom floors, and connected to the skylight through the central atrium, where natural ventilation were maximised. The space is defined by the surrounding special programme classrooms, and extends itself to a green roof deck at the south side, which is also the pivot point of the site arrangement. The deck connects to the main school entrance, the outdoor fields, and different parts of the building at different heights by stairs, ramps and bridges. Such a "Platform", consisting of indoor space and outdoor deck, not only generates and amplifies energy of interactions, but also adds visual characters to the exterior building appearance because of the application of distinctive materials and space modules. A series of green technologies are proposed in this project, such as geothermal system, storm water management, green roof, permeable landscape, passive ventilation, maximised natural daylight, and recycled material.

设计师希望在这个小学设计中着眼于对于"教"与"学"这种生活方式对于空间的需求，尝试提供学生和老师、学生和学生之间充分而富有层次的交流的机会和场所。

设计起始于对交流空间的行为和空间模式的研究和分析。为了寻求最合理的空间功能布局，最终将一个共享的交流"平台"设置在二层，它像三明治一样被一层和三四层的普通教室夹在中间，最大程度上带来该空间使用的易达性和必达性。而各个年级交叉，教学形式相对自由，师生和学生之间交流互动最为频繁的专业功能教室则成为这个交流"平台"的功能载体。整个建筑活力最强，能量最集中的空间通过一个中庭在顶部获取自然光和加强自然通风，同时它延伸出室外，和位于其南侧的一层绿色屋顶平台相通，成为连接建筑各部分和教学楼前后景观的一个中心枢纽。设计中倡导运用一系列的绿色环保措施，主要包括地源热泵、绿色屋顶、可渗透景观、自然通风和采光最大化等。

Feng Jicai Literature and Art Academy at Tianjin University

天津大学冯骥才文学艺术研究院

Location: Tianjin
Area: 6,000m²
Architect: Zhou Kai
Photographer: Yang Chaoying, Wei Gang, Zhou Kai
Completion Date: 2005

地点：天津
建筑面积：6,000 平方米
建筑师：周恺
摄影：杨超英、魏刚、周恺
建成时间：2005 年

Tianjin, China
中国，天津

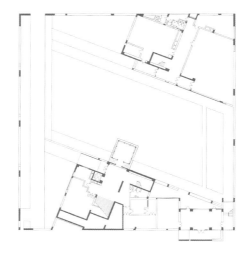

Feng Jicai Literature and Art Academy at Tianjin University is located in the main campus of the university. It is constructed on an upright and foursquare foundation, with the main stadium in the east, the teaching and laboratory building in the south, the saddle-shape gymnasium in the north, and the biggest Youth Lake in the west.

At the very beginning of design, Mr. Feng Jicai proposed that the academy should be filled with oriental features for the correspondence to the study orientation, and thus how to apply the contemporary vocabulary to create a prescribed atmosphere turns to be the focus of the design.

The plan proposes carrying out from the foundation, enclosing the site with a square yard and then imbedding in functional modules to form a united structure. The oblique volume being hung in the upper space divides the yard into the north part and the south part. The wall with the same height of building functions as the façade of the whole and the spatial border of the yard. It corresponds to the scale of the surrounding constructions and cooperateds with the water-planted trees to create an atmosphere of quiet and modern academy.

The main space of the building is unfolded along an axis of east-west direction, which faces the Youth Lake in the northwest. The zigzag path connects different spatial points, which correspond with the lake scenery. This construction has maintained the original trees in the yard, with the water flowing along the north and south yards, finally connecting them together. The grey concrete wall is sealed in the lower part and with colourful perforations in the upper part, isolating the noisy outside and also forming the spatial border of the yard.

天津大学冯骥才文学艺术研究院，选址于天津大学主教学区。基地形状方正，东侧是主体育场，南侧是教学实验楼，北侧是马鞍形体育馆，只有西侧与校园内最大的青年湖相邻。

设计之初冯骥才先生就提出研究院要具有东方意境以期与学院研究方向对应，所以如何以当代语汇营造特定的场所意境便成为设计的焦点。方案从基地出发，方形院落围合场地，以功能体块嵌入其中，共同形成统一的整体。院中斜向架空的建筑体量将院落分成南北两院，与建筑等高的院墙既作为整体的外立面，也作为限定院落的空间界面。其外与周边建筑尺度呼应，其内则以水体树木共同形成静逸的现代书院意境。

建筑主要空间沿东西走向的斜轴展开，斜轴从院内对向西北侧的青年湖。转折向上的行走路线，欲扬先抑，将不同的节点空间串联起来，重要节点空间与湖景形成对话。建筑院内结合环境保留原有树木，架空下方穿过的水面在南北院落形成关联。灰色调凿毛混凝土院墙下部围实，上部透空，既遮蔽了外部的干扰也形成了院落的空间限定。

Tangshan City Hall
唐山市城市展览馆

Location: Tangshan, Hebei Province
Area: 5,900m²
Architect: Urbanus Architecture and Design
Photographer: Yang Chaoying
Completion Date: 2008

地点：河北，唐山
建筑面积：5,900 平方米
建筑师：都市实践设计有限公司
摄影：杨超英
建成时间：2008 年

Hebei, China
中国，河北

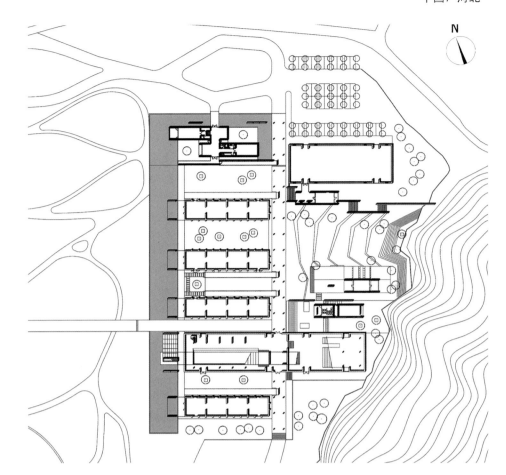

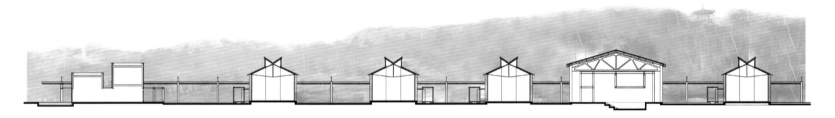

If the great 1976 earthquake in Tangshan is an unintentional destruction in its architectural history, then the Urbanization Movement deliberately denies the contemporary architecture of Tangshan. The original flourmill in Tangshan is a victim. When the plant was relocated, its dozens of homely warehouse were planed to be completely turned down and turned into the City Park. For a city where a majority of the buildings are only 30 years old, the four warehouses built during the Japanese invasion should be reserved and created into a core of museum group that is built at the foot of the mountain. The six parallel buildings are perpendicular to the mountain and lead the mountain to the city in a rhythm. The new construction gives prominence to this form, and tries to make the original building and mountain to be the focus. The new construction with rich functional areas, such as the reception room, bookstore, café, and gift shops, makes much space for the mountain with small volumes and parallels to the original building. Materials only refer to the transparent metal grill and preservative wood in order to strengthen the spirit and natural feeling of the site. Against the alternative but also simple materials, the intact walls of the old warehouses reveal its inner beauty. Along the park side, s steal porch is added to each warehouse so as to inject the closed warehouse with a sense of openness. The reflection of the porch in the pool has highlighted the beauty of the old building. The "X"-shaped steel structure roof has transformed the closed room into a bright, perfect, and standard showroom. Currently, this museum complex is defined as the city exhibition hall which tells the story about the city. What's more, in the park, Taobao village, games circle and other public facilities are also created to encourage the public to participate more easily. Landscape design is to use the natural beauty to decorate the building, to create a relaxing cultural environment, thus creating a sense of intimacy. In such a city, this park will provide the public with a good place to read more about the history of Tangshan.

如果说1976年的大地震对唐山的建筑史是一次无意的摧毁，那么，今天的城市化运动则是对唐山平庸的现当代建筑史的一次有意的抹杀。原唐山面粉厂便是这样的牺牲品。当工厂外迁后，它的数十座相貌平平的仓库，由于几乎毫无美学价值，计划中被彻底推掉，变成城市公园。对于一个大多数建筑只有30余年岁数的城市，厂区中四幢日伪时期建的旧库房似乎很值得保留。在多方努力下，它们和另两幢80年代建的粮仓得以保存，并以此为核心，形成一个山脚下的博物馆群。这六栋平行的建筑恰巧垂直于山体，它们有节奏地将山引到城市。新加建的部分更加突出了这种天作之合，所以内容非常有限，尽可能使原建筑和山体成为视觉的主体。用来丰富功能活力的新建筑，例如接待室、书店、咖啡、礼品店等，尽可能用小体量来让出山体，并平行于原建筑。材料使用上也很节制，只用通透的金属格栅和防腐木板，来强化场所固有的工业化精神和自然的面貌。在这些另类、却也朴素的材料映衬下，原封保留的旧仓库的墙面透出了内在的美。沿公园一面，每个仓库增添一个钢结构门廊，让封闭的仓库具用一种开放性。这些门廊落在反射水池上，使旧建筑的美进一步放大。"人"字形仓库的屋面用"X"形钢结构来代替，形成的侧高窗使原先封闭的室内变成明亮的、非常完美和标准的展示厅。目前这个博物馆群被定义为城市展览馆，向市民讲述城市的故事。从有生机的公园活动的角度，公园内还规划了淘宝村、游戏圈等公众更易于参与的内容。景观设计是用自然美来装点朴素的建筑，以形成一种轻松的人文环境，从而创造一个与一般市民没有距离感的公共空间。在一个似乎一切都很平淡的城市，人们在这个公园里可以不经意地读到唐山沉淀的历史和值得关心的历史残片。从这些历史中，找回对自己城市的信心。

Qinhuangdao Bird Museum
秦皇岛鸟类博物馆

Hebei, China
中国，河北

Location: Qinhuangdao, Hebei Province
Area: 2,000m²
Architect: Yu Kongjian, Xiang Jun, Zhangyuan / Turen Landscape Planning Co., Ltd, The Graduate School of Landscape Architecture, Peking University
Photographer: Yu Kongjian
Completion Date: 2009

地点：河北，秦皇岛
建筑面积：2,000平方米
建筑师：俞孔坚、向军、张媛 / 土人景观规划设计研究院、北京大学景观设计学研究院
摄影：俞孔坚
建成时间：2009年

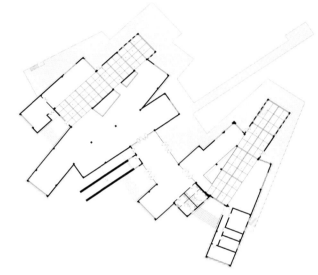

The building, inspired by the "anchored fishing boat along coast", was composed of a series of intersecting striped spaces to reflect the balance between wind power and holdback power. Moreover, the design takes advantage of sea wind and expresses the poetic scene of the beautiful site.

The structure was just one floor and partially lifted to correspond with the surrounding trees. Roof terrace was proposed and the outdoor passages could directly lead onto it. Thus, visitors can reach during the closing time of the museum.

The main structure was constructed by reinforced concrete. White rostone and wooden panels were employed to clad the exterior walls. The two different materials alternated to form an interesting scene: the façade facing the road expresses itself as white stone while the façade facing the sea shows the refined textile of wood. In addition, the white stone façade was simple and elegant with diaglyphed sea plants patterns; the wooden façade highlighted naturalness and intimacy.

Good ventilation and natural lighting were the two main concepts in the interior design. In order to realise this, large French windows were installed at the southern and northern sides as well as several skylights. At the main entrance, natural light was brought in. Moreover, two light courts were created to invite more light in as well as setting plants and holding outdoor exhibition.

建筑从"秦皇岛外打渔船"在海边的停泊状态获得灵感，由条状虚、实空间穿插交错构成。这种肌理反映了风力和牵制力之间的一种平衡状态。是对场地海风的利用和诗情的表达。

建筑高度以一层为主，局部高起，可以和周围树木相互掩映。建筑还设计了屋顶平台，并有专门的户外通道，一直达到屋顶，以便观光者在闭馆时也可使用屋顶。建筑采用钢筋混凝土结构，标准柱网。外墙由两种材料：一种是白色人造石（混凝土）挂板，有阴刻海草图案，隐约可见，整体白色，素雅。另一种是木板，自然亲切。两种材料交互使用在不同的穿插建筑体上，从而形成了临路一侧为白色人造石，而临海一侧为木纹的"阴阳"建筑。表达了城市与自然的交互关系。

建筑强调自然通风和自然采光。除南北两头大面积落地窗外，还有许多天窗。主入口尤其强调天光的利用。此外，利用建筑穿插块之间的三角交错带，留出两个采光天井，内可种植物或布置户外展览。

Tangshan Museum
唐山博物馆

Location: Tangshan, Hebei Province
Area: 24,444m²
Architect: Wang Hui, Cheng Zhi, Du Aihong, Chen Chun, Hao Gang, Wei Yan, Zhang Yongjian, Chen Lan, Zheng Na, Yang Qing, Wu Wenyi, Liu Yinyan, Liu Nini
Photographer: Chen Yao, Hao Gang
Completion Date: 2011

Hebei, China
中国，河北

地点：河北，唐山
建筑面积：24,444 平方米
建筑师：王辉、成直、杜爱宏、陈春、郝钢、魏燕、张永建、陈岚、郑娜、杨勍、吴文一、刘银燕、刘妮妮
摄影：陈尧、郝钢
建成时间：2011 年

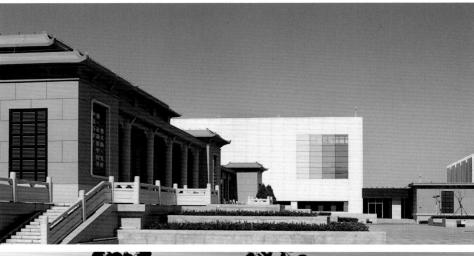

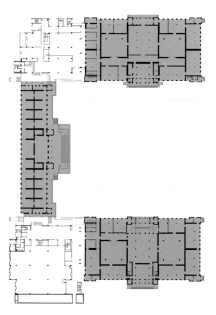

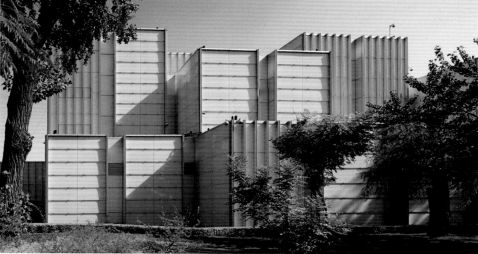

130 / NORTH CHINA

The original Tangshan Museum is composed of three old "Long Live Exhibition Centres" which are allocated in a triangle pattern. In addition to highlighting the ancient history of Tangshan, the Ping opera as well as the shadow puppets exhibition, this renovated museum also provides plentiful exhibition spaces. Additionally, a full range of integrated features of collection, research, education, communication and appropriate facilities have been also provided, including forum, bookstore, library, and restaurant. This new museum has integrated these three detached buildings into a whole by an organisation of organic dynamic lines and also interpenetrated a series of non-exhibition halls, so as to make itself a modern museum. As an important infrastructure of Tangshan and a part of the open public place of Lantau Peak, Tangshan Museum also plays a key role in the chain of city's culture and leisure and entirely opens for free. It will be the most beautiful urban living room, providing a perfect place for the citizens to relax and communicate as well as entertain guests. As for dealing with the volume of the old and new buildings, the extension part not only highlights the old building body successfully, but also makes up for the maladjustment of the building size.

From the aspect of construction materials, the modern material of super white colour ceramic glaze glass has been boldly used, which not only enlarges the era of distance of the two buildings, but also reduces the volume size of the new building, highlighting the old building. As for the relationship with the environment, the extension part has harmonised the ambivalent relations between the museum and the mountain. In the process of renovating the old square favoured by the citizens, the design creates some semi-intimate spaces for small groups, which makes people feel senses of domain and belonging.

This project explores the strategy of organic maintaining the urban spirit and city memories of the second or third-tier cities and achieving the perfect transformation in the current urbanised conditions.

唐山博物馆前身便是三座呈"品"字形的旧"万岁馆",经历了大地震的考验,是唐山的建筑古董。改造后的新博物馆以唐山古代史、唐山评剧和皮影展览为亮点,兼有大量的其他展陈空间。并具备收藏、研究、教育、交流等全方位的综合功能及相应的配套设施,包括讲坛、书店、图书馆、餐厅等。设计通过有机的动线组织,把三座离散的建筑组织成一体,同时在这个展线中穿插进许多非展陈性的厅堂,实现了一个现代化的博物馆功能。唐山市最重要的基础设施之一,唐山博物馆是凤凰山开放的公共空间的一部分,又是整个城市文化、休闲链上的一个关键节点,向市民免费开放,成为供市民休闲交往,自豪地招待来客的场所,是唐山最美丽的城市客厅。

在新旧建筑体量处理上,加建部分成功地使旧建筑主体突出,并弥补了其尺度失调的不足。

在材料处理上,大胆地采用了超白彩釉玻璃这一现代材料,既拉大了新旧建筑之间的时代距离,又弱化了新建筑的体量,同时突出了旧建筑。在和环境关系上,加建部分协调了博物馆和公园山体之间的矛盾关系。在改造被市民们热爱的旧广场时,用城市客厅的概念,在开敞的公共广场中植入半私密的、适合小群体活动、细腻的个人化空间,让人在城市尺度上有领域感和归属感。

这个设计探讨了如何有机地保持当前二三线城市的城市精神和城市记忆,有效地实现其在当前城市化条件下的华丽转身。

Wonder Mall
万象城

Location: Shijiazhuang, Hebei Province
Area: 170,000m²
Architect: amphibianArc
Photographer: Nicholas May, Zhou Ruogu
Completion Date: 2005
Awards: 2006 China Innovation Award of Best Commercial Real Estate; Listed as "Major Project" of Hebei Province "Great Changes of Every Three Years" Plan; Listed as "Offering Project" for the 60th Anniversary of National Day; One of the "Top Ten Best Public Architecture" of Heibei Province 2010; Mall China "City Advancement Award" 2010

Hebei, China
中国，河北

地点：河北，石家庄
建筑面积：170,000 平方米
建筑师：amphibianArc 设计院
摄影：尼古拉斯·梅、周若谷
建成时间：2005 年
获奖："2006 年度中国最佳商业地产创新奖"；被列入河北省市"三年大变样"重点工程；被列入 60 周年国庆献礼项目；荣获 2010 年河北省"十佳公共建筑"称号；"中购联中国购物中心 2010 年度城市推动奖"

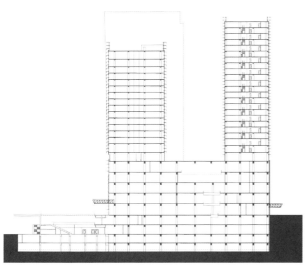

Wonder Mall, a mixed-use development with a shopping mall at the podium as its anchor, is a celebration of China's contemporary urban life that has been partly defined by its pursuit and consumption of fashion. While fashion is ever-changing and multiple in expressions, Wonder Mall strives to embody the essence of fashion by articulating fashion as an act of wearing and accessorising oneself. It is a conscious act of covering the naked body with items, either utilitarian or frivolous, which create identity and self-expression. The architecture of Wonder Mall takes place at the building's surface where the naked curtain wall as body skin is covered with weaves of metal louvres, the Prêt-à-porter of architecture. With the architectural solution to fulfill the client's vision of creating a locus for fashion and urban amenities, architects employed an intricate louvre system as an agent of free expression that will capture contemporary Chinese enthusiasm toward fashion. The louvre system at the same time functions as shading device to mediate sun light and heat gain, making the building more sustainable. The twisting geometry of the louvre, from vertical to horizontal and vice versa, allows the louvre panels on different elevations to respond to changing angle of the sunlight.

万象城是一个综合性项目，购物中心位于基座结构内。其设计目标旨在宣扬中国现代城市生活——一定程度上取决于对时尚的追求以及对时尚品的消费。时尚一直处于变化之中，万象城的设计意在抓住时尚的主旨，将其转化成穿衣装扮的一种方式。穿衣装扮是自然而然的意识行为，无论从功能性还是美观性层面来讲，因此这便构成了所谓的"时尚的主旨"。这一建筑在设计上犹如穿衣一般，裸露的幕墙被比作成"未穿衣服的身体"，之后采用金属窗"包裹"。这一构思满足业主需求的同时，更实现了遮阳功能——防止阳光的灼晒以及减少热量的获得，使得整幢建筑更加环保。此外，金属窗蜿蜒的几何造型更使得不同立面的窗板随着太阳照射角度的变化而发挥着"自身的作用"。

Complex of Hebei Education Press

河北教育出版社综合楼

Location: Shijiazhuang, Hebei Province
Area: 16,800m²
Architect: Zhang Yonghe, Xu Yixing, Liu Xianghui, Zhang Lufeng, Wang Hui, Yu Lu, Dai Changjing / Atelier FCJZ
Completion Date: 2003

地点：河北，石家庄
建筑面积：16,800 平方米
建筑师：张永和、许义兴、刘向晖、张路峰、王晖、于露、戴长靖 / 非常建筑
建成时间：2003 年

Hebei, China
中国，河北

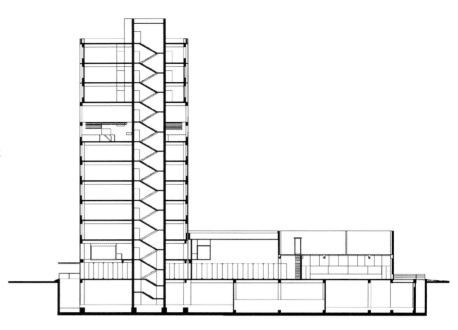

The complex includes various functional spaces, offices (both for the press and for renting), conference and exhibition centre, art gallery, hostel, restaurant, café, bookstore, culture salon, canteen, indoor basketball gym and so on.
The challenge was how to combine all these spaces in only one building. However, the architects started from the opposite, planning all the functions first and then defined the final architectural form on the basis of that. Eventually, all these spaces were separated into three groups according to different functions. Three structures were proposed, respectively separated in function but combined in form.
The structure accommodating the press was clad with wooden partition panels, while the exterior walls of the other two were of concrete block. In addition, public spaces are often created between buildings in the urban context. In this project, a vertical garden was devised among the three structures, expanding in parallel along the roof. It was also used as fire escape staircase and connected with the water park to the south visually.

河北教育出版社综合楼功能包含河北教育出版社自用办公、出租办公室、会议展览中心、美术馆、旅馆、餐馆、咖啡馆、书店、文化沙龙、职工食堂、室内篮球场等。
如此复杂的使用功能似乎不应简单得在一栋建筑中切割出来。而应首先策划／组织功能，在其基础上确定建筑形态。最终，功能被明确划分为三组，一栋建筑于是相应拆解为彼此相对独立又有关联的三栋建筑。
三栋建筑组合的形成，将设计的思路再次引向城市。河北教育出版社，在其外部做了木隔扇表层；其他两栋建筑均为混凝土砌块外墙。在城市中，建筑之间往往形成公共空间。在河北教育出版社的三栋建筑之间设置了公共绿地：一个垂直的花园，在办公与出租办公楼之间的屋顶上水平展开，同时也是建筑的消防梯系统，在视觉景观上与南侧水上公园连在一起。文化性的美术馆也是这个建筑内部公共空间的一部分。

Tangshan Exhibition Centre
唐山会展中心

Location: Tangshan, Hebei Province
Area: 30,000m²
Architect: Atelier 11, China Architecture Design & Research Group / Cui Kai, Xu Lei, Yang Jinpeng, Wang Yuhang
Photographer: Zhang Guangyuan
Completion Date: 2004

地点：河北，唐山
建筑面积：30,000 平方米
建筑师：中国建筑设计研究院拾壹建筑工作室 / 崔恺、徐磊、杨金鹏、王宇航
摄影：张广源
建成时间：2004 年

Hebei, China
中国，河北

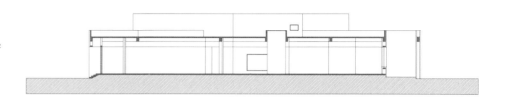

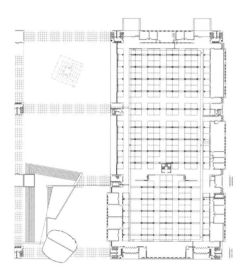

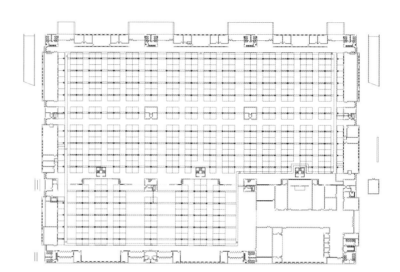

The city of Tangshan is one of the most important industrial centres in Northern China. Its economy depends largely on coal mines, heavy industries and ceramics manufacturers. The local government therefore has decided, during their urban renewal process, to construct a new exhibition centre. The centre would provide space for ceramics exhibition and other economic exchange programmes near the main entrance of the city. The project site is divided by the main boulevard, and a park is planned to its north. Based on the current condition, architects have decided that the exhibition centre should become an extension of the park with an overpass crossing the main boulevard allowing pedestrians to walk through. This gesture corresponds well with the conception of a park like exhibition centre, since it will provide popularity needed for exhibition buildings. The supporting structure uses pre-stressed concrete box girder for the large span roof, providing the needed spaces for HVAC ducts and other equipment. The parcelled volume created by this structure helps to divide the internal functional space, possible expansion in the future and generating some ceremonial interior spaces.

Architects also tried to apply the expressivity of the large span structure to terraces, ramps and other elements on the exterior. The material used for the façades and other exterior elements are red earth brick, concrete and steel, referring to the industrial heritage of Tangshan. In this project, the considerations for all spatial, formal, structural and mechanical requirements have been the core of the design process, and all of them have been accomplished with success.

唐山在近代是北方的一个工业重镇，以煤矿、重工业、陶瓷为主导产业，因此在城市更新的过程中，政府决定在主要的城市进入通道边建造一座会展中心，为陶瓷博览会和其他的经济交流项目提供场所。建筑的用地沿道路展开，跨过了一条城市道路，北侧则是规划的一个公园。因此设计师决定了建筑应当成为公园的延续，把日常人流引入到建筑平台上，并跨过城市道路。这种做法形成会展公园的概念，而在实质上则解决了会展建筑经常遇到的日常人气的问题。支撑结构采用了钢筋混凝土预应力箱形梁作为大跨度屋面的承载，同时为风管和桥架提供了敷设空间。由此产生的若干空间模块，既有利于内部功能的分割和未来的扩建，也形成了具有某种仪式感的内部空间。

在建筑的外部，大跨度的结构形成了建筑最富表现力的形式母题，平台、坡道等细节的处理上也延续这种力度的表达。建筑外部采用红色的陶土砖，和混凝土、钢材形成质感上的组合，也强调了建筑所在城市的特征。在这个项目里，对空间、形式、结构、设备的统一考量成为设计过程中的核心路径，并且最终实现了预期的目标。

Ming-Tang Hot Spring Resort

河北茗汤温泉度假村

Location: Bazhou, Hebei Province
Area: Approx. 12ha. (first phase)
Architect: C T Design + Associates
Photographer: Tony
Completion Date: 2010

地点：河北，霸州
建筑面积：12公顷（一期）
建筑师：CT设计联盟
摄影：托尼
建成时间：2010年

Hebei, China

中国，河北

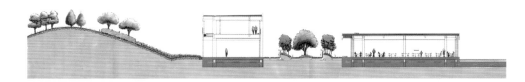

Bazhou has been planned and developed as a hot spring town in Hebei Province, China. There are six pieces of land allotted to six investment groups in the new development area and the project is one of the six hot spring resorts. How to use these varied experience for resort is the one of main points for design. Topography also works as base to create three landscape typologies which will create multiple landscape experience – the hilltop as grassland, the hillside as forest, and the low land as hot spring. Hotel and villas will be set at different positions and zoning as groups with different landscape themes such as hot spring, forest, and lake. For a hot spring resort, water is the main subject and is used as main landscape element. From hot spring to spa, from dam to waterfall, and from water courtyard to surrounding lake, the architects intend to create more chances for guests to experience hot spring in many different ways.

"Architecture in Nature, Nature in Architecture" is the basic concept for the hotel. In this way, the architects intend to create weak and humble architecture which is harmonious and consistent with the surroundings. It is hoped that this resort could be an environment which people can experience by not just only vision, but hearing, smell, and touch as well. You can hear voice of water, birds, and wind going through bamboo, and can feel hot spring and smell flower as well.

凭借天然的地理优势及特有的温泉资源，霸州经过规划和开发，将成为河北省的温泉城。在新开发地带，共有六个地块被分配给六家投资集团，该项目是六个温泉度假胜地之一。如何利用这些丰富的度假村体验是设计的一个重点。该项目以地形为基础，创造出三种景观类型，为人们带来多重景观体验——山顶为草场、山坡为森林，而低地则为温泉。酒店与别墅被设置在不同的方位，根据温泉、森林、湖泊等不同的景观主题各自组团分区设计。作为一个温泉度假胜地，水源当然是主题，也是这里主要的景观元素。从温泉到水疗馆，从堤坝到瀑布，从庭院到周围的湖泊，设计师们力图为客人创造更多的机会以不同的方式去体验温泉。

"自然中的建筑，建筑中的自然"是设计师为该酒店确定的基本设计概念。设计师力以这种方式创造出外观含蓄、简朴的建筑，与周围环境保持和谐一致。他们希望这个度假村不仅能够满足人们的观感，还能从听觉、嗅觉和触觉上给人带来全新的体验。在这里，人们能够倾听到潺潺的流水声、鸟儿的吟唱以及风吹过竹林飒飒作响，感受沐浴在温泉中的惬意、清嗅花朵的芳香。

Shanxi Grand Theatre
山西大剧院

Location: Taiyuan, Shanxi Province
Area: 72,100m²
Architect: ARTE-Charpentier / Zhou Wenyi, Pierre Chambron
Photographer: Shen Zhonghai, Zhou Wenyi, Pierre Chambron
Completion Date: 2012

地点：山西，太原
建筑面积：72,100 平方米
建筑师：夏邦杰建筑设计公司、周雯怡、皮埃尔·向博荣
摄影：沈忠海、周雯怡、皮埃尔·向博荣
建成时间：2012 年

Shanxi, China
中国，山西

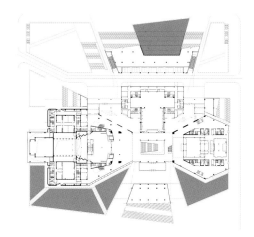

Shanxi Grand Theatre is located in the new district of Changfeng in Taiyuan, in the heart of a green island. As a visual focus, it should be transparent to extend the landscape. Furthermore, it will get together with the green island and other cultural buildings to shape an orderly special enclosure and form a united but distinct visual relationship. It will be conceived as a window openning to the Fenhe River or a viewing frame that connects the nearby Fenhe River with the distant Western Hill.

The main building is based on the layered platform with the height of 57.5 metres and the length of 210 metres, forming a visual effect of entering into the art world step by step and also providing perfect viewing platforms for the green island and the Fenhe River. Both sides of the volume are respectively occupied by a 1,600-capacity performance space and a concert hall with 1,200 seats. The platform below is the places for 600-seat hall, exhibition hall, ticket lobby, shared hall, souvenir shop, bookshop, coffee house and other supporting facilities. The whole building seamlessly integrates with the city square and the landscape of Fenhe River. The portal part and the entry platform can be used as a stage opening to the city, which will be paved with red carpet and used for holding various famous film festival activities and other culture and art events.

The whole building is thus designed like an urban sculpture, whose folding surface gives a sense of rational geometry, grand and majestic, just like others historic buildings in local, representing the profound culture of Shanxi. The light color stone curtain wall of the façade and the top plays with the partial transparent glass curtain, reflecting the sunlight from the sky in a sensitive and varied way.

Located near the Fenhe River and in the middle of the city's axis, Shanxi Grand Theatre is an art gateway opening to the city and leading people into a cultural palace. It is an open stage for the public and also a gateway of city with the solemn city hall square in the west and the energetic green land in the east. It plays a role of bridge and viewing frame between the east-west mountains and the Fenhe River.

山西大剧院位于太原长风文化商务区内，汾河畔的文化绿岛中央，它既是视觉的焦点又必须是通透的，使景观得以延伸。它与绿岛其他文化建筑一起形成有秩序的空间围合，使其相互之间拥有既统一又富有变化的视觉关系，建筑师把它设想成一个对汾河敞开的窗口，一个取景框，它将近处的汾河和远处的西山一并框入构图中。

主体建筑高57.5米，长210米，位于层层平台上，形成渐渐进入崇高艺术世界的效果，也为绿岛和汾河提供观景平台。门式空间两侧分别为1600座主剧场和1200座音乐厅，平台下布置有600座小剧场和展览大厅，以及售票厅、共享大厅、纪念品店、书店、咖啡厅等配套设施。建筑与城市广场，汾河景观之间形成流动的、共融的、浑然一体的关系。在举行大型露天活动时，门式空间和入口平台可以作为向城市敞开的舞台，这里将铺上红地毯，成为举办国内外著名电影节等文艺活动的场所。

整体造型拥有现代雕塑般的力度，折叠的界面形式具有一定的严谨的理性几何性，如同山西古建筑那么舒展大度，以神似的方式，现代的手法象征着山西凝重渊远的文化。表皮的处理上达到一种近乎纪念性的纯净效果，主体立面和顶部采用浑然一体的浅色石材幕墙，与局部透明玻璃幕结合，形成强烈的虚实对比的效果，在一天中不同的时间与光照条件下，产生不同的光线效果。

屹立在汾河之畔，城市轴线的中心，它是一扇向城市敞开的艺术之门，引导人们进入文化的殿堂，是市民共享的公共活动舞台。它也是一座城市之门，其西侧是庄重肃穆的市政广场，东侧是郁郁葱葱的文化绿岛公园，它在之间形成一种沟通，成为太原市东西山脉和汾河的取景框。

Yungang Grottoes Museum
云冈石窟博物馆

Location: Datong, Shanxi Province
Area: 10,000m²
Architect: Cheng Dapeng
Photographer: Cheng Dapeng
Completion Date: 2009

地点：山西，大同
建筑面积：10,000 平方米
建筑师：程大鹏
摄影：程大鹏
建成时间：2009 年

Shanxi, China
中国，山西

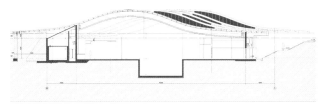
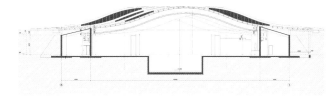
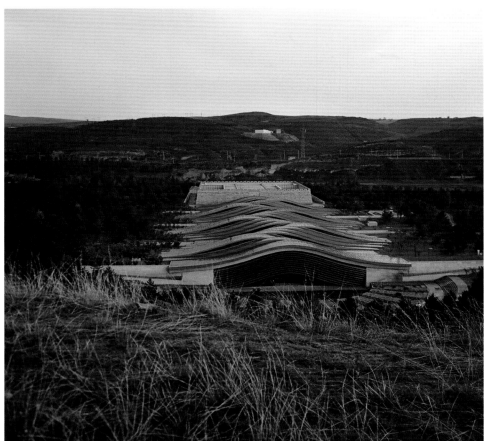

142 / NORTH CHINA

Yungang Grottoes Museum is located on the west side of the Yungang Grottoes in Datong, within the scenic spot of Conservation Zone, being one of the subprojects at the whole renovation area of Yugang Grottoes. The whole museum is formed by several arch-shaped strips in 4 metres wide by 40 metres long each. The exhibition hall is a single layer long-span space, which is 6 metres deep underground while the highest elevation above ground is 5.8 metres. The southern performing arts centre is recovered at the site of the ancient city. It has an underground layer which is as high as 3.6 metres, with 7.0 metres height of the cornice on the ground floor.

The new building must have a clear and positive attitude to the improvement of the scenic spot. Hence, the strategy of constructing this museum is applying modern techniques to integrate the museum into the entire scenic area. In the process of designing, the large dimension volume of an area of 6,000 square metres with the width of 50 metres is almost hidden under the ground. Additionally, strengthening with its 300-metre-long roof, the building expresses a simple and smooth architectural language, as a fine rhythm into the beauty of land. With the "openness and vast" artistic conception, it expresses its respect for the profound history and brilliant artistry over 1,500 years.

This proposal aims to create a new architectural landscape, bringing a bright future for the distinctive cultural heritage and recalling its beauty of delicacy and eternity to the whole world.

位于大同市云冈石窟景区内西侧，风貌保护区范围内，是整个云冈石窟景区改造的子项之一。博物馆以宽4米，跨度40米为一个单元的拱形交错排列而成。陈列馆为单层大跨度空间，地下6米，地面以上最高点标高5.8米。南面演艺中心为在原古城遗址上复原。地下一层，层高3.6米，地上一层檐口标高7.0米。新的建筑必须以一种清晰的积极态度参与整个景区的改善。所以，建在风貌保护区的云冈石窟博物馆，是使这个现代技术支撑的当代建筑，与整个景区"意与境"融合。在设计中，不仅仅将面宽50米，占地6000平方米的庞大体量建筑基本隐藏在地下，而且将绵延300米长的屋面语言强化，以其简朴单一的变化韵律融入云冈地文之美，以"空旷"的意境表达对辉煌了1500多年历史的石窟的沉默的虔诚。

设计师希望以一个1500年后新的建筑景观的清晰存在，重塑或唤回在岁月的摧残之下，云冈石窟所依存的精致之美，空旷之美，沧桑之美。

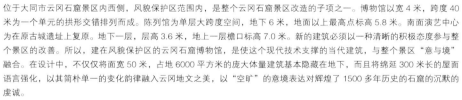

Dalai Nur Nature Reserve Education Centre
达里诺尔自然保护区宣传教育中心

Location: Chifeng, Inner Mongolia Autonomous Region
Area: 1,500m²
Architect: Zhang Yonghe / Atelier FCJZ
Cooperated architects: Beijing Yishe Architectural Design Consulting Co., Ltd.
Completion Date: 2005

地点：内蒙古，赤峰
建筑面积：1,500 平方米
建筑师：张永和 / 非常建筑
合作设计：北京意社建筑设计咨询有限公司
建成时间：2005 年

Inner Mongolia, China
中国，内蒙古

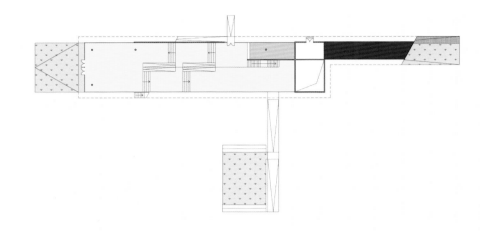

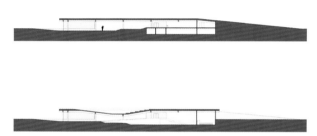

The building is a visitor centre of the Nature Reserve, housing exhibition hall and screening rooms that feature the introduction of the regional geology and geography, flora and fauna as well as the tradition. In order to minimise the impact of the building on the grassland environment, the designers conceived it as an integral part of the prairie terrain, treating it like an extension and rising part of the ground. In order to limit the height of building and thus make it seamlessly integrate into the grasslands, the basement is designed to sink into the earth; for this reason, the inner space is formed at different levels, connecting all of the routes for visitors around the inner space into an undulating and changing loop. Also, the rising part above ground is partly covered with the soil slope. The roof connects with the ground has planted the same vegetation as that on the prairie, attracting people or cattle or sheep unwittingly to ascend from the prairie to the roof, sightseeing or grazing. Therefore, Dalai Nur Nature Reserve Education Centre can be considered as a distinct building with a little appearance and a practice of designing starting from an artificial terrain.

这幢建筑是保护区用来接待游客的，内部设有介绍区内地质地理、动植物以及风土人情的展览、电影放映等使用功能。为了尽量减少对草原环境的影响，设计师把建筑看作是草原地形的一个组成部分，即将建筑处理为地面的延伸和隆起。为了限制建筑的高度，使之更加融合在草原之中，建筑的底层是沉入地下的；从而也形成室内地面在一系列不同标高上，将整个游客在建筑内的路线连成一个起伏变化的环路。在同一原则指导下，建筑突出地面的部分也局部用土坡覆盖。与地面相连的屋顶种植了与草原一样的植被，人和牛羊将会从草原不知不觉地走上屋顶，或观光或放牧。因此，达里诺尔宣教中心可以认为是一栋几乎没有外表的建筑，同时也是以人造地形为设计出发点的又一次尝试。

Site Entrance of XANADU
元上都遗址大门

Inner Mongolia, China

中国，内蒙古

Location: Plain Blue Banner, Inner Mongolia Autonomous Region
Area: 409.95m²
Architect: Li Xinggang, Qiu Jianbing, Sun Peng, Yi Lingjie, Zhang Yuting, Zhao Xiaoyu
Photographer: Qiu Jianbing, Zhao Xiaoyu, Li Ning
Completion Date: 2011

地点：内蒙古，正蓝旗
建筑面积：409.95 平方米
建筑师：李兴钢、邱涧冰、孙鹏、易灵洁、张玉婷、赵小雨
摄影：邱涧冰、赵小雨、李宁
建成时间：2011 年

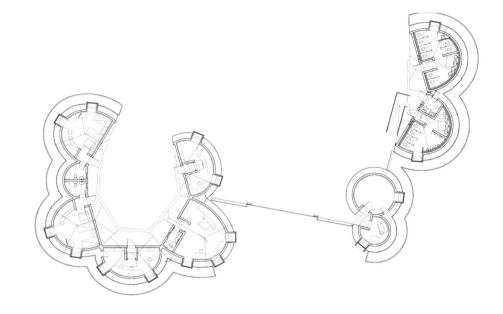

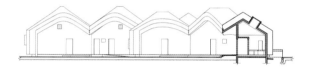

146 / NORTH CHINA

 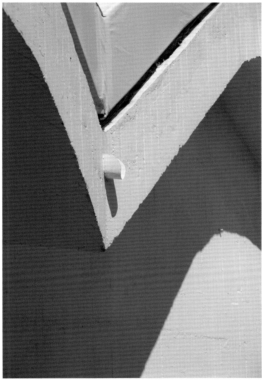 

Site of XANADU, the capital site of Yuan Dynasty, belongs to Important Historical Monuments under Special Preservation of China. Created by Chinese northern nomadic nation, this capital on grassland is located in the territory of Wuyi Pasture, Plain Blue Banner, Xilin Gol League, Inner Mongolia Autonomous Region, and on the north shore of alluvial plains of Lightning River (Luan River). There are magnificent mountains, boundless prairies and beautiful views. A new site entrance is needed to apply for World Cultural Heritage, and to solve the scenery ticketing, tourist services and other supporting functions.

The entrance is located in the southeast of the Site as the status quo, the stone lintel carved "Site of XANADU" and the stone map of XANADU are set along the axis connected to the Site, while the architecture, sculpture of Genghis Khan, battery car parking and other entities are set at east of the axis to allow for the landscape visual corridor.

The entrance's functions include two parts, visitors' toilets and a small amount of management rooms. Specifically they are ticket office, public toilets, security control room, offices, dorms and storage room, etc.; building area is about 410 square metres.

This group of white, pitched roof and small circular and elliptic buildings enclose two courtyards, internal and external, respectively for the use of staff and tourists. According to functional requirements, these scattered buildings have different sizes and heights, forming some interesting dialogue between them.

The inner surfaces are finished with bare concrete, while the outer ones use white translucent PTFE membrane, triggering the imagination of Mongolian yurts and bringing the feeling of temporary buildings on the grassland, to reduce the interference to the Site environment. Modulation tubes hidden in the gap between the membrane and exterior wall will shimmer with white light at night, making the buildings more light and even seem ready to move out. That coincides with the characteristics of nomadism of the grassland, and expresses the respect for the Site as well.

元上都遗址是中国元代都城遗址，属全国重点文物保护单位。由我国北方骑马民族创建的这座草原都城，位于内蒙古自治区锡林郭勒盟正蓝旗五一牧场境内、闪电河（滦河上游）北岸冲积平地上，山川雄固，草原漫漫，风景优美。为申报世界文化遗产需建设一处新的大门，以解决景区售票及游客服务等配套功能。

大门基地选址于遗址东南方向现状入口处，将题有"元上都遗址"的门楣和刻有元上都遗址地图的石块沿基地与遗址相连的轴线设置，而将建筑、成吉思汗雕塑、电瓶车停车场等实体偏于轴线东侧，以留出面向遗址的景观视觉通廊。

大门功能包括游客卫生间和少量管理用房两部分，具体为售票处、公共卫生间、警卫监控室、办公室、宿舍和储藏间等；建筑面积约410平方米。

这组白色坡顶的圆形和椭圆形小建筑，围合成对内和对外的两个庭院，分别供工作人员和游客使用。根据功能需求，这些小建筑大小不一高低错落，相互之间的群体关系形成了有趣的对话。

建筑内界面采用清水混凝土，外界面采用白色半透明的PTFE膜，引发蒙古包的联想，带来草原上临时建筑的感觉，降低对遗址环境的干扰。膜与外墙之间空隙里隐藏的灯管将在夜晚发出白色的微光，更显轻盈，似乎随时可以迁走一样，暗合草原的游牧特质，同时表达了对遗址的尊重。

Ordos Art Museum
鄂尔多斯美术馆

Location: Ordos, Inner Mongolia Autonomous Region
Area: 2,700m²
Architect: Xu Tiantian, Guillaume Aubry, Chen Yingnan/DnA _ Design and Architecture
Photographer: Zhou Ruogu
Completion Date: 2007

地点：内蒙古，鄂尔多斯
建筑面积：2,700 平方米
建筑师：徐甜甜、纪尧姆·奥布里、陈英男 / 徐甜甜建筑事务所
摄影：周若谷
建成时间：2007 年

Inner Mongolia, China
中国，内蒙古

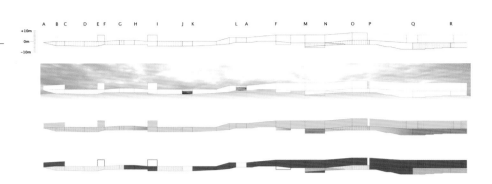

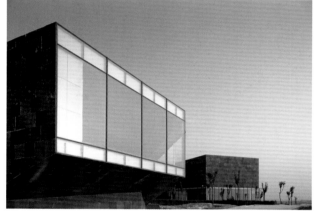

Museum of Fine Arts has two function lines: the public exhibition line and the internal data line.

For the exhibition line, circulation commenced in a linear, low-level entrance along the dune slope of the natural terrain of sand to reverse, and lift in the high-level to overlook Kaokaoshina reservoir, and then turn around, use its own building as the structure carrying, winding its way back to the ground exhibition hall – it is the ending climax of public exhibition streamline and continues downward to be a start to internal data streamline – as the entrance and exit correspond with each other, the entire public circulation presents an "8"-shaped continuous line.

The architectural form then winds extending on the sand dune, interactive with the terrain to form a half-enclosing yard or plaza. This flow line space has ups and downs along the way, according to the height and with the surrounding terrain to form intersecting surfaces in different sizes, heights and scales. The combination of landscape, space and terrain along the road, or to the best endpoint-glass curtain overlooking the reservoir, or the introduction of the courtyard landscape through side walls of continuous glass, or the enclosure contents of the exhibition to the top of the window. The light is so diverse, lighting and space interwoven to create a unique rhythm. In this rhythm, the art scenery within the architectural space and the outdoor natural scenery alternately present, which makes the visitor's psychological experience an elastic conversion between artistic context, architectural scale and natural beauty.

美术馆有两条功能主线：公共展览线路和内部资料线路。

展览的交通以线性展开，水平低矮的入口顺沿沙丘坡地的自然地形扭转，在高处则挑起远望考考什那水库，继而反转，以自身建筑作为结构承载，蜿蜒回落到地面展厅——既是公共展览流线的结束高潮，或又向下延续开始内部资料流线——作为出口与入口相互对望，整个公共交通呈现一个"8"字型的连续线路。

建筑形体因此在沙丘上蜿蜒延展，与地形相互作用形成半围合的院落或广场。这个流线空间一路上跌荡起伏，根据高度和周边地势形成大小高宽尺度不同的横截面。沿路结合景观、空间和地势，或以尽端点式玻璃幕远眺水库，或以侧墙连续玻璃引入院落风景，或以顶部开窗围合展览内容。光线也因此多样化，光影和空间交织出特有的节奏。在这个节奏里，建筑空间内的艺术情景和外部自然风景交替呈现，参观者的心理空间在艺术情境、建筑尺度和自然风光之间弹性转换。

Ordos Museum
鄂尔多斯博物馆

Location: Ordos, Inner Mongolia Autonomous Region
Area: 27,760m²
Architect: MAD Architects
Photographer: Iwan Baan
Completion Date: 2011

地点：内蒙古，鄂尔多斯
建筑面积：27,760 平方米
建筑师：MAD 建筑事务所
摄影：伊万·班
建成时间：2011 年

Inner Mongolia, China
中国，内蒙古

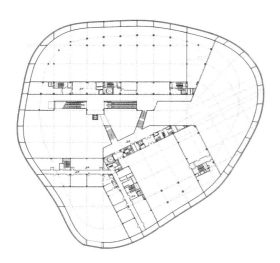

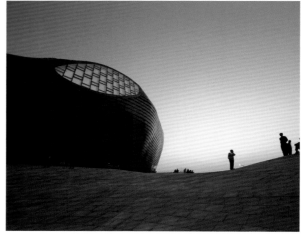 

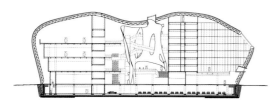

150 / NORTH CHINA

Located in Ordos, Inner Mongolia, the Art and City Museum is a crossroads for a community working to interpret its local traditions in a new urban context.

The project was first envisioned six years ago, in the then-desert wilderness of Inner Mongolia, when the municipal government of Ordos commissioned MAD to design a museum for the unbuilt metropolis. Amidst the controversy surrounding the planned city, it became evident that the museum for Ordos must navigate the many contradictions that emerge when local culture meets with visions of the future city. Inspired by Buckminster Fuller's "Manhattan Dome", MAD conceived of a futuristic shell to protect the cultural history of the region and refute the rational new city outside. Encapsulated by a sinuous façade, the museum sits upon sloping hills – a gesture to the recent desert past and now a favourite gathering place for local children and families. Upon entering the atrium, a brighter, more complex world unfolds. A canyon-like corridor connects the east and west entrances, allowing the space to become an open extension of the outer urban space. Visitors meander through the space as if in the future – yet eternal – Gobi desert.

The completion of the museum offers a moment of pause in a city which has seen no end to construction. In this vital space where the past and the contemporary are joined, people can meet with art and with each other, giving new spirit to this young community.

MAD 设计的鄂尔多斯博物馆近日落成，它好像是空降在沙丘上的巨大时光洞窟，其内部充满自然的光线，正将城市废墟转化为充满诗意的公共文化空间。六年前还是一片戈壁荒野的内蒙古鄂尔多斯新城今天充满争议，而争议本身已经将其置于更广泛的中国当代城市文化反思的焦点，它让公众重新理解地方传统和城市梦想的关联和矛盾，同时，也迫使我们理解那些被边缘化的地方文化所爆发出的对未来深切的渴望。2005 年，在一片荒野上建立一个新城区的城市规划图制订后，MAD 受到鄂尔多斯市政府的委托，为当时尚未成形的新城设计一座博物馆。受到巴克明斯特·富勒（R. Buckminster Fuller）的"曼哈顿穹顶"的启发，MAD 设想了一个带有未来主义色彩的抽象的壳体，在它将内外隔绝的同时也对其内部的文化和历史片段提供了某种保护，来反驳现实中周遭未知的新城市规划。博物馆飘浮在如沙丘般起伏的广场上，这似乎是在向不久前刚刚被城市景观替代而成为历史的自然地貌致敬。市民们在起伏的地面上游戏玩乐，歇息眺望；甚至早在博物馆还未完工时，这里就已经成为大众、儿童和家庭最喜爱的聚集场所。在步入博物馆内部的一刹那，好像进入了一个明亮而巨大的洞窟，与外界的现实世界形成巨大反差的峡谷空间展现在眼前，人们在空中的连桥中穿梭，好像置身于原始而又未来的戈壁景观中。在这个明亮的峡谷空间的底层，市民可以从博物馆的两个主要入口进入并穿过博物馆而不需要进入展厅，使得博物馆内部也成为开放的城市空间的延伸。

内部的流线是一条游动在光影中连续的线，时而幽暗私密，时而光明壮观，峡谷中的桥连接着两侧的展厅，人们在游览途中会反复在穿过空中的桥上相遇。明亮的漫射天光使得博物馆大厅完全采用自然光照明。

Daihai Hotel
岱海宾馆

Location: Neimenggu, Inner Mongolia Autonomous Region
Area: 23,000m²
Architect: Beijing New Era Architectural ltd.
Completion Date: 2003

地点：中国，内蒙古
建筑面积：23,000 平方米
建筑师：北京新纪元建筑工程设计有限公司
建成时间：2003 年

Inner Mongolia, China
中国，内蒙古

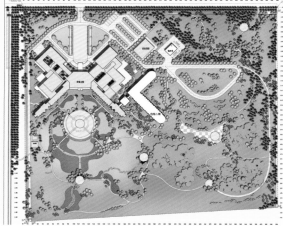

152 / NORTH CHINA

The biggest wealth of this project is the beautiful natural environment around the hotel; it gathered natural landscape such as lake, mud flat, wet land, grassland and mountain forest, and culture landscape like cultural relics of the ancient human beings and ancient Great Wall. This project covers an area of 20 hectares, but the Phase I project only has 2,000 square metres. For this reason, the architects insist on the following practice in the overall layout:

First, make Phase I project away from the lake as far as possible and make it close to the North Road around the lake, in order to emphasise the integration of building and local natural environment, and give maximum protection to the ecological environment.

Second, the whole building is only built in two floors, and three floors in some parts. The height of the building is 10 metres, so it won't give an oppressive feeling to the south lake scenic area. It is forming an abundant external outline with the different heights of buildings and being in harmony with nature. Walking around the lake, you can still feel the original forest skyline.

Third, the layout is a butterfly-like axis net and opens to the lake, letting the main rooms such as the lobby, guestroom, restaurant, meeting room, and swimming pool lounge totally have the lake view.

Fourth, Inner Mongolia aloha customs and bonfire characteristics are combined to design the welcome plaza and bonfire plaza, and an axis relationship with hotel lobby is made.

Fifth, design a patchwork yard between each part of functional area in the central building and sidewalk. It enhanced the enjoyment and vigour of space and solves the problem of natural lighting and ventilation. Make the building keep good permeability and avoid the fully distributed layout.

Sixth, try to arrange the Phase I project near the west to reserve good condition for the next phase. Before development of Phase II project, this area can be reserved as plant nursery and cultivation.

宾馆周围优美的自然环境是本项目最大的财富，它集中了湖畔、滩涂、湿地、草原、山林等自然景观和古人类文化遗址、古长城等人文景观，而同时本项目用地达20公顷，一期工程又仅仅有2000平方米，因此在总体布局上我们坚持了如下做法：

第一，一期建筑尽量远离湖面，靠近环湖北路，强调建筑融入当地的自然环境，对所处的生态环境给予最大保护；

第二，整个建筑只涉及2层，局部3层，高度10m，使得建筑不对南面湖边景区产生压迫感，同高低错落的建筑形成丰富的外轮廓，与周围的绿荫自然融为一体，在湖边活动的人群依然感受到原始的山林天际线；

第三，平面布局呈45°轴网蝶形，面向湖面敞开，使得大堂、客房、餐饮、会议、泳池、休息厅等主要使用功能房间百分之百观赏到湖景；

第四，结合内蒙古迎宾的习俗及篝火特点，设计迎宾广场和篝火广场，并和宾馆大堂成轴线关系；

第五，在建筑中心部位各功能区和走道之间设计错落的庭院，增加了空间的趣味性和活力，同时又解决了自然采光、通风问题，使建筑物保持良好的通透性，又避免了完全分散式布局的特点；

第六，将一期建筑尽量靠近西边布置，为后续建设预留宽松条件，二期开发前，此预留地可作为苗圃、种植等绿化用地。

Baotou Children's Palace and Library 包头市少年宫、图书馆

Location: Neimenggu, Inner Mongolia Autonomous Region
Area: 52,046m²
Architect: China Architecture Design Institute
Photographer: Zhang Guangyuan
Completion Date: 2012

地点：中国，内蒙古
建筑面积：52,046平方米
建筑师：中国建筑设计研究院
摄影师：张广源
建成时间：2012年

Inner Mongolia, China
中国，内蒙古

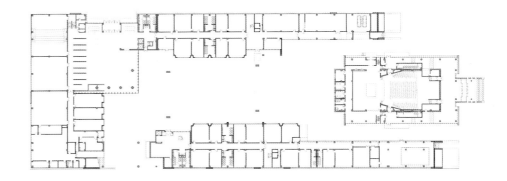

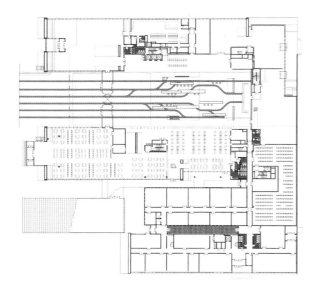

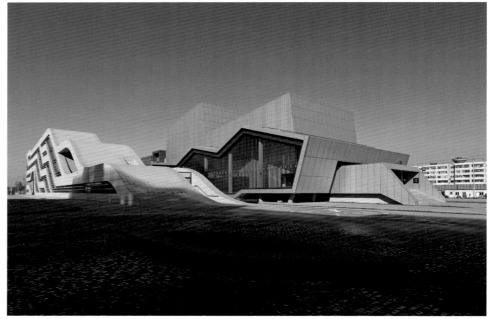

The architects are inspired by the special terrain of the rolling Inner Mongolia grassland. The site measures more than 400m from east to west and looks like it grew out from the land. The curved back set design of the children's palace and library fully encourages the interaction between the users, public space and the buildings. The courtyard extending from indoor to the outside maximise the sun exposure. The open staircase and corridor, together with streamlining arrangement of the greenery area and paving reaching for the interior of the two buildings, enhance the visual continuity in space and provide a vague space of drama.

A variety of materials are used in the new library and palace facades, including stone, compound ecological panels, aluminium extrusions, and glass brick embedded concrete panels. All materials perform well with each other in the same space and the integrity and harmony of the two buildings are enhanced. At the same time, they are independent from one another for the different functions: the library uses mainly centralised large space but the children's palace comprises small units. Different forms and features emerge in the connection with the central public square.

In the children's palace, a circular communication area is formed by the C shaped courtyard and the central square. By directing the eyes of the visitors with objects of different height, colours and shape on the circular platform, different game areas are available for playing. The central square serves as an empty space to emphasise the volume of the library. At the central square, you will find yourself facing the gentle slope of stone at different angles on the west facade of the library, as it indicates the public nature of the architecture to the city in a most welcoming way.

设计的灵感来自于内蒙古大草原的起伏延绵的特殊地貌特征，场地自东向西形成了400余米巨大的跃动曲线，整个建筑群就像从地面生长出来一样。图书馆和少年宫曲线型的退台设计，最大程度的促进了使用者、公共广场和建筑之间的互动。由室外延伸至室内的中庭最大程度地导入阳光，开放式的楼梯和连廊以及延伸进两栋建筑内部的流线型绿化区域和铺装部分，增强了空间上的视觉连贯性，为人们提供了一个充满戏剧感的模糊空间。新的图书馆、少年宫外立面运用了多种材料：石材、复合生态板、铝型材、预嵌玻璃砖混凝土挂板等，各种材料特征被高度协调在一个语境下，加强了之间的整体性与交融感。同时它们又彼此独立，根据其各自不同的功能，图书馆集中式大空间少年宫多为单元式小空间，结合中央公共广场产生了不同的形式特征。对于少年宫来说，内部C形院落与中心公共广场组成了一个环形交流空间，并且在环形的空间界面上通过不同高度、不同色彩、不同块体的造型进行视线引导，产生出适合儿童嬉戏游玩的各个平台。而中心广场作为一个虚空间正好能衬托出图书馆的体积感，站在中心广场之上迎面而来的是图书馆西侧主立面不同角度的石材缓坡，它们以一种热情开放的姿态呈现着，向城市传递着强烈的场域公共性。

Number Two Middle School at North District, Shenyang

沈阳市第二中学北校区

Location: Shenyang, Liaoning Province
Area: 87,000m²
Architect: Lv Panfeng / New World Architecture Design Co., Ltd., Shenyang
Photographer: Lv Panfeng
Completion Date: 2006

Liaoning, China
中国，辽宁

地点：辽宁，沈阳
建筑面积：87,000 平方米
建筑师：吕攀峰 / 沈阳新大陆建筑设计公司（中国）
摄影：吕攀峰
建成时间：2006 年

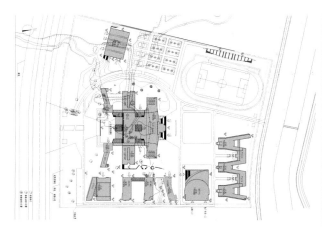

Number Two Middle School at North District is located at New District at the North of Shenyang, on the south bank of Pu River, with New Town Road at the south, South Pu Road on the north, and North Dili Street on the east. An architecture for 48 classes of junior school students is sited on a plot of 158,700 square metres.

The design principle of this project is to take advantage of the site and make some reasonable changes, illustrating the frozen music with plain notes. The surrounding natural environment is to be kept as much as possible.

The site is a sloping plot with the south end higher than the north. In the middle there is a terrace with a height difference of three to four metres, which is well adopted in the one-storey central library. The library connects the single buildings together, forming a two-storey circulation system linking interior spaces on different heights. Varied heights and interior/exterior spaces, as well as tens of existing trees, provide a beautiful environment for the teachers and students at Number Two Middle School.

Façades of the teaching building are made of 300mm-wide building blocks, either grey-painted or clad with grey stone. Fixed vertical louvres are adopted on façades of the stadium in order to reduce the penetrating sunshine for a better interior lighting effect.

In classrooms in the main teaching building, columns are arranged in a grid system of 9.0mX7.8m. The floors are made of high-tensile thin plates, with less weight and thickness than cast-in-situ ones, thus reducing project cost. The columns in the library are regularly distributed, and a special cross roof girder is adopted. A semi-exterior courtyard is designed in the centre of the library, which has a steel-structured roof with glass skylights to ensure natural lighting and ventilation for the semi-underground library.

沈阳二中北校选址于沈北新区蒲河南岸，新城路北侧、蒲南路南侧、地利北街西侧。用地面积158700平方米，是48个班的高级中学。

该方案设计的指导原则是"型有其用、变有其理"，用朴实的音符诠释凝固的音乐，充分利用现有地势并尽力保留现状环境。

基地内地势走向由南向北逐渐降低，基地中部有3至4米落差的台地。设计中充分利用了这一地势特征，将3至4米的台地断面设计为一层高的中心区图书馆。图书馆将各单体教学楼有机结合在一起，形成的双层水平交通系统联系着位于不同标高的室内外空间。多变的地势、丰富的室内外庭院空间同极力保留的几十棵原状树为二中北校区的师生提供了优美的教学及生活环境。

校区建筑外墙为300毫米厚建筑砌块，外饰面采用灰色涂料及灰色石材。体育馆的外墙采取了固定垂直百叶式的构造方式，用来削弱直射阳光对室内照明的影响，减少室内运动场地的眩光。

主教学楼教室柱网为9.0米×7.8米，楼板采用了高强薄壁管楼板结构，减小了现浇板厚及自重，从而降低了工程造价。图书馆柱网比较均匀，在屋面梁特别采用十字梁布置；图书馆中心部位设置半室外庭院，屋面为钢结构加上玻璃采光顶，用来为半地下图书馆提供自然采光及通风。

Northeast Yucai Bilingual School
东北育才双语学校

Location: Shenyang, Liaoning Province
Area: 90,000m²
Architect: Ji Peng / New World Architecture Design Co., Ltd., Shenyang
Photographer: Ji Peng
Completion Date: 2006

地点：辽宁，沈阳
建筑面积：90,000 平方米
建筑师：计鹏 / 沈阳新大陆建筑设计有限公司（中国）
摄影：计鹏
建成时间：2006 年

Liaoning, China
中国，辽宁

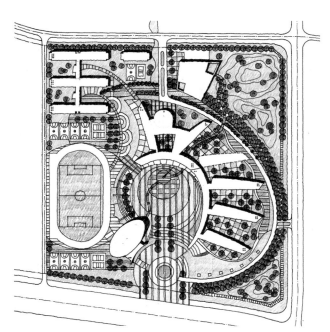

In the design of Northeast Yucai Bilingual School, the architects are glad to make some change, different from traditional educational architecture. The change occurred from design to construction, and even would continue with future development of the school. The architects used a romantic, elegant and Chinese-styled architectural language in the design. Grey and white constitute the main colour scheme, with punctuating black metal and shining glass curtain walls. Stone pillars and wainscots enrich the texture of the buildings. All the walls clad with tiles are well calculated, having accurate amounts and sizes of tiles even for a most complicated façade.

Besides, a garden-like environment is also a key point in the design. The existing fertile soil on the site provides a good planting condition. The architects chose various proper arbors, shrubs and grasses. The height difference enriches the school environment, and students can explore the mystery of nature in several small gardens in break time or when doing morning reading. Water is indispensable in gardens, so the architects made a small stream, which immediately enlivens the ambience. Fishes swimming in the water, birds flying in the sky, breezes stroking your face… This is heaven for a memorable childhood. A wooden bridge is set up above the stream, and bamboo boats are ready for a row. In winter when the water is frozen up, you can even skate.

在东北育才双语学校整个的设计过程中,设计师高兴地看到了一种变化——明显区别于以往教育建筑的分格。这种变化不但伴随着设计建造的始末,而且也将伴随着学校日后的历史变迁。设计师把一种浪漫典雅又富有中式设计韵味的建筑处理手法贯穿于整体设计中。灰、白成为校园的主要基调,黑色的金属、晶莹的玻璃幕墙点缀其间,一些体现石材自然肌理的石柱、石墙裙所带来的质感让人体会到一种文化底蕴流淌在建筑中。所有的建筑墙面都进行了精心的排尺,包括面砖的大小组合、分析的宽窄疏密,即使是一块复杂的立面单元,也能精确地标明所需的贴砖数量。

另外,亲近自然的花园式校园环境设计也是这一过程中的重要环节。基地内现有肥沃的土壤提供了适宜的种植条件,在设计中考虑了各种适合的乔木、灌木和草场,这种高度上的良好搭配丰富了地面植被的层次感,同时,其中还自由设置了多个小面积的植物标本园,可以让学生们在课间时、晨读中就能直接探求生命的深奥,体会自然的选择。花园中不能缺少水的点缀,于是设计师引入了一条河流,打造曲水流觞的意境。浅浅潺潺,清澈见底,水中荡漾微波,鱼儿跳跃,鸟儿掠过水面,清风拂面,年少的憧憬伴着这诗一般的仙境生长着,就像白天鹅在水面游动一样,在平淡中升华着自己的理想。踏着水面上的木板桥,划着竹木小舟在河水中畅游一番,冬天水冻住了还可以滑冰,这可以说是年少成长的一个真正天堂。

Luxun Academy of Fine Arts
鲁迅美术学院

Location: Dalian, Liaoning Province
Area: 139,000m²
Architect: Auer+Weber+Assoziierte, Munich
Photographer: Auer+Weber+Assoziierte, Munich
Completion Date: 2006

地点：辽宁，大连
建筑面积：139,000 平方米
建筑师：奥尔 + 韦伯 + 合伙人建筑设计有限公司
摄影：奥尔 + 韦伯 + 合伙人建筑设计有限公司
建成时间：2006 年

Liaoning, China
中国，辽宁

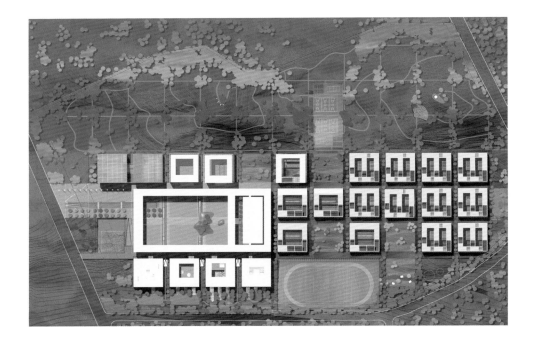

160 / NORTHEAST CHINA

Within an expansive open landscape, a memorable place for the arts shall be achieved, which obtains its appearance and its energy from a closed architectural ensemble, implanted in the landscape. The rolling topography is emphasised as a memorable landscape element through the overlaying of a "stamped" grid corresponding precisely with the four points of the compass, with modular fields of identical size.

The various city components with their respective functions and their identical compact dimensions are inserted into the system of streets, lanes and squares, whose modular dimensioning also permits adjustment to changing programmatic requirements without breaking through the organisational structure. Only the central square – the "agora" – extends beyond the single module as an outdoor reference and focus point for the campus, as does the museum in accordance with its heightened significance.

The spatial experience of the "City of Arts" arises from the dialogue-like tension between the fundamental space-defining architectonic elements and their relationship to landscape by means of enclosure, opening and framing.

鲁迅美术学院选址在一片开阔的景观内，封闭式的建筑群营造了一个永恒的艺术圣地，并被赋予了独特的外观及活力。延绵起伏的地形决定了建筑的造型，整个地块被分成体量相同的四个部分，逐级上升的建筑结构被堆砌在一起。

街道、幽径、广场等不同城市元素 "压缩" 之后被全部引入进来，发挥着各自的功能，模块般的造型使各自部分可根据需要加以调整，同时无需改变整体结构。中央广场是唯一一个超越了模块功能的结构，延伸出去并成为整个校园的核心。

"城市艺术" 这一空间体验理念源于不同结构之间的 "对话"，通过封闭结构、开放式结构以及不同框架结构的运用进而决定了建筑元素的确立以及其与周围景观的关系。

Kindergarten for the Dalian Software Park

大连软件园幼儿园

Location: Dalian, Liaoning Province
Area: 4,600m²
Architect: Charles Debbas / Debbas Architecture
Photographer: Shu He
Completion Date: 2010

地点：辽宁，大连
建筑面积：4,600 平方米
建筑师：查尔斯·德巴斯 / 德巴斯建筑设计公司
摄影师：舒赫
建成时间：2010 年

Liaoning, China
中国，辽宁

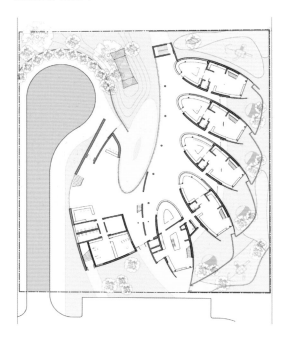

Response to the local climate was an integral part of the initial schematic design. Both the shape and location of the large, bow-shaped concrete façade of the administration wing serve to deflect cold, winter winds around the school while the classrooms are open at the south and east to take advantage of the sun for day lighting and passive heat gain on colder days.

Exterior materials consist mainly of architecturally form-worked concrete, pre-finished wood composite wall panels and boards, tempered tension-mounted insulated glass curtain walls and a sculptural zinc-coated steel roof over the administration wing. Interiors are to be warm, with most floor and ceiling surfaces finished in natural woods. Composite materials will be used in wet areas while carpet will be used in more intimate and sound-sensitive areas.

Each classroom is contained inside a classroom "pod" that consists of two split-level classrooms as well as a dedicated ground floor educational room adjacent to the main hall. The pod concept takes its inspiration from flowering seedpods that gently release their fragile sheltered seeds allowing the wind to carry them to eventually take root, blossom and renew the cycle of life.

Overall, the concept was to create a fluid and spiritual environment distant from the more mechanistic, rigid and prosaic late-modern designs that have come before it. The architecture tries to invigorate a child's sense of wonder and generate unique memories without resorting to theatrical stage design or applied nostalgic detailing.

设计师在着手项目之初便将当地气候条件作为整体理念的一部分，因此无论在建筑造型还是外观上都着重考虑这一元素。行政管理楼弓形的混凝土外观用于抵御冬季的冷风，教室全部朝向东、南方向，充分利用自然光线。

外观材质主要包括混凝土、预制木材复合板、压力绝缘玻璃幕墙以及锌覆层钢材屋顶。室内设计主要突出温暖的氛围，地面与天花板多以天然木材装饰。其中，易潮湿地区主要采用复合材料装饰，而较为私密及噪音敏感区则采用地毯装饰。

教室包含在不同的"豆荚"结构内，其中每个"豆荚"内设置上下两间教室以及与一层大厅先练的教育空间。建筑造型源于"开花的结荚植物"，缓和地释放出成熟的种子，让它们随风散落，然后扎根于土壤中，开花结果，诠释出生命的循环。

总之，设计的任务即为打造一个流畅并具备自身灵魂的建筑环境，远离那些机械化造型、毫无特色的结构。这一建筑能在一定程度上激起孩子们的好奇心，无需通过精心的细节刻画便可给他们留下独特的回忆。

Howard Johnson Parkland Hotel
百年汇豪生酒店

Location: Dalian, Liaoning Province
Area: 30,000m²
Architect: gmp Architekten – von Gerkan, Marg and Partners
Photographer: Hans-Georg Esch, Jan Siefke
Completion Date: 2009

地点：中国，大连
建筑面积：30,000 平方米
建筑师：gmp 建筑设计事务所
摄影：汉斯–乔里·埃施、简·赛风
建成时间：2009 年

Liaoning, China
中国，辽宁

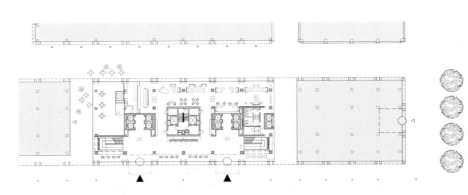

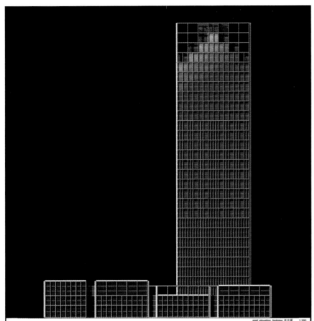

The hotel is part of the Dalian Financial Centre ensemble. The approximately 300 hotel rooms and suites in the 5-star hotel are accessed by means of a 30-metre-high and elongated atrium. The reception, lobby, lounge and restaurants with show cooking zones as well as a conference area with ocean view are located at the base of the building.

The two-storey entrance hall is completely clad with beige-coloured natural stone from France. It is the primary material in the hotel in combination with stucco lustro wall coverings, palisander wood, matte suede leather and metallic screens. The reception desk is covered with brown suede leather and functions as the focal and distribution point to the lifts and the hotel room levels. Passing through the entrance hall, which has been conceived as a flowing room, the hotel guest is directed to the lobby lounge, which has been outfitted with a cocktail bar. The generous and comfortable lounge seating accounts for early evening relaxation.

An elongated and slender atrium vertically connects all floors from the fifth to the twelfth level and serves as a means of access to the hotel rooms.

Each room has either a balcony or a so-called "French window" and can be naturally ventilated. The bathroom has been inserted into the room as a "wooden box".

百年汇豪生五星酒店是大连金融中心整体规划的一部分，共包括300间客房，入口为30米高的中庭。大堂、接待台、休息区、开放式餐厅及会议中心位于建筑底层，可欣赏外面的海景。

两层高的入口大厅全部采用来自法国的米色天然石材饰面，其与来自德国的全新墙面装饰（stucco lustro）、黑黄檀木、亚光绒面革以及金属屏风作为主要装饰材料。接待台构成了空间的焦点元素，指向通往不同楼层的电梯，台面采用褐色毛绒皮革装饰。穿过入口大厅，客人便可到达休息区，内设有鸡尾酒吧。宽大而舒适的座区提供了一个理想的休息场所。

Maritime Museum of Arts
海中国·美术馆

Location: Dalian, Liaoning Province
Area: 4,200m²
Architect: URBANUS/ Wang Hui, Tao Lei, Zhao Hongyan, Du Aihong, Hao Gang, Chen Chun, Liu Shuang, Zhang Yongjian, Zhang Yongqing
Photographer: Yang Chaoying
Completion Date: 2008

地点：辽宁，大连
建筑面积：4,200平方米
建筑师：都市实践 / 王辉、陶磊、赵洪言、杜爱宏、郝钢、陈春、刘爽、张永建、张永清
摄影：杨超英
建成时间：2008年

Liaoning, China
中国，辽宁

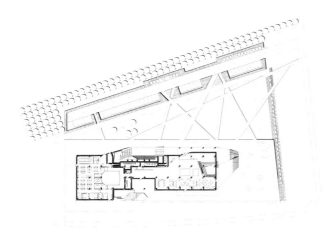

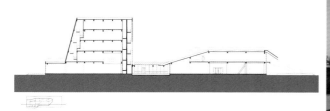

Maritime Museum of Arts at Dalian Development Zone developed by China Resources Land is a large-scale estate project on a land of sea reclamation. The Sales Centre would stand alone facing the endless sea for a long time, so the architects tried hard to make the architecture "survive" with energy. Among several initial proposals, the architects and the client both chose the one that translates architecture into landscape, i.e., the architecture is conceived as an artificial hill made of a series of wooden terraces on the sand beach.

The architecture has its entrance to the north while facing the sea to the south. From the parking lot, the obliquely crossing floating bridges bring visitors to the porch, where outdoor staircases leading to the first floor are placed on both sides. The staircases also serve as viewing platforms, expecting dramatic events to happen in front of the entrance. As you enter the entrance, you would encounter an unfolding scroll painting of sea at the bottom, communication spaces with views to the sea in the middle, and the model zone ahead – a two-storey-high space. The spaces are not static, as on the left a dramatic terraced stage links the ground and first floor. Numerous floating "silver fishes" are used to further enhance the dynamism of the spaces. These swirling fishes also bring the eyes to the bridge on the first floor on the right. The bridge leads to the outdoor terrace on the first floor, where you can meet people who step up from the southern and northern staircase, and then move forward to the second floor terrace. This terrace is the best place to enjoy a sun bath and views to the sea, and also a good place to hold a rooftop party. Without a doubt, the outdoor terraces bring not only dynamism, but also opportunities for more activities.

The architecture becomes a scenery spot with the scenery of the sea. The developer named it "Maritime Museum of Arts" probably because they didn't want to mix it with conventional estate developments. In museums you enjoy pieces of arts with your eyes, but here, you have to experience the architecture with your whole body, particularly with your ears to hear the stories and events that happened here.

华润置地在大连开发区的"海中国"项目是一个填海而出的百万平米大盘。它的销售中心在相当长的时间内要独自面临浩瀚的沧海，怎样的建筑才能让这海边的一粟能存活呢？面对初期的几种构思，建筑师和开发商都一致地选择了把建筑变成地景的方式，即建筑是由沙滩上一片木平台演绎而来的假山。

建筑南面是大海，入口朝北。从停车场出来，斜交的浮桥把人吸引到入口门廊。门廊两翼是通向二层的室外楼梯，这些楼梯又构成了室外的看台，等待着门前戏剧的发生。进了大门，底景是长卷般展开的海，中景是看海的洽谈空间，前景是两层挑高空间下的模型区。这个空间不是静的，左手边一个戏剧化的阶梯讲坛把一层和二层串连起来。室内设计师用漂悬在空中的无数银鱼把动感弥漫到整个空间中，这些漩涡状的游鱼进而把视线引到入口上空右手二层桥。这座桥通向二层的室外平台，在这里可以和室外南北两侧拾级而上的人会合，再折身走向三层的平台。三层的平台是享受海景和日光浴的绝佳之地，也是举办屋顶派对的好场所。不言而喻，室外的平台不仅仅带来了运动，还带来了活动。

这栋建筑成为海边风景中的风景。开发商大概不舍得把这栋建筑和一般的房子混淆，将之命名为"海中国美术馆"。美术的体验要靠眼睛，但体验这栋建筑需要身体。当然还需要耳朵，去听她的身上曾经发生过的故事和事件。

Museum of Culture, Fine Arts and Science, Changchun

长春美术馆、科技馆及文化馆

Location: Changchun, Jinlin Province
Area: 107,500m²
Architect: gmp Architekten – von Gerkan, Marg and Partners
Photographer: Marcus Bredt
Completion Date: 2011

地点：吉林，长春
建筑面积：107,500 平方米
建筑师：gmp 建筑设计事务所
摄影：马库斯·布赖特
建成时间：2011 年

Jilin, China
中国，吉林

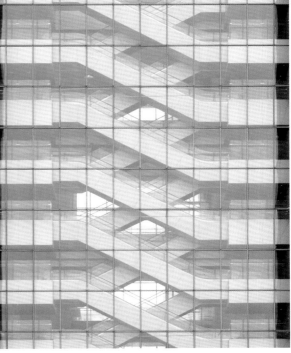

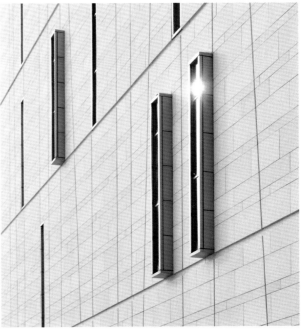

Three stone cubes house three museums arranged round a central entrance building like the sails of a windmill. The configuration of the buildings clearly expresses the functional distinction between the museums. Each of the three cubes has a sculptural feel – lofty, individually tailored air spaces act as central lobbies and set their stamp on each museum.

The lobby for the Museum of Culture is derived from the shape of a deep canyon. The space of the Museum of Art features clear diagonals, as in a modernist painting. The Museum of Sciences has a central foyer that cuts rationally into the solid mass in right angles, almost as a symbol of technology and the scientific approach.

Functionally, all three museums are similarly organised: a square route leads visitors from exhibition area to another on all levels. Crossing the central space by means of bridges in four places in each museum facilitates orientation in the building. In addition, visitors get impressive views both in the central halls of individual museums and of the landscape and museum park that surrounds the Museum of Culture, Fine Arts and Science.

What's more, the façades of the three museunms have the same colour and material palette. The museum laid the foundation for a new neighbourhood in the southeast of the city. The three museums have the same site development idea: sun-screened exhibition spaces arranged along the visitors' tour.

长春美术馆、科技馆及文化馆由三个石材外观的立方体形状结构组成，围绕着中心入口建筑排列，犹如风车的风叶一般。简约的造型清晰地体现了场馆的不同功能，其中每个"立方体"都如同一座雕塑，中心大厅设计独特，从这里可以走入不同的场馆。

文化馆的大厅设计从"峡谷"中获得灵感；美术馆大厅突出清晰的斜线结构，好似一幅现代风格画作；科技馆的大厅直接通向内部，犹如一条科技通道。

从功能上看，三个场馆布局简单——中央通道通向各层之间的不同展区，上层通过"桥"连通，使得整个建筑格局更加灵活。此外，在这里，参观者可以看到不同场馆的中央大厅以及周围的景观。

更值得一提的是，三座场馆在外观色彩及材质上完全相同，不同的是其各自的表达方式以及立面细节处理。它们更源于统一设计理念：展区沿着便于参观者行走的方向展开。这一建筑为长春东南新城的发展奠定了基础。

Shanghai World Financial Centre
上海环球金融中心

Shanghai, China
中国，上海

Location: Shanghai
Area: 381,600m²
Architect: Kohn Pedersen Fox Associates PC, East China Architectural Design & Research Institute
Photographer: Mori Building Ltd, Tim Griffith
Completion Date: 2008

地点：上海
建筑面积：381,600 平方米
建筑师：美国 KPF 建筑师事务所、华东建筑设计研究院有限公司
摄影：日本森大厦株式会社有限公司、蒂姆·格里菲斯
建成时间：2008 年

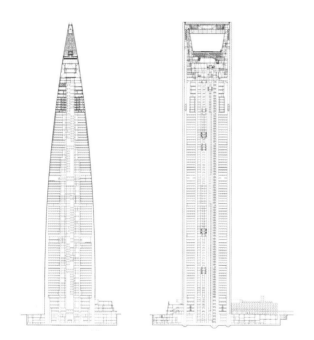

170 / EAST CHINA

A square prism – the symbol used by the ancient Chinese to represent the earth – is intersected by two cosmic arcs, representing the heavens, as the tower ascends in gesture to the sky. The interaction between these two realms gives rise to the building's form, carving a square sky portal at the top of the tower that lends balance to the structure and links the two opposing elements – the heavens and the earth.

Soaring 101 storeys above the city skyline, the Shanghai World Financial Centre stands as a symbol of commerce and culture that speaks to the city's emergence as a global capital. It features the highest occupied floor and highest public observatory in the world, and was recently recognised by the Council on Tall Buildings and Urban Habitat as the Best Tall Building in the World 2008.

The elemental forms of the heavens and the earth are used again in the design of the building's podium where an angled wall representing the horizon cuts through the overlapping circle and square shapes. The wall's angle creates a prominent façade for the landscaped public space on the tower's western side, and organises the ground level to provide separate entrances for office workers, hotel guests and public access to express lift service for Sky Walk visitors. The wall is expressed in Jura yellow limestone, and the base of the tower is clad in Maritaca green Brazilian granite with a split-face finish, which contrast beautifully with the metal of the circular wall and diaphanous glass skin enveloping the retail volume.

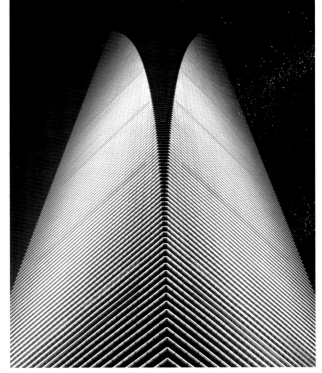

方形棱柱体在中国古代象征"地",半弧则象征着"天"。这一设计便从中获得灵感,两个巨大的半弧结构在顶端交叉,形成了中空方形棱柱体,奠定了整幢大厦的基本造型,更将"天"与"地"两个对立的元素联系起来。
大厦高达101层,是上海经济与文化不断崛起的象征。作为全球可居住的最高楼层高度及最高公共观景台建筑,上海环球金融中心曾被"国际高层建筑与城市住宅协会"评为"2008世纪最高建筑"。
此外,"天"与"地"的造型在建筑基座的设计中再一次被运用,角状墙壁象征着穿过层叠的圆弧与方块造型结构的水平切口。此外,墙壁呈现一定角度在西侧公共区处形成了独特的外观,并打造了不同的入口,分别共大厦内公司员工、酒店客人及公众使用。墙壁上黄色石灰岩与大厦基座绿色大理石同环形墙壁上金属材质及商业空间玻璃外观形成鲜明对比。

Atlas of Contemporary Chinese Architecture

You You Grand Sheraton International Plaza
由由喜来登国际广场

Shanghai, China
中国，上海

Location: Shanghai
Area: 200,000m²
Architect: B+H Architects
Photographer: Kerun Ip.
Completion Date: 2007

地点：上海
建筑面积：200,000 平方米
建筑师：B+H 建筑设计公司
摄影：科伦·伊普
建成时间：2007 年

You You Grand Sheraton International Plaza spans two city blocks. The development consists of three towers: hotel, residential, and office. The towers share a common form but are individually distinguishable through variations in the floor-plate, cladding choices and height. The development meets the region's strict building codes. For example, all rooms have access to natural light and ventilation, and all primary rooms such as master bedrooms and living rooms face south.

The west block is comprised of two towers, a 33-storey hotel and a 29-storey residential building, which are connected by a mixed-use podium at the base. Amenities on the ground floor include two restaurants, a café, retail space and private lobbies for both the hotel and residential towers. The shape of the podium forms a courtyard that is beautifully landscaped to create a welcoming oasis to the public and a calming retreat for guests and residents.

The east block consists of a 23-storey office tower that sits atop a 5-storey podium composed primarily of retail space. The two blocks are connected by a pedestrian bridge that joins the conference centre to the office tower.

由由喜来登国际广场横跨两个街区，共由三幢建筑构成：酒店、住宅区及办公楼。三幢建筑呈现统一造型，但在底板、覆层材料、高度上带有差异。这一项目在设计上严格遵循这一区域特色，如所有的房间全部利于光线照射及自然通风，而卧室及客厅灯主要空间全部朝向南侧。

西侧由33层的酒店及29层的住宅楼构成，之间通过多功能基座结构连通。基座结构内布置着餐厅、咖啡馆、商店以及分别通往两幢大厦的大堂。基座的独特造型使其构成一个美丽的庭院，为公众、酒店客人以及住户提供了一个完美的"休憩绿洲"。

东侧是一幢23层的办公楼，"耸立"在5层高的基座上。两个街区之间通过步行桥连通。

Wanxiang Plaza
万向大厦

Location: Shanghai
Area: 42,000m²
Architect: gmp Architekten – von Gerkan, Marg and Partners
Photographer: Hans Georg Esch
Completion Date: 2010

地点：上海
建筑面积：42,000 平方米
建筑师：gmp 建筑设计事务所
摄影：H·G·埃施
建成时间：2010 年

Shanghai, China
中国，上海

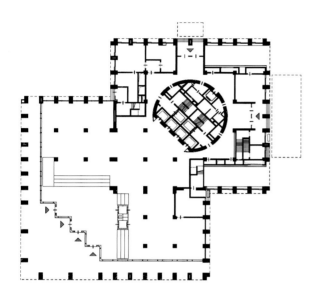

The site in Pudong, Shanghai, on which the regional headquarters of the Wanxiang Holding Group Corp is built, is impressive with its unique location on the Huangpu River. One of the building's distinctive features is its façade arrangement: typologically it is a façade faced in natural stone that has a consciously intense sculptural depth and tall window openings extend over two storeys each. On the outside they have vertical metal grilles, behind which on the inside are placed ventilation windows.

The sculptural form of the façade is continued on the building's base by the deeply recessed windows and a high colonnade, which leads into the three-stoery high main entrance. This is carved into the building's base with set backs and faces south, around a corner. The structure's nineteen storeys reach a total height of 79.40 metres including the roof level. A three-storey atrium connects the levels with one another and, in so doing, creates a pleasant spatial expanse. The circular core, which is written in the square base of the tower, takes up the necessary infrastructure of the office floors. White natural stone is the material used throughout on the façade and the floors of the lower base levels. This is contrasted on the building's interior with hunter green wall surfaces of glass and warm precious woods.

万向集团大厦选址在黄浦江沿岸，其主要特色之一即为外观设计。自然石材打造的表皮呈现一定的进深，高大的窗户延伸两层的高度；外侧为垂直的金属条格子结构，后面安装着通风窗，营造出雕塑般的造型。

表皮雕塑般的结构同样被运用到基座的设计中，"内陷"的窗户以及通向主入口的高高石柱廊奠定了整体造型。主入口高达三层，朝向南面。整幢建筑共18层，高为79.4米。三层高的中庭将不同的楼层连通，营造强烈的空间感。环形中心结构内装置着所有办公楼层的基础设施，天然石材构成主要表皮材质，与室内绿色玻璃墙饰以及暖色木材形成鲜明对比。

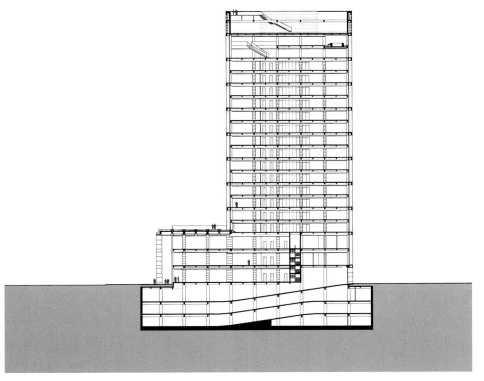

Riviera TwinStar Square
浦江双辉大厦

Location: Shanghai
Area: 200,000m²
Architect: Arquitectonica
Photographer: Rogan Coles
Completion Date: 2011

地点：上海
建筑面积：42,000 平方米
建筑师：Arquitectonica 建筑室内设计规划公司
摄影：罗根·科尔斯
建成时间：2011 年

Shanghai, China
中国，上海

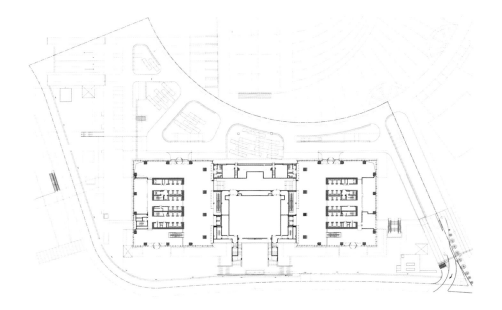

176 / EAST CHINA

The design of the office towers forms a gateway in the shape of a ship aligning the axis of slipway and leads to the Celebration Park on the waterfront. The spectacular site fronts the Huangpu River and the historic Bund, and the slipway will be restored for use as a new museum.

The towers will be occupied by two different bank headquarters, so architects designed the symmetrical towers that are all about balance, order, and stability. Their interdependence sends the message of union and respect for each other at once.

As they face each other, their façades curve dramatically to form an imaginary space that frames the skies and the city skyline. Their nautical symmetry will convey memories of the ships that were once launched from the now relocated Shanghai Shipyards. Architects want the towers to glow from within but designed the inner curves at the centre to be lit to emphasise the distinctive shape and monumental scale of the space.

双辉大厦呈现船舶的造型，沿着船台轴线"排列"，犹如一直通往江边"革命公园"的入口。处于江边及外滩的独特地理位置赋予其极大的优势，而船台结构将被保护起来用作新博物馆。

大厦业主为两家银行总部，设计中强调对称性、平衡性、顺序性以及稳定性。相互依赖的造型传达出两个结构共为一体的信息，同时又相互独立。

两幢建筑对面而立，弯曲的立面构成了一个"梦幻"区域，仿佛将天空和城市天际线"框入"其中。与船舶设计相关的对称性让人不禁联想到上海船厂中停泊的轮船。设计师将照明设施全部设置在室内空间，中心区域内的曲线结构在灯光的照射下使得建筑整体造型以及其宏伟空间规模更加突出。

BEA Financial Tower
东亚银行金融大厦

Location: Shanghai
Area: 70,000m²
Architect: TFP Farrells
Photographer: Paul Dingman Photo, Zhou Ruogu
Completion Date: 2009

地点：上海
建筑面积：70,000 平方米
建筑师：TFP 建筑设计事务所
摄影：保罗·丁曼摄影公司、周若谷
建成时间：2009 年

Shanghai, China
中国，上海

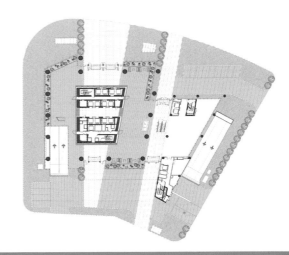

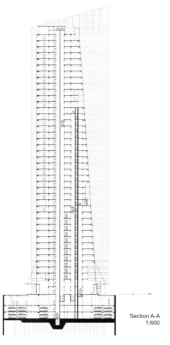

Section A-A
1:600

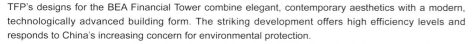

TFP's designs for the BEA Financial Tower combine elegant, contemporary aesthetics with a modern, technologically advanced building form. The striking development offers high efficiency levels and responds to China's increasing concern for environmental protection.

Although there are view corridors of the river and the Bund, the site is set back from the waterfront and has to compete with prominent high rises, notably the Jin Mao Building, currently the tallest skyscraper in China. TFP proposed a structure that was layered into three principal forms. A central circulation and service core is flanked by two floor plates with the west wing of the building rising above the other two components. The creation of this stepped effect brings a level of clarity and directness to the building's massing. Each element functions independently but is bound into a singular composition by complementary materials and modularity. This adds significantly to BEA Financial Tower's instant-recognition factor and enhances both the view potential and the building's silhouette on the skyline.

The façades react differently to the environment through orientation, materials and technology within the building envelope. Fluctuations in heat gain and loss are limited, the building's sustainability is maximised, and operational efficiency is improved. Following detailed analysis into solar insulation, four types of cladding were established, each of which is designed to deal with a specific environmental aspect. To minimise excessive solar gain and building heat load on the south-west and south-east elevations, the percentage of glazed areas is reduced, horizontal shading devices are provided and low-emissive glass is used.

这一建筑集典雅、现代美学及融合先进技术的造型于一身，达到高水准要求的同时，更重要的是满足了中国当前时代日益关注的绿色理念要求。

建筑选址在远离河岸处，与中国目前最高的建筑，金茂大厦相"抗衡"。三种造型层叠在一起，中央通道及服务中心通过楼板"挂"在西侧结构上，似乎从其他两个结构上"冉冉升起"一般。阶梯状的造型突出了层次感及简约性。相互补充的材质以及模块结构的运用使得每一种元素在行使自身功能的同时，又依附于整体。这一设计增添了建筑本身的可识别性并增强了其潜在的"能量"及形象。

外观通过朝向、材质以及内部技术的运用可应对不同的环境变化，热量的获取及损失的变化被缩小，可持续性被放大，从而实现了能源节约。通过对太阳绝缘材料的仔细研究，四种覆层结构被运用，分别应对不同的环境条件。为减少西南及东南两侧太阳热量的获取，玻璃覆层的运用被大量减少，遮光结构以及低反射玻璃被采用。

City Hall of the Shanghai Nanhui District
上海南汇区政府办公楼

Location: Shanghai
Area: 100,860m²
Architect: gmp Architekten – von Gerkan, Marg and Partners
Photographer: Marcus Bredt, Berlin
Completion Date: 2008
Award: 2005 Competition 1st prize

地点：上海
建筑面积：100,860 平方米
建筑师：gmp 建筑设计事务所
摄影：马库斯·布赖特
建成时间：2008 年
获奖：2005 年竞赛一等奖

Shanghai, China
中国，上海

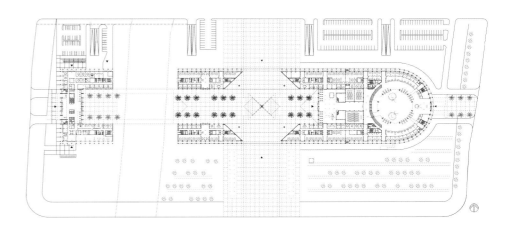

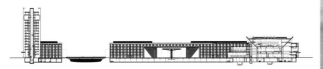

The Administrative Office Centre is embedded in the first green belt, which surrounds the city centre of Lingang New City. A river, which flows through, divides the complex into two separate areas. The prominent position of the plot by one of the main access roads and the important function for the district and the city demands a similarly exposed as well as strict architectural composition for the two individual buildings, which are grouped together to form a superordinate ensemble.

The high-rise at the west side of the site – facing the main entrance axis of the city – is designed as a landmark for Lingang New City. Tow linear office buildings are forming an elongated block. At the eastern end they are joined to a round closure. In between the inner courtyard with its water basins and green trees two pedestrian bridges connect both parts of the complex, leading over the river.

The official access leads from the yard into the representative main lobby. Starting here, a cascade-like staircase leads up to the central zone, which is crowned by the conference hall with a capacity for up to 1,000 people. The public entrance for the inhabitants of the city and the Nanhui District is facing east – to the centre of Lingang New City and the lake.

南汇区政府选址在临港新城中心区周围的第一条绿化带内，毗邻该区的主干道。独特的地理位置以及南汇区的重要功能要求打造一个结构严谨的建筑体。为此，设计师将两幢独立的建筑统一在一起，满足了这一需求。

该场址西侧的建筑朝向城市主入口轴线，被作为临港新区的地标。两幢线型办公楼在视觉上被拉伸，并在东侧交汇形成一个圆形体量。中心庭院内设计着水池，种植着绿色，两座步行桥将两部分建筑连通。

入口设置在庭院内，可通往各自的大堂。此外，瀑布般的楼梯一直通往中心区内。这里布置着可容纳1000人的会议大厅。供大众使用的入口朝向东侧，与临港新城中心及人工湖相望。

Liantang Town Hall, Qingpu
青浦练塘镇政府

Location: Shanghai
Area: 200,000m²
Architect: Arquitectonica
Photographer: Rogan Coles
Completion Date: 2011

地点：上海
建筑面积：42,000 平方米
建筑师：Arquitectonica 建筑室内设计规划公司
摄影：罗根·科尔斯
建成时间：2011 年

Shanghai, China
中国，上海

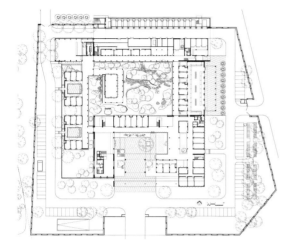

The site of the town hall is located at the centre of the new town area and a canal on the western side connects with the old town area. A major courtyard is located in the centre of this building which has a multiple courtyards structure, and smaller courtyards of different functions are scattered around it. These independent courtyards form different spatial features in a harmonious way, which create a pleasant working environment. A three-storey main office building in two rows is located near the southern square. The ground floor of the southern row is partly elevated, which makes the entrance courtyard between these two rows widely open to the square. The cloister on the second floor, which features a double-storey height, functions as the main building's spatial frame. Meanwhile, four independent offices units and three aerial gardens in the southern row create small pockets of space. In the east is the community service centre. In the north is the conference centre and a cafeteria, with a veranda opposite the main courtyard in the south. In the west are the functional offices directly under the town government, which form three interlocked courtyards connected to the canal through a pierced tracery wall. A library and a recreational room are located in the major courtyard like a bright pavilion, separating the main courtyard into two distinct areas. The building presents an image of continuous courtyards with white walls and single slope roofs with black tiles as its main characteristics. The delicate details require appropriate materials and tectonics with local style, as well as feasible creativity.

镇政府的建设场地在新镇区中心的一片旷地上，西侧的河道向北与老镇区相连。整个建筑是一个四面围合的多重院落结构，围绕着内向的大尺度主庭院铺展；在其周围是一系列大小不一的分属不同功能空间的独立小院，这些庭院由不同的经营位置生成不同的空间性格，层层相应，左右逢源，形成宜人的工作环境。南侧临入口广场为三层两进的办公主楼，南进底层局部架空，使两进间的入口前庭向广场开放。上部两层高的环通敞廊形成主楼的空间骨架，而南进上部则由四个独立办公单元与三个空中花园间隔形成小尺度的空间肌理。东侧是社区服务中心；北部为会务与后勤服务中心，南面对着内院是连续的两层通高檐廊；西侧布置政府直属职能部门，组成三个串联的一层小院，通过镂空花墙与西侧的河道景色相钩连。而图书室和文体活动室如同一个明轩坐落在主庭院内，将院子分为东西两个性格不同的部分。建筑总体呈现为单坡顶为主的连绵的白墙黛瓦院落，细部设计中出于自身内在结构与功能的需要有选择地融入体现地方特征的材料和构造方式，并挖掘其创新的可能性。

Jinqiao Office Park
金桥开发区研发楼

Location: Shanghai
Area: 26,955m²
Architect: AS Architecture-Studio
Photographer: AS Architecture-Studio
Completion Date: 2009

地点：上海
建筑面积：26,955 平方米
建筑师：法国 A.S. 建筑工作室
摄影：法国 A.S. 建筑工作室
建成时间：2009 年

Shanghai, China
中国，上海

The building should keep the connection with block G2 and isolate the noise coming from Jinqiao Road as well as fully use the landscape along the band of Majiabang. Under all these conditions architects design the "U" shape scheme with an open to the south. The garden is surrounded by the building on three sides with the southern opening to block G3. The east part of the building is raised to block the noise from Jinqiao Road, while the height of the west part is reduced for direct sunshine. The vegetal roof protects the roofing and interior space. The arch-shaped raising roof covered by massive greensward becomes a symbolic ecologic architecture of this area.

The treatment of the façade is critical for the appearance of the building. Usage of biotechnology promises the quality of the workspace, access of the light and high level of comfort. The south and north façades are permeable glass walls. The horizontal and vertical sunshades on the south and north façades promise sufficient light to enter into the building and provide with a comfortable nature environment through the exchange of interior and exterior sights. Along with the change of the curve of the façade, the length of the outstretching board of the vertical sunshade is gradually reduced from the north to the west and finally vanishes in the middle part of the building where the sunlight can't reach.

建筑需要保持与G2地块良好的联系，同时处理好金桥路较大的噪音，以及利用好马家浜的优美的河岸景观。在这样的前提下，设计师提出了北向开口的U形的方案。

庭院三面由建筑围合，北面向G3地段敞开。东面的建筑高起，阻挡了金桥路的噪音，西面的部分降低，使阳光能够照射进来。屋顶草皮保护了屋面和室内空间，给相邻的办公楼提供了良好的景观。大面积的草皮覆盖在弧形的屋顶上面，与地面脱开一定的距离，形成一个标志性的逐渐上升的生态建筑。

建筑立面的处理是建筑外形的关键部分，它运用生态技术决定工作空间的质量，光线的质量及舒适的程度。建筑的南北表面是一个通透性的玻璃墙面，在水平和垂直的遮阳板的作用下，能够保证充足的光线和良好的室内外景观的交流和舒适的自然环境。随着立面曲线的变化，水平遮阳板的伸出长度从北向西侧逐渐缩小，并且消失在建筑的中段没有阳光的部分。

Cobest Design & Manufacturing
科倍集团

Shanghai, China
中国，上海

Location: Shanghai
Area: 17,500m²
Architect: ABSTRAKT Studio Inc. / Voytek Gorczynski Architect, OAA / Zhong Ke Design Institute
Photographer: Hong Wu
Completion Date: 2006

地点：上海
建筑面积：17,500 平方米
建筑师：ABSTRAKT 设计工作室、Voytek Gorczynski 建筑师事务所、中科设计院
摄影：洪武
建成时间：2006 年

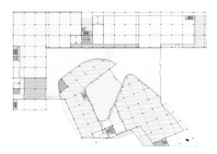

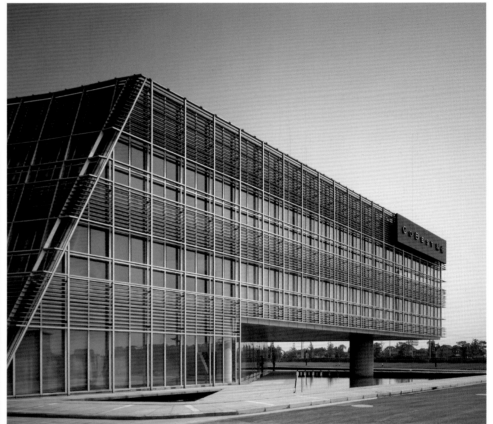

The architectural concept was derived from programmatic juxtaposition of office and manufacturing components, with interstitial space to be an ambiguous interior-exterior landscaped public "condenser". The Chinese character for the "Moon" in the version of standard Chinese script was the origin of the figurative plan for the compound. The rectilinear part of the building contains Design and Technical Centre, Canteen and Dressmaking Shop. More organic free-flowing form houses offices. The office component contains large auditorium space in addition to the training facilities. Residual space between these two elements is composed of partially climate controlled interior and partially exterior landscaped area which provides backdrop for fashion shows and social gatherings.

Originally the interstitial space was to be covered with glass roof. In the design process the glass roof has been reduced to only cover the entrance link between the office and design/manufacturing area. The building containing offices is clad in clear glass and metal curtain wall with dense sunshade layer placed on the south-facing exterior. The volume housing Design, Manufacturing and Technical Centre is clad in local stone. Vehicular traffic has been designed with separate drop-offs in front of the main components of the complex as well as with convenient access to the on-grade parking.

这一项目的建筑理念源于将办公空间与生产车间的独特排列方式，整体格局呈现"月"字型。建筑直线型的部分包括设计及技术中心、食堂及剪裁车间；弯曲的部分设置着办公区，包括礼堂以及培训室等；两部分之间的剩余部分分为温度宜人的室内空间以及室外景观空间，用作服装展示及公共集会。

最初，中间部分打算运用玻璃屋顶遮蔽，但设计过程中决定减少玻璃的使用，因此仅将办公区与设计及生产车间的入口上加盖玻璃顶。办公区部分外观采用玻璃及金属幕墙，南侧安装遮阳板；设计、生产及技术中心部分外观采用当地石材饰面。此外，交通通道经过精心设计，建筑主结构前方设计着独立的车辆入口以及通往停车场的便捷入口。

Henkel Asia-Pacific and China Headquarters
汉高亚太中国总部

Location: Shanghai
Area: 23,000m²
Architect: HPP International Planungsgesellschaft mbH
Photographer: H.G. Esch
Completion Date: 2007

地点：上海
建筑面积：23,000 平方米
建筑师：德国 HPP 国际建筑规划设计有限公司
摄影：H·G·埃施
建成时间：2007 年

Shanghai, China
中国，上海

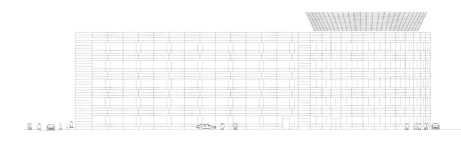

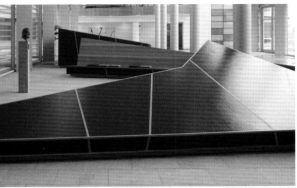

The new administration and research centre stands in a green field and in its basic form represents an asymmetrically curving "U". The design for the new building is based on the concepts of "skin" and "mantle", as well as the reflection of nature within architecture. As the official city flower of Shanghai, the magnolia was the inspiration for various design ideas. From a functional standpoint, the new headquarters mirrored Henkel's identity as an international and innovative company. The designers' guiding principles were environmental compatibility as well as the most modern building standards with respect to sustainable building. In addition, the values of Chinese culture were to be integrated into the design. The building is an earthquake-proof construction with a concrete and steel framework. The most conspicuous sign of the influence of "Feng Shui" is that there are no right angles in the entire building.

With five storeys and a basement, the new building has a gross surface area of 23,000 square metres and approximately 500 work stations. A continuous construction grid of 5.5 metres and façade grid of 2,75 metres ensure the greatest flexibility as well as variable office and laboratory layouts.

汉高亚太中国行政研发中心坐落在一片绿地中，整幢建筑呈现出非对称的"U"形。设计灵感源于上海市花"木兰花"，以"表皮"及"幕墙"为主题，同时在建筑内部彰显自然本色。从功能角度出发，这一设计旨在体现出汉高作为国际公司的特色。此外，设计中遵循可持续发展理念，并将中国文化融入其中。建筑采用钢筋混凝土框架，具备抗震功能。为体现中国传统的风水理念，建筑结构中完全摒弃了直角造型。

建筑共为5层，包括大约500个工作区。外观采用5.5米的格子结构拼接而成，以确保室内格局的灵活性。

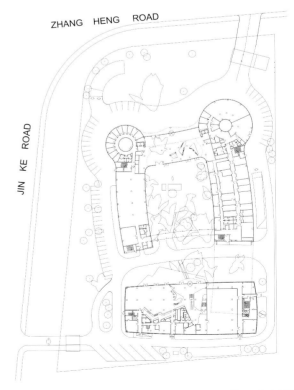

Shanghai International Cruise Terminal

上海国际港客运中心

Location: Shanghai
Area: 263,448m²
Architect: SPARCH
Completion Date: 2010

地点：上海
建筑面积：263,448 平方米
建筑师：思邦建筑设计咨询有限公司
建成时间：2010 年

Shanghai, China

中国，上海

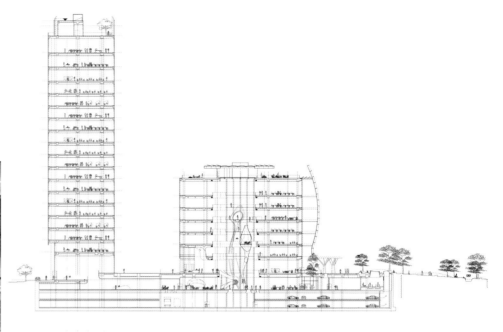

The design of the architecture for the cruise terminal site considered the Herculean scale of the cruise ships that will dock alongside. The brief required that 50% of the total construction area be placed underground, including the cruise terminal passenger facilities (planned by Frank Repas Architects), thus freeing up most of the site as a green park terracing down to the water's edge. SPARCH's challenge was how to deal with the "under world" as well the architecture rising out of it. Their solution was to create ambiguity as to where the ground plane is, by opening up a honeycomb of sunken courtyards. The buildings appear to disappear into these sculpted holes, providing abundant opportunities to explore connections between the ground and "lower ground" levels. The concept also explored the idea of ripples in the landscape being amplified into standing crystal waves that wrap over the buildings. This augmented over time into a second skin that protects the commercial office spaces from their due south orientation, and is populated with semi outdoor balcony spaces overlooking the Huangpu River. The river front faces the city, and illuminates at night into a herring bone array of delicate curved masts that tie the pavilion buildings together. An intriguing gap appears in the middle – a glazed table top supports amorphous pods on cables. One, two and four storey pods contain cafés, bars and restaurants, hovering over a public performance space below. There is a symbiosis between Shanghai's fun-loving desire for diversity, and SPARCH's approach to design, that has made this architecture a reality.

上海国际港客运中心的建筑设计已经考虑到了今后码头船运的巨大吞吐量。在项目任务书中更是要求50%的面积需要作为地下空间，包括为客运旅客设计的设施，然后敞开所有的场地作为绿色公园，层层叠叠地向黄浦江边推进。这里SPARCH的挑战是如何处理"地下世界"以及地上的建筑。对此，设计方案是运用向上开放的一系列蜂窝式下沉广场来创造模糊地平面的效果。建筑如同是从这些蜂巢状的洞口中生长出来的一样，从而为人们提供了更多的机会来探索地面和多层地下空间的联系。层层线性起伏的景观设计同样被引伸到建筑的幕墙设计中去。他们为大楼表面提供了第二层幕墙，让朝南的商业办公空间免受强烈光线的影响，同时在两层幕墙之间设置了室外阳台空间以供人们可以俯视黄浦江景。入夜，大楼表面的鲱鱼骨头般曲度精美的排排桅杆被城市的灯光所照亮，并映射在江面上。而在其中两栋大楼之间的空隙处矗立着一个巨大的玻璃"桌子"，桌子里面悬吊着数个不规则造型的吊舱。这些吊舱悬浮在一个公共演出空间之上，位于桌子内部的一、二、四层，功能分别为咖啡馆、酒吧和餐厅。SPARCH的设计和上海的多元娱乐需求融为一体，最终将这一奇特的建筑变为现实。

Maritime Museum
中国航海博物馆

Shanghai, China
中国，上海

Location: Shanghai
Area: 46,400m²
Architect: Meinhard von Gerkan / gmp Architekten – von Gerkan, Marg and Partners
Photographer: Hans-Georg Esch, Jan Siefke
Completion Date: 2009

地点：上海
建筑面积：46,400 平方米
建筑师：迈因哈德·冯·格康 / gmp 建筑设计事务所
摄影：H·G·埃施、简·赛风
建成时间：2009 年

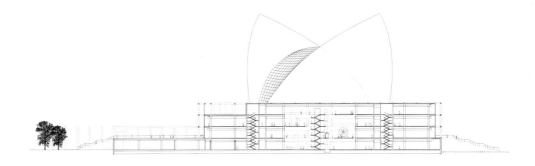

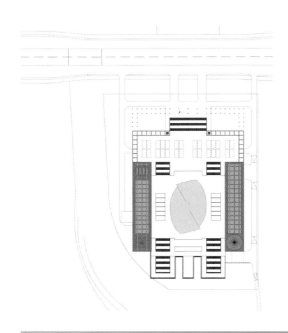

192 / EAST CHINA

The 46,400m² large museum, intended to house an exhibition on China's naval history, lies in the middle of Lingang New City approximately 60km south of Shanghai.

In perfect symmetry, two parallel four-storey-high wings define the rectilinear plan and the functional layout of the museum quarters. The museum is organised around a raised courtyard on a one-storey-high pedestal in which all the functions of the museum are accommodated. From the land and the waterfront side large sweeping staircases lead up to the pedestal. The rectilinear geometry of the façade design, the arcades and the "air-beams" – all clad in a light-grey granite – literally harbours the museum on the lake shore.

The analogy of a ship in a port is developed further in the abstract symbolism of the two "sails" located almost in the middle of the museum square. The 58-metre-tall concave grid shells lean against each other and touching only at one intersection point approximately 40m above ground. This masterpiece of engineering allows the two-way curved "sails" to rest on only two base points, approximately 70m apart. Like a huge fishing net, a pre-tensioned cable net structure clad in glass forms the façade between these two "sails". The enormous cathedral like interior space allows the display of a fully rigged junk.

In the flat coastal topography the Maritime Museum can be spotted for miles due to its expressive silhouette – a feature that will undoubtedly make it one of the most successful landmarks of Lingang New City.

中国航海博物馆选址在临港新城中心，位于上海以南60公里处，其设计目的旨在展现中国的航海历史。

两个四层楼高的侧翼结构平行而立，奠定了整体建筑的直线造型及内部的功能格局。此外，主体结构围绕着一层高的基座上的庭院展开，内部布置着全部功能空间。从陆地及河岸一侧，宽大而威严的楼梯一直通向基座。线型外观、拱廊以及张弦梁结构全部采用浅灰色花岗岩饰面，保护着建筑免受湖水的"侵袭"。

主体结构上方正中央的"船帆"造型进一步诠释了港口中"轮船"形象，高达58米，相互倾斜，仅在大约距离地面40米高的交叉点处"相遇"。独特的工程学原理使得"船帆"结构坚固地屹立在两个大约相距70米的基点上。"船帆"之间预应力索网结构犹如一张巨大的渔网，采用玻璃材质覆面。此外，室内空间格外开阔，用于展示不同时期废弃的船只造型。

博物馆建在平坦的地势处，其独特的造型在几公里外都可见，名副其实地成为临港新城地标建筑之一。

Museum Shanghai-Pudong
上海浦东博物馆

Shanghai, China
中国，上海

Location: Shanghai
Area: 41,000m²
Architect: gmp Architekten – von Gerkan, Marg and Partners
Photographer: Christian Gahl
Completion Date: 2005
Award: Competition 2002 1st prize

地点：上海
建筑面积：41,000 平方米
建筑师：gmp 建筑设计事务所
摄影：克里斯琴·加尔
建成时间：2005 年
获奖：2002 年竞赛一等奖

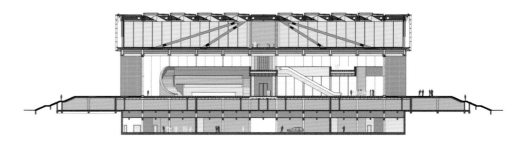

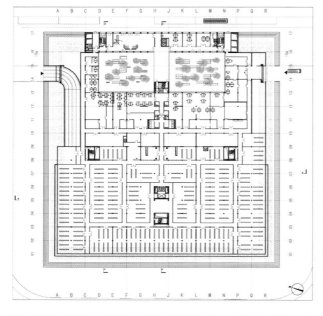

The Shanghai-Pudong Museum is one of the most important urban projects in this new district. It is supposed to document and archive the district's history and development comprehensively. At the same time modern multifunctional and open exhibition spaces are created to inform the public with a permanent exhibition and special exhibitions about selected topics.

Three elements form the building complex: the square-shaped horizontally orientated glass body with exhibition halls, a much broader, 4 meters high base with surrounding stairs, which accommodates the archives, and a bar-shaped building on the eastern side for the administration. The base as one of the main architectural features of the museum lifts the main building with the exhibition halls above the level of the surrounding streets and emphasises the central importance of the complex. Simplicity and reduction of the materials dominate the clear cube.

The façade of the upper, closed part of the main building not only serves as weather protection but also as a communication surface. It is made of two parallel façade-layers. The outer layer consists of glass and the inner one of room-high closed wall panels. These elements can be rotated along their longitudinal axis and can be opened or closed, according to the particular requirements of the exhibition concept, so that views from the inside to the outside and vice versa are generated.

作为浦东新城城市规划重要项目之一，浦东博物馆主要用于记录及存档该区的历史及发展情况。同时，一些现代风格的多功能展示空间用来举办永久性展览以及关于某一主题的展览，供公众参观。

整幢建筑有三个结构组成：方形的全玻璃大厅、更加宽敞的4米高的基座（四周设计着楼梯，用于存储博物馆档案文件）以及东侧长条形的结构（用作办公管理区）。基座结构构成主要特色之一，将展览大厅提升到四周街道高度之上，强调了这一建筑的重要地位。较少材质及装饰元素的运用，突出了简约风格。

主体建筑上方全封闭结构外观为双层，其中外层由玻璃材质打造，内层由封闭墙板构成，不仅可以抵御恶劣的天气，更传达出交流信息的特色。更值得提出的一点是，其外观可以沿着水平轴线旋转，根据展览的需求开启或闭合，促进了室内外空间的融合。

Shanghai Museum of Glass
上海玻璃博物馆

Location: Shanghai
Area: 5,500m²
Architect: logon | urban.architecture.design
Photographer: Jan Siefke
Completion Date: 2011

地点：上海
建筑面积：5,500 平方米
建筑师：罗昂建筑设计咨询公司
摄影：简·赛风
建成时间：2011 年

Shanghai, China
中国，上海

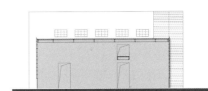

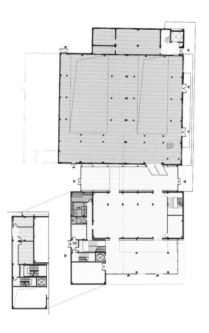

The Shanghai Museum of Glass multi-functional design combines exhibitions with hot glass shows, DIY workshops, lectures, libraries, restaurants, coffees, events, shops, public space and so on. Its sustainable adaptive reuse design and modern feel incorporate old and new ideas making it the first of its kind in China. The Entrance Plaza is the face of the museum enabling immediate recognition and recall for visitors where it guides people into the museum, hot glass show and surrounding areas. The new entrance building stands on the Entrance Plaza acting as a welcoming platform for the museum. Contrasting the dark façade with the bright lobby interior creates a unique first impression for visitors to the museum. The façade is made from U shaped glass imported from Germany, sandblasted and enameled to reveal transparent glass-related words in various languages. Behind the glass façade is an LED backlight that allows light to glow through each word on its black background; the final effect is breathtaking by night. The Shanghai Museum of Glass will educate and entertain thousands of visitors whilst adding value to the local district government and people for years to come.

上海玻璃博物馆的多功能设计方案囊括了展览空间、热玻璃演示、DIY工坊、演讲空间、图书馆、咖啡馆、活动空间、商店、公共空间等。这个可持续的改造设计，将新旧特色融于一体，赋予项目现代感，是中国"第二类"博物馆的先驱。入口广场是博物馆的门面，具有高识别度，引导游客进入博物馆、热玻璃演示中心及园区其他地方。矗立于入口广场上的全新建筑迎接着来往的游客。明亮的博物馆大堂与黑色外立面形成鲜明对比，给游客以耳目一新的感受，令人印象深刻。外立面采用德国进口的U形玻璃，经过喷沙和涂层处理勾勒出和玻璃有关的多国文字。玻璃立面背后的LED灯管点亮了黑色背景上的文字，营造出令人眩目的效果。上海玻璃博物馆将为游客提供集娱乐教育于一体的体验场所，也将为宝山区政府和人民带来更多福祉。

Zhujiajiao Museum of Humanities and Arts

朱家角人文艺术馆

Location: Shanghai
Area: 1,818m²
Architect: Zhu Xiaofeng / Scenic Architecture Office
Photographer: Iwan Baan
Completion Date: 2010

地点：上海
建筑面积：1,818 平方米
建筑师：祝晓峰 / 山水秀建筑事务所
摄影：伊万·班
建成时间：2010 年

Shanghai, China
中国，上海

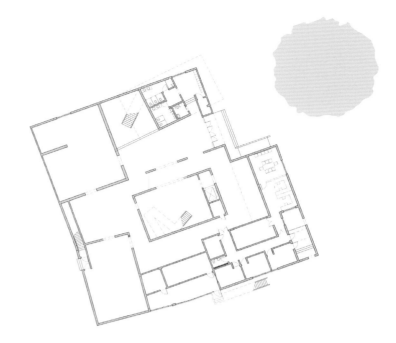

As the most integrally preserved canal-town in Shanghai, Zhujiajiao attracts an increasing number of visitors every year with its authentic tradition of eastern China. The site, located at the entry of the old town, faces two 470-year-old ginkgo trees. This 1,800sqm museum will house paintings and other art works related to the history of Zhujiajiao. The design approach is to delineate an art-visit experience that is rooted in Zhujiajiao. The architecture will be the carrier of this experience.
In the spatial allocation, the central atrium becomes the heart of the circulation. On the ground floor, the atrium brings natural light into the surrounding galleries through carefully positioned openings. On the first floor, a corridor around the outskirt of the atrium links several dispersed "small-house" galleries and courtyards, which absorb surrounding sceneries and provide diverse spaces for small exhibitions and events. This building-courtyard layout makes a clear reference to the figure-ground texture of the old town, and orientates the visitors to wander between the art works and the real sceneries with an experience of intimate interactions between matter and thought. A reflecting pool, laid in the east courtyard on the first floor, accomplishes an ultimate collection by bringing the reflection of the ginkgo tree into the museum.

作为上海保存最完整的水乡古镇，朱家角以传统的江南风貌吸引着日益增加的来访者。人文艺术馆位于古镇入口处，东邻两棵470年树龄的古银杏。这座1800平方米的小型艺术馆将定期展出与朱家角人文历史有关的绘画作品。设计师希望在此营造一种艺术参观的体验，它将根植于朱家角，而建筑是这一体验的载体。
在空间组织中，位于建筑中心的室内中庭是动线的核心。在首层，环绕式的集中展厅从中庭引入自然光；在二层，展室分散在几间小屋中，借由中庭外圈的环廊联系在一起，展厅之间则形成了气氛各异的庭院，收纳着周围的风景，为多样化的活动提供了场所。这种室内外配对的院落空间参照了古镇的空间肌理，使参观者游走于艺术作品和古镇的真实风景之间，体会物心相映的情境。在二楼东侧的小院，一泓清水映照出老银杏的倒影，完成了一次借景式的收藏。

SPSI Art Museum
上海油雕院美术馆

Location: Shanghai
Area: 3,000m²
Architect: Wang Yan / Architects Ring
Photographer: Lv Hengzhong
Completion Date: 2010

地点：上海
建筑面积：3,000 平方米
建筑师：王彦 / 上海绿环建筑设计事务所
摄影：吕恒中
建成时间：2010 年

Shanghai, China
中国，上海

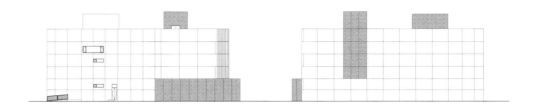
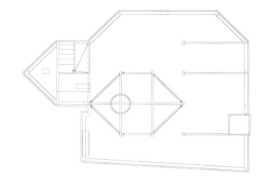
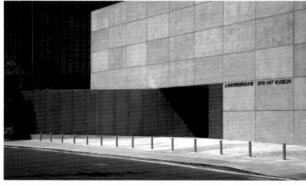
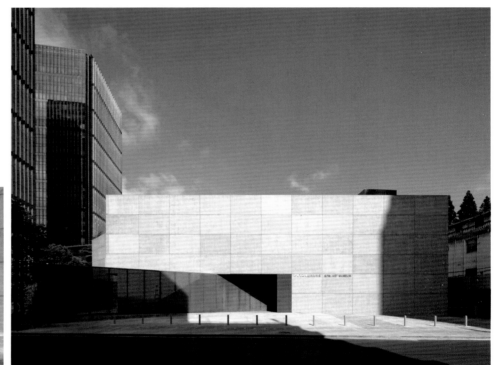

The thinking of architectural design of SPSI Art Museum is to start from environment and architecture itself, and to find the result during the process of answering the basic architectural questions. The effect of its volume, space, material and structure has touched us deeply, which turns out to be a kind of pure power. Like a polygon form stone lies the Museum along Jinzhu Road, calm and angular, simple but powerful, which rescues people from decorating façades in neighbourhood. The wall between Museum and its neighbour East Centre had to be remained. A piece of freestanding wall cannot form the feeling of volume, which is uncoordinated to the Museum main volume. The architects decide to continue the wall back into the main body of Museum, so that it creates an impressive entrance space and at the same time a nice inside garden, which supplies good view for the VIP room on the first floor. Half-transparent weaved stainless steel panels make the inside garden and outside street both visible.

Smoother the surfaces are, clearer the volume become and purer the space appear. To the interior, smooth wall inside-surfaces create pure interior space, while to the city, smooth wall outside-surfaces definite clear open spaces. So the architects abandon all decoration on both sides of wall, try to create the atmosphere of pure spaces, in order to highlight the art works at the same time.

SPSI美术馆设计完全由建筑本身出发，在回答建筑基本问题的过程中，逐渐接近设计师想要的答案。建筑本身的体量、空间、材质、构造展现出了一种质朴纯净的力量。美术馆如一块被切割成几何多边形的坚硬石块横卧于金珠路旁，它沉稳而棱角分明，简洁而厚重有力的体量让人们从相邻建筑的装饰化立面感受中解脱出来。美术馆南侧与东银中心接壤，由于地界划分的需要，分界墙必须保留。单片矗立的墙无法形成体量感，与建筑主体极不协调。因此设计将围墙折向延伸，顺势直接斜插入建筑主体，巧妙地形成了入口空间，同时围合了一个内院，为接待厅提供了室外景观。半透明的不锈钢编织网并不隔断内院与墙外绿地之间的视觉联系，院中情景隐约可见。

墙面越平滑，体量越清晰，空间越纯净。对于室内来说，平滑的墙面围合纯净的室内空间，对于城市来说，平滑的外墙限定了明确的室外空间。所以无论内外墙面的处理都力求平滑，去掉一切不必要的装饰，最大限度地营造出干净的空间体验，同时也突出了艺术作品本身。

Shanghai Museum of Contemporary Art
上海当代艺术馆

Location: Shanghai
Area: 3,900m²
Architect: Atelier Liu Yuyang Architects
Photographer: Atelier Liu Yuyang Architects
Completion Date: 2005

地点：上海
建筑面积：3,900 平方米
建筑师：刘宇扬建筑事务所
摄影：刘宇扬建筑事务所
建成时间：2005 年

Shanghai, China
中国，上海

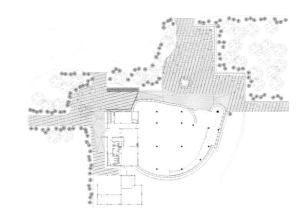

The design of the Shanghai MOCA was not much about establishing a new manifesto, but instead to exploit the notion of duality, mediating between an existing form and new intentions. The original structure was an un-utilised but structurally intact glass and concrete building. A series of geometric glass volumes were introduced to replace the main entrance and to extend part of the third floor, dissolving the predictable form of the original glass pavilion. The diagonally-laid "Mongolian Black" stone cladding over the existing concrete building gives a much subdued yet differentiated expression, highlighted by the deep-recessed stainless steel window frames that are intentionally mis-aligned. The resulting work could neither be defined as a new building, nor a mere addition.

The new programme mediates between the requirements for art exhibition – the need for a generic white box, and the desire for an appropriate architectural expression – one which celebrates the intrinsic quality of architecture, be it tectonic, spatial, structural, or material. The design of the ramp is one such example. Thanks to an elegantly resolved structural model, the ramp connects the two principal exhibition floors at a maximum span with a minimum usage of steel (just under 3 tons), producing a sweeping curvature that "dances" through the existing reinforced concrete columns in different tangential relationships, and allows for a circumscribed and ascending viewing of large-scale installations placed in the centre of the main exhibition space. The new museum mediates between landscape and city. Because of its unique location within the People's Park, visitors are required to meander through the "garden of the proletariats" before arriving at the institution that is paradoxically more akin to the consumerist nature of the city outside. The largely unobstructed glazing of the glass pavilion and the roof deck of the third floor bring together a view of the park with the images of the city. Here the former function of the building as a flower pavilion is somewhat re-incarnated; instead of housing the flowers, the new building affords the visitors a view of the Park's lush vegetation from its galleries, café, and sun deck.

"上海当代艺术馆"的设计并不想要一个"宣言式"的建筑，而更多的追求是其建筑语汇的"双重性"（duality）。它是一座介于现有和新造的建筑。首先，在原入口和三楼加建的部分，新的设计加入了数个比较突出的几何玻璃体，来"融解"旧玻璃房原有的外形。在原水泥房的部分，外墙以斜着排列的"蒙古黑"花岗岩配合了同样斜列的不锈钢窗框，体现出另一种含蓄却与众不同的外观。这结果既不是仅仅的装修，也不完全是单纯的加建，而是在保留原结构体的条件下做彻底的改头换面。

新"艺术馆"的功能必须兼顾艺术展览空间——所谓的白盒子——的要求，又欲求体现其建筑本质上的意义——无论是建构、空间、结构、材料等方面。其中，斜坡桥算是一个例子。借由电脑结构模型的辅助，斜坡桥的设计用最少量的钢材（不到3000kg）和最大的跨距（近30米）连接了两个主展览层。弧状的钢结构体从不同切面穿过有如树林般的混凝土柱，提供了参观者从不同定点与高度观看大型艺术装置的可能性。新"艺术馆"是一座介于园林和都市的建筑。由于它特殊的地理位置，人们先穿越了这绿意盎然的人民公园，再抵达这座更近乎于当代城市脉搏的机构。从展馆的主玻璃体内和三楼的大阳台向外望，参观者得到了园林与都市的双重印象。在此，原"花卉馆"的功能得到了再生。人们不再来看馆里的花，而是从这里向外看到所有公园里的花草。

Zhou Chunya Art Studio
周春芽艺术工作室

Shanghai, China

中国，上海

Location: Shanghai
Area: 1,46m²
Architect: Tong Ming, Huang Xiaoying
Photographer: Tong Ming, Huang Xiaoying
Completion Date: 2010

地点：上海
建筑面积：1,460 平方米
建筑师：童明、黄潇颖
摄影：童明、黄潇颖
建成时间：2010 年

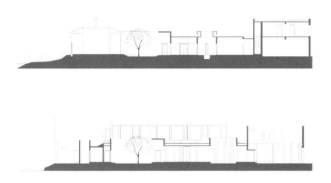
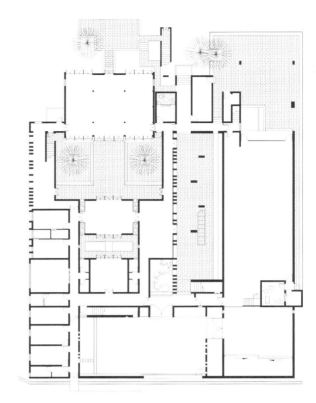

The site is surrounded by water on three sides and with the beautiful scene of Southern village as the main context. Architects particularly devised a pure concrete structure and exterior cladding for this project in order to seek balance between context and building. At the same time, the simple design can fully expresses the quality and atmosphere of an art studio.

The structure was divided into two parts based on the owner's work and living habits as well as the special site. The eastern volume is two-storey high and expands along the river in a narrow-striped form, accommodating work area and living space. It is relatively enclosed to ensure privacy.

Part of the western structure roof was used as terrace and linked with the work area and living space. Such a design not only enlarged the public space and created more usable area, but also formed an open square to enjoy the beautiful landscape.

In addition, it always rains in spring, and thus open drain system was adopted. Through the skylight, the draining rain water can be seen, which highlights the local feature.

工作室的基地三面环水，周边是典型的江南水乡风貌。为了适应这种自然而野趣的环境，建筑选择采用清水混凝土的结构及外表，这也相应能够体现艺术工作空间的品质与氛围。

设计的主要出发点在于艺术家本人的工作及其生活特征要求，同时结合基地现场特征，将工作室在总体上分为东西两个部分。东侧沿河为两层楼高度的长条体量，以容纳艺术家的工作空间及居住空间，总体上对外相对封闭，以保证私密性的要求。

由于东西两部分建筑存有高差，西侧一层部分的屋面则被利用设计为屋顶平台，并经由各种交通联系与艺术家的工作空间和生活空间联络起来，从而扩大了公共面积，提供了多重的使用可能性，另一方面也为工作室提供了一个开敞的四周景观。

江南春季多雨，屋面的雨水排放采用明沟系统，并与采光天窗的改造结合起来，使得雨水的流淌为人所见，以显示地方的特征。

Beicai Culture Centre

北蔡市民文化中心

Location: Shanghai
Area: 8,187m²
Architect: KH Architect
Photographer: KH Architect
Completion Date: 2011

地点：上海
建筑面积：8,187 平方米
建筑师：夏恩尼曦（上海）建筑设计事务所
摄影：夏恩尼曦（上海）建筑设计事务所
建成时间：2011 年

Shanghai, China

中国，上海

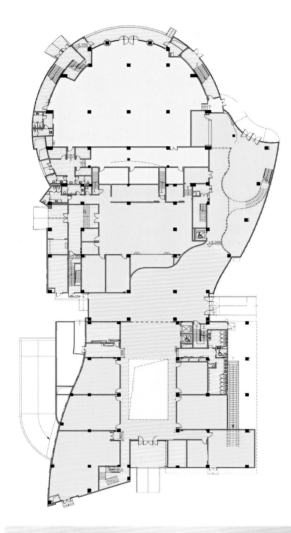

The streamlined spatial layout is the feature of the architectural design: the main circulation extends from square, through theatre entrance, exhibition corridor, central entrance, courtyard to the landscaped square in the southern part of the site. The east elevation is installed with transparent glass wall, combined with the exterior green belt and the interior tall atrium completely open to the south, to link the outside and inside together visually.

The landscaped circulation along the eastern part of the site runs from the extended landscape square, along which people can move to the centre of the site and the entrance/exit of the building as well as the green area aside the river. The western side open to the town roads is designed with entrance/exit for traffic, separating pedestrian from traffic. The southern part is mainly the green area with landscape architecture, riverside deck and sport facilities to connect closely with the riverfront landscape area in a natural way.

The façade design emphasises the organic integrity of the building to break the limits of narrow site. In addition, the roof is planted with grass to improve the ecological environment of the site.

建筑设计上采取流线型布置，主要动线由广场开始经剧场入口、展示廊道、中心入口、中心中庭流向基地南面绿化广场。建筑东侧采用全透视大挑高玻璃墙设计，与室外绿化带形成协力，再转到中心全南向开敞的挑高中庭，室内外空间浑然一体。

沿基地东侧绿化动线，顺着延伸的绿化广场设置，人流也可沿此动线，逐渐深入基地内部，到达各出入口及基地南面的沿河绿化区。而基地的西侧由城市道路开口，布置客货车流和后勤出入口，形成人流车流分侧出入，最大程度地缓解相互干扰。基地南侧布置大面积集中绿地，并辅设景观小品、亲水平台和部分露天健身设施，以便与日后建成的滨河景观绿化带自然衔接，充分利用滨河绿化带的日照和景观资源。

立面设计上强调建筑物的有机性，以克服基地过于局促所带来的限制，并刻意在屋顶设计大面积屋顶绿化以改善基地的生态环境，以弥补建筑覆盖率较大而造成的不足。

Jintao Village Community Pavilion
金陶村村民活动室

Location: Shanghai
Architect: Zhu Xiaofeng, Ding Penghua / Scenic Architecture
Photographer: Zhu Xiaofeng
Completion Date: 2010

地点：上海
建筑师：祝晓峰、丁鹏华 / 山水秀建筑事务所
摄影师：祝晓峰
建成时间：2010 年

Shanghai, China
中国，上海

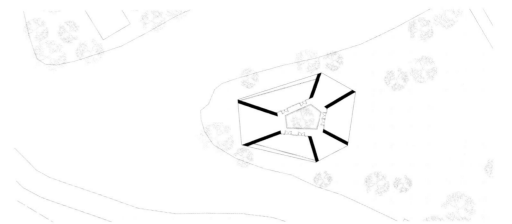

The site chosen for the new community pavilion is located at a public property by the "T" shape river cross. Since the environment has the spatial quality of openness and aggegation, the architect designed a hexagonal ring building for the village residence to rest, communicate and have recreation activities. Three interior spaces contain a recreation room, a teahouse, and a stage facing the grain-sunny ground; whereas three semi-outdoor spaces face three different sceneries: the concrete bridge to the northwest, the river cross to the southwest, and the little stone bridge to the southeast. These six spaces are defined by six radial bearing walls, and surround a courtyard that serves as both the spatial centre and the destination of drainage from the roof.

The concrete podium, blue brick clad bearing masonry walls, light-weight wooden partition window between walls and aluminium clad steel roof constitute the main features of the building. Moreover, components are buried in advance at the top of constructional column in the bearing wall to connect with the roof, creating "reinforced concrete structure". The logic of architectural structure and the style of southern village are taken into consideration in terms of selecting materials. The special method of wedging walls between podium and roof makes the building a prototype.

新建村民活动室的选址是村头三岔河口旁的一块集体用地。由于四周环境开阔、有聚合焦点的场所感，因此设计了一座六边形的环状建筑。三个功能空间容纳了活动室、茶室和一个面向谷场的小舞台；另外三个半室外空间则分别面向三幅风景：西北方向的水泥桥、西南方向的河口以及东南方向的石板小桥，成为村民休息纳凉和聊天的场所。这六个空间由六片放射状的承重墙划分，当中围着一个天井，它是空间聚集的中心，也承担了屋顶排水的收集工作。

混凝土基座、青砖表面的承重砌体墙、墙之间的轻质木窗隔断、铝板包敷的钢结构屋顶由木板条吊顶、小青瓦做屋顶。承重墙内的构造柱顶部放预埋件，和钢结构屋顶连接，形成了"钢混结构"。材料的选择兼顾了建构的逻辑和江南村落的风貌，而墙体夹在屋顶和基座之间的做法则暗示了成为空间原型的企图。

Culture and Sports Centre, Anting
安亭镇文体活动中心

Location: Shanghai
Area: 16,171m²
Architect: Zhang Bin, Zhou Wei / Atelier Z+
Photographer: Atelier Z+
Completion Date: 2010

地点：上海
建筑面积：16,171平方米
建筑师：张斌、周蔚 / 致正建筑工作室
摄影：致正建筑工作室
建成时间：2010 年

Shanghai, China
中国，上海

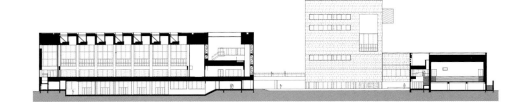

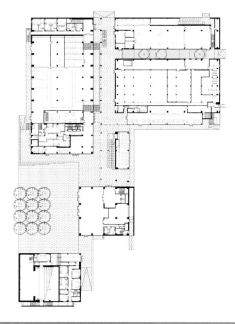

The mixed-use project is located at an irregular site in northern Anting town along a river, almost surrounded by factories and warehouses except a supermarket to the northwest. The entire programme is composed of a swimming gym, ball room, culture centre and cinema to satisfy the basic concept of relative independence, simplifying spatial complexity and reducing outdoor space. The four separate structures differ in size, height and dimension connected through a stilt terrace and can be freely accessed through large ramps both on the western and northern sides. The main functional spaces inside each structure are allocated on the first floor or above.

The abstractness of each structure and the richness of the entire programme are being reinforced through the application of unitary and understated material. Large expanse of stone wall seemingly arranged in a free way creates a steady background and at the same time controls the ribbon windows through which light and scenery penetrate inside. The aluminium plate clad on the mobile roof and landscape courtyard as well as the vertical greenery is extremely eye-attractive and makes visitors feel the tension and dialogue among separate structures. Moreover, the design separating the entire programme into different parts endows each structure with a clear function.

这是一个功能极端复合的项目，位于安亭镇区北部一块临河的三角形不规则基地内，周围基本都是厂房和仓库，唯西北角临近一处大型超市。顺应内部使用状态的相对独立性、简化空间复杂程度及减小外部尺度的意图，整个建筑被分为游泳馆、球类健身馆、文化馆和电影馆四个大小、高度和尺度各不相同的独立体量。四个单体建筑的主要使用空间都在二层及以上，它们被一个紧凑的架空公共活动平台联系成一体，并能从西侧和北侧通过宽大的坡道自如上下出入。

个体的抽象性和总体的丰富性通过统一而克制的材质使用而得到强化。大面积的石材墙面构成沉稳的背景，看似自由分布，实则控制室内采光和视野景象的横向长窗，以及在活动屋顶、景观庭院处出现的预涂装铝板表皮和编织金属网垂直绿化成为强烈的认知焦点，使得使用者在不同的位置可以体验到个体之间的张力和对话，同时单纯化的体量特点能很好地反映出建筑内部的使用状态。

Hanbiwan Holiday Villa
涵璧湾花园

Shanghai, China
中国，上海

Location: Shanghai
Area: 50,510m²
Architect: Zhang Yonghe / Atelier FCJZ
Photographer: Zhang Yonghe / Atelier FCJZ
Completion Date: 2010

地点：上海
建筑面积：50,510平方米
建筑师：张永和 / 非常建筑
摄影：张永和 / 非常建筑
建成时间：2010年

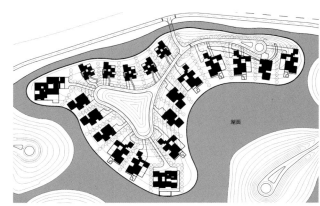

Hanbiwan Holiday Villa is located in the Qingpu District of Shanghai (it is originally a fishpond, with nearly 43 hectares of natural lake during the breeding season and a large number of waterfowl). The site is located at Zone B (north of the island) with a total of 20 units, ranging from 514 to 1,022 square metres.

Architects hope to integrate architecture and environment. The environment is natural as well as cultural – the water is the main element of the local natural environment; the traditional style of South China serves as the main design concept. At the same time the construction of the contemporary architecture can not be bound to the traditional simple repetition.

Thus, the design starts from three words – "division", "couryard", and "garden".

Division: The building re-split the functions, so that it is more like a collection of several buildings; one can make the rooms more transparent, get good ventilation, improve lighting quality, and adapt to local wet and rainy climate. Landscape and outdoor space combine harmoniously.

Couryard: Split between rooms, the formation of a number of different couryards enclosed, or semi-enclosed to provide residents with outdoor living space.

Garden: View from the road to the water is introduced into the leisure life, and echoes the experience of traditional gardens in South China; each villa is both a house and a miniature garden; people here are both residents and visitors.

涵璧湾花园地处上海青浦区，定位是度假别墅，用地原为养鱼塘，拥有近43公顷的天然湖面，在繁殖季节，有大量水鸟在此栖息，生态环境良好；地块位于园区北侧B岛，一共20栋，5种户型，地上规模514～1022平方米不等。

设计师希望建筑和环境融为一体。这个环境既是自然的也是人文的：水是当地自然环境中的要素，江南建筑传统是这里人文环境的主线。同时当代的生活方式和建造条件也注定建筑不可能是传统的简单重复。

由此，设计围绕三个关键词展开——"分"、"院"、"园"。

分：化整为零，将建筑内各功能进行拆分重组，使一个建筑更像若干个建筑的集合；由此，可以使更多的房间前后通透，得到良好的通风、采光质量，适应当地阴湿多雨的气候，同时与室外空间与景观更紧密地咬合。

院：拆分重组后的房与房之间，形成多个不同尺度的围合、半围合院，提供给住户可居的室外空间；

园：从路到水的景观系列引人进入休闲生活的状态，也呼应了传统江南园林的体验；至此，每个别墅既是一个房，还是一个微缩的园；对它的使用既是住，也是游。

World Expo Village
世博村

Location: Shanghai
Area: 550,000m² and 160,000m² garden
Architect: HPP Architects
Photographer: HPP Architects
Completion Date: 2010

地点：上海
建筑面积：550,000 平方米及 160,000 平方米景观
建筑师：HPP 国际建筑规划设计有限公司
摄影：HPP 国际建筑规划设计有限公司
建成时间：2010 年

Shanghai, China
中国，上海

The occasion for the master plan is the world's fair, so an employee and visitor district is part of the design. However, the requirement that the project be sustainable has necessitated that the project requires the design concept retain its validity after the period of the World Expo. The master plan thus makes possible the continued use of the district after the world's fair as a new, dynamic quarter that harmonises with the preexisting urban structure of Shanghai. The planning of the individual buildings allows for maximum flexibility with a view to future marketing, because the building structure is set up in such a way that with few exceptions it can be converted and inserted into the private real estate sector. This will occur above all thanks to a conveniently organised construction grid and structures whose use is easily adapted. Modular planning, a high degree of prefabrication of the façades, and the deployment of energy-saving technical systems characterise the qualitative requirements of the individual buildings.

The architectural image of the World Expo Village has a European character. The guiding concept is based on the structures and three-dimensionality of the greatest European cities. While Shanghai's surroundings are heterogeneous and marked by isolated developments, the idea of the coherent structure and the unifying stone façades of historical European city districts was one of the foremost influences on the design. In the overall cityscape of Shanghai, the perforated stone façades of the World Expo Village transmit stability and calm while also contributing something sustainable and energetic to the city.

项目的规划设计是为了世博会，因此设计中需要考虑员工和访客区域。然而，项目的可持续性要求使得世博村在世博会之后可以另作他用。因此世博村总图的设计使得世博村区域在世博会之后依旧是一个全新的、和上海城市周边肌理相和谐的活力社区。由于建筑设计使得世博村可以很方便地改造成为私有的居住社区，因此每个建筑的规划设计都可以很灵活地面向世博会之后的市场需求。当然，合理的结构和施工体系也是使得上述改造得以实现的重要保证。模数设计、立面的高度预制、节能环保措施的使用保证了每幢建筑的品质。

世博村的建筑概念具备欧洲特色。总的理念是基于欧洲知名城市的结构以及三维特性的。然而，上海的周边环境却是独立发展并且复杂多变的，欧洲历史名城协调一致的建筑结构和风格统一的石材立面，是整个设计中最有影响的设计理念。就上海整个城市风景线而言，世博村洞口式的石材立面表达了稳重与平静，然而又不失城市的可持续性发展与活力。

Shanghai Oriental Sports Centre
上海东方体育中心

Location: Shanghai
Architect: gmp Architekten – von Gerkan, Marg and Partners
Photographer: Bildnachweis, Marcus Bredt
Completion Date: 2011
Award: Competition 2008 1st prize

地点：上海
建筑师：gmp 建筑设计事务所
摄影：Bildnachweis、马库斯·布赖特
建成时间：2011 年
获奖：2008 年竞赛一等奖

Shanghai, China
中国，上海

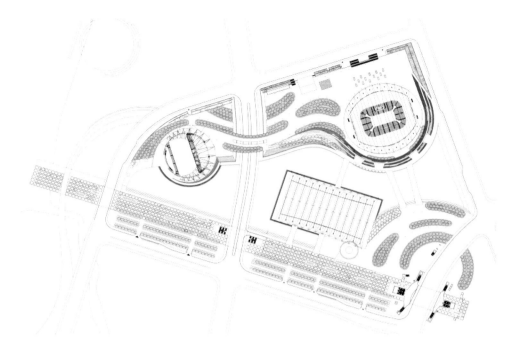

The Shanghai Oriental Sports Centre (SOSC) was built on the occasion of the 14th FINA World Swimming Championships in Shanghai. It consists of a hall stadium for several sports and cultural events, a natatorium (swimming hall), an outdoor swimming pool and a media centre. In keeping with a sustainable urban development policy, the SOSC was built on former industrial brownfield land along the Huangpu River. The individual venues are designed so that after the Swimming Championships, they can be used for a variety of other purposes.

Water is the overarching theme of both the park and the architecture of the stadiums and the media centre. It is the connecting element between the buildings, which stand on raised platforms in specially constructed lakes. Thus the round stadiums have a curved lakeside shore round them, while the rectangular Natatorium has a straight lakeside shore. Design affinities and a shared formal idiom and use of materials give the three stadiums structural unity. The steel structures of broad arches with large-format triangular elements made of coated aluminium sheet form double-sided curved surfaces along the frame of the sub-structures, thus evoking sails in the wind.

上海东方体育中心专为2011年第十四届国际泳联世界锦标赛设计。综合体包括一座可举办多种体育赛事以及文化活动的多功能体育馆，一座游泳馆，一座室外跳水馆和一个媒体中心。在可持续性城市发展的方针政策下，体育中心取代了黄浦江沿岸的工业区旧址，为这一地区重新注入活力。中心内的每座体育场馆在设计理念上均考虑到了赛后的多样化利用问题。

水体景观是整个体育公园、体育场馆和媒体中心设计理念的核心主题。建筑体坐落于从人工湖中升起的数个岛屿之上，水面如同纽带般将其连为一体，在圆形的体育馆外留下蜿蜒的柔和边界，而在游泳馆外形成笔直的边缘。建筑体结构上的异曲同工，相仿的造型手法以及类似的材质选用刻画出建筑群整体的简洁感与秩序感。大跨度的钢拱结构，呈三角形的大块涂层铝合金板覆盖于龙骨结构之上，形成双向弯曲的拱面，从远处眺望令人联想起千帆竞发的壮观景象。

Renovation of Qingpu Stadium and Training Hall

上海青浦体育馆、训练馆改造

Location: Shanghai
Area: 8,100m²
Architect: Hu Yu Studio
Photographer: Hu Yu Studio
Completion Date: 2008

地点：上海
建筑面积：8,100 平方米
建筑师：胡越工作室
摄影：胡越工作室
建成时间：2008 年

Shanghai, China
中国，上海

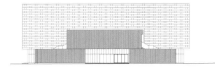

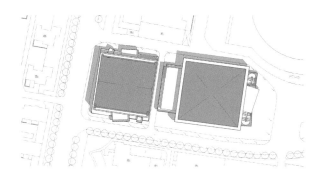

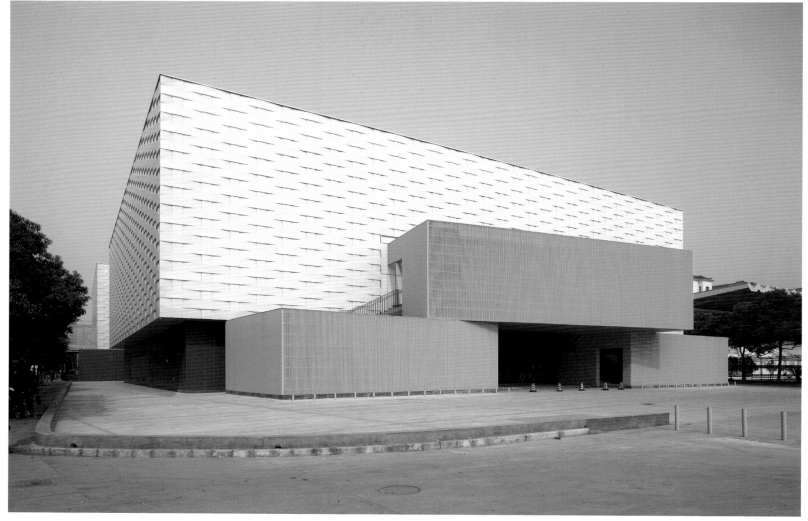

218 / EAST CHINA

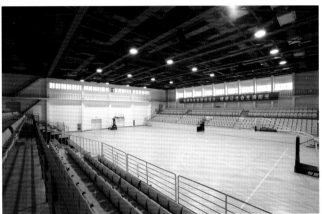

Qingpu Stadium and Training Hall are located on the northeast side of the two urban roads' intersection in the old city of Qingpu, Shanghai, covering an area of 8,100m^2.

The stadium and the training hall were built in the early 1980s, and both fail to meet the need of the fast-developing city as a result of old age, worn-out structures, dated facilities, and great defects in the elevation design. The government intends to reinvent the original buildings with this renovation, and improve the internal facilities to provide a space for public sports and fitness. The make-over focuses on altering the setback structure, by adding an envelope enhancing the geometric relations of the structure, and enlivening the structure moderately. For the inside of the stadium, the function of holding smaller games will be reinforced, the stands transformed, subsidiary spaces and audience service facilities added. The architects overcomed many unfavourable conditions including the loss of original design and small budget, and creates an "open-ended coat" of original polycarbonate plate woven wall for the buildings, using new materials, new technologies and new construction techniques. It not only ensures the natural lighting inside, but also creates a unique work of architecture. Its neutral texture reflects the spirit of the times, and coordinates the urban surroundings. Metal grills and perforated aluminium sheets are employed in the project to enclose the outdoor staircases, the entrance marquee, outdoor air-conditioning units, and old walls that are originally defects of the building. This gives the building a new look while retaining the original façade and building components.

The renovation has dramatically changed the urban landscape of the area, making the stadium and training hall a popular public sport and fitness arena in Qingpu, Shanghai.

上海青浦体育馆、训练馆位于上海青浦区的旧城内，两条城市道路交叉路口的东北侧。建筑面积 8100 平方米。

两个馆均建于 1980 年代早期。由于年代久远，原有建筑、设施破旧，建筑立面造型存在较大缺陷，不能满足快速发展的城市的要求。政府希望通过这次改造彻底改变原建筑的面貌，同时改善其内部设施以便为市民提供一个运动健身的场所。改造中通过改变建筑体型——增加一层外皮、强化体型中几何逻辑关系和适度地活跃体型来完成建筑物外部改造。对体育馆内部完善其举行小型比赛的功能，改造看台、增加附属用房和观众服务设施。

本项目设计克服了原始设计资料缺失、造价低廉等不利条件，合理巧妙地应用新材料、新技术以及新的构造做法，采用独创的聚碳酸酯板编织外墙，给建筑物穿上了一层"开放式的外衣"，不仅保证了建筑内部的自然采光效果，还创造了独具一格的建筑形象。其中性的质感富有时代气息，对城市环境起到整合的作用。同时本项目利用金属格栅和穿孔铝板将原建筑中在造型上存在缺陷的室外楼梯、入口雨罩、空调室外机和旧墙包裹起来，使建筑在保留原有外墙和建筑构件的同时焕然一新。

本工程建成后，使这一地区的城市面貌发生了很大变化，同时它已经成为上海市青浦区一处深受市民欢迎的体育健身场所。

Kindergarten in Jiading New Town

嘉定新城幼儿园

Location: Shanghai
Area: 6,600m²
Architect: Atelier Deshaus
Photographer: Shu He
Completion Date: 2010

地点：上海
建筑面积：6,600 平方米
建筑师：大舍
摄影：舒赫
建成时间：2010 年

Shanghai, China
中国，上海

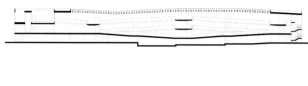

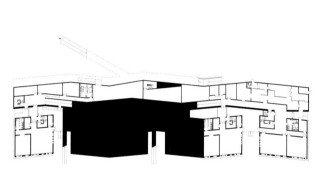

220 / EAST CHINA

The kindergarten is located in the Jiading New Town in the northern suburb of Shanghai; it is neither countryside nor urban in the traditional sense. Facing to the ambiguous and uncertain surrounding, the architecture emphasises the self-improvement, directly intervening into the site, and making a very clear juxtaposition of the architecture and the site. Compared with the open environment, the kindergarten acts as an introversive area. The architecture is divided into two general parts: one is the rational and efficient area, which is composed of 15 classrooms and a number of playrooms; the other is the intentionally enlarged transportation space, which is an atrium with ramps connecting with different storeys. This atrium contributes to an emotional and entertaining spatial experience that is beyond the common daily experiencing. Beside the transportation function, it is the vagueness and uncertainty that also provide a number of possibilities in spatial utilisation. The external form and the façade represent the inevitability to the internal relation as well. Apart from the two separated shapes of this architecture, the constantly transformed height among different floors is represented on the façade as well. Open activity spaces are created on elevation places, and the courtyards are extended along the vertical direction instead of the horizontal direction in the traditional method. Thus, the courtyards and activity spaces for children act as an important part of the façades. Everyday, kids and teachers shuttle between the architecture and the site, between the two separated parts of the architecture. They are shielded by the architecture and experiencing the inside and outside, the balance of passion and reason.

嘉定新城幼儿园地处上海北部郊区嘉定新城，这里一片空旷，不同于传统意义上的城市。面对基地周边的暧昧不定，建筑强调自我完善，以完整而有力的体量直接介入场地，与之形成泾渭分明的并置关系。场地依旧保持开放，而建筑则比较内向，给人以庇护。建筑本身也由两部分并置而成：一部分是由15个班级单元及不同的活动室聚合而成，理性而富有效率；建筑的另一部分则是个刻意放大的主交通空间——一个充满了连接不同楼层坡道的中庭。中庭提供了超越日常经验的空间体验，感性且充满情趣。除了交通功能外，中庭的模糊和不确定性为承载不同的功能提供了更多的可能性。建筑的外在形体与立面处理也是对这种内在关系的直接反映，除了形体被清晰地表达成两个不同部分的并置外，建筑内部不同空间的高差变化最终也反映到立面上。与此同时，设计在立面上还有意设置了一些内凹的户外活动空间，令传统意义上的沿水平方向展开的庭院组织模式转化为沿垂直方向展开，"庭院"及其幼儿的活动由此也成为建筑立面的一部分。当老师与孩子们每天穿梭于建筑与场地之间，穿梭于建筑的两个部分之间，在受到建筑庇护的同时，也取得内与外、情与理的平衡。

Xiayu Kindergarten, Qingpu
青浦夏雨幼儿园

Shanghai, China
中国，上海

Location: Shanghai
Area: 6,328m²
Architect: Atelier Deshaus
Photographer: Zhang Siye
Completion Date: 2004

地点：上海
建筑面积：6,328 平方米
建筑师：大舍
摄影：张嗣烨
建成时间：2004 年

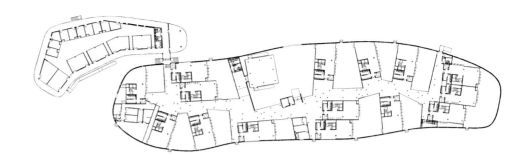

222 / EAST CHINA

Xiayu Kindergarten lies on the edge of the Qingpu New Town. Qingou is one of the several sub-districts around Shanghai, which still preserve some traditional buildings. The eastern elevated highway is the potential source of exhausted gas and noise, but it also provides the possibilities of viewing the building of various eyesight, altitude and speed in process of passing by. The river provides the fine landscape, and it also makes the architects think about the way of ensuring the children's security and the figure of the building by the river. In the design of Xiayu Kindergarten, the architects emphasise the difference between inside and outside. The inner region is protected, while the outer environment is filtrated. The kindergarten contains 15 classes, and each one has its own living room, dining room, bedroom and outdoor playground. After placing all the functions in a linear way in the narrow site, the architects found that a soft curve form could suit the site better than straight lines. So they separate the 15 classrooms and teacher offices into two curve clusters that are wrapped by solid material and void respectively. A painting finished wall clarifies all the classrooms while offices and special classrooms are fenced by elevated U-shaped glass. In design of classroom, the architects arrange all the living rooms on the first floor with outdoor playgrounds, and leave brilliant coloured bedrooms on second floor. To emphasise floating and uncertain feeling they detach the classroom floor from the roof of first floor. It's this uncertain and isolation on proper scale that leads to a seemingly random convergence condition and produces spatial tension. Every three bedrooms are linked by raised wooden walkways. Architecture volume is scattered by the tall trees dotted in those courtyards, while the final architectural figure is full of vigour because of the trees. Thus the architecture and tall trees bring out the best in each other and cohabit harmoniously in the narrow riverside.

夏雨幼儿园的基地位于青浦新城区的边缘，所在区域已经丝毫感觉不到地方建筑所能给予的影响，倒是基地东侧的高架高速公路及西侧的河流对设计产生了决定性的影响。

因此，幼儿园的设计强调"内""外"有别，内部领域是受保护的，而外部环境是被筛选的。

幼儿园总共有15个班级，每个班都要求有自己独立的活动室、餐厅、卧室和室外活动场地。就既定的基地而言，一个柔软的曲线型边界可能会比直线更容易与环境相融合，于是15个班级的教室群和教师办公及专用教室部分被分为两大曲线围合的组团，分别围以一实一虚的不同介质，班级教室部分的曲线体是落地的实体涂料围墙，办公和专用教室部分是有意抬高并周边外挑的U形玻璃围墙。

在班级单元的设计上，活动室因为需要和户外活动院落相连而全部设于首层，卧室则被覆以鲜亮的色彩置于二层，卧室间相互独立并在结构上令其楼面和首层的屋面相脱离，强调其飘浮感和不定性，这种不定性以及恰当尺度的相互分离导致一种看似随意的集聚状态，空间产生张力。每三个班级的卧室以架空的木栈道相连。

当高大的乔木植入各个院落，建筑在空中被化解，而最终的建筑形象也因这些树木而生机勃勃，两者相得益彰，共同栖息在这狭长的小河边。

Botanical Gardens "Chenshan"
辰山植物园

Location: Shanghai
Area: 58,600m²
Architect: Auer+Weber+Assoziierte
Photographer: Auer+Weber+Assoziierte
Completion Date: 2010

地点：上海
建筑面积：58,600平方米
建筑师：奥尔＋韦伯＋合伙人建筑设计有限公司
摄影：奥尔＋韦伯＋合伙人建筑设计有限公司
建成时间：2010年

Shanghai, China
中国，上海

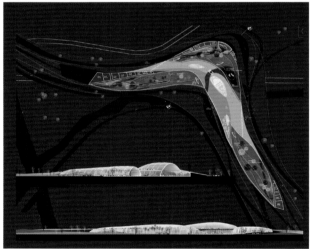

The architectural elements are integrated into the continuum of the band-like "ring of gardens". The buildings strengthen the idea of the garden and formulate spots of concentrated experiences within the garden band; they are embedded into the undulating landscape and become a part of it. The dynamic forms and arrangement of the buildings and the change of materiality between concrete, as a supporting element of the landscape and glass that acts as a filling between the openings of the ring, integrate the architecture within the landscape design. The emphasis of the architectural programme, for instance central entrance building, greenhouses and research centre, are integrated into the continuum of the "garden ring" and positioned according to their particular function and relevance: central entrance building with visitor centre, education centre, exhibition area and administration on the south; research centre located between the research/experiment area and the botanical garden on the north constellation of greenhouses by the Shen Jing He canal, corresponding to the panoramic outline of the Chenshan Mountains in the northeast.

建筑延续了园区内带状结构，强化公园设计理念的同时将活动体验集中。建筑在造型、排列以及材质的变化（混凝土以及玻璃）上使其与景观设计融为一体，成为整个园区的一部分。建筑主要功能结构，如中央入口大楼、研发中心以及花房分别根据各自的功能及相互之间的联系排列：中央入口大楼包括游客中心、教育中心、展区以及行政区，布置在南侧；研发中心介于研发试验区及植物园之间设计在北侧；花房设置在东北侧。

Sino-French Centre, Tongji University

同济大学中法中心

Location: Shanghai
Area: 13,575m²
Architect: Zhou Wei, Zhang Bin / Atelier Z+
Photographer: Zhang Siye
Completion Date: 2006

地点：上海
建筑面积：13,575 平方米
建筑师：周蔚、张斌 / Z+ 建筑工作室
摄影师：张嗣烨
建成时间：2006 年

Shanghai, China
中国，上海

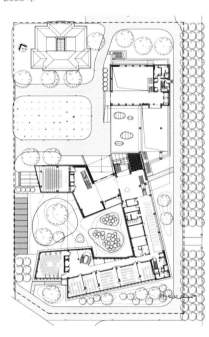

Sino-French Centre at Tongji University is located at the south-east corner of the campus, with 12.9 Building, the oldest existing building in the campus, and 12.9 Memorial Park on its west side, tracking field on its south side, and Siping Road on its east side.

The programme is composed of three parts, college, office and public gathering space. Two similar but different zigzag volumes, occupied by college and office sector respectively, overlap and interlace each other, and then they are linked together by the volume of public gathering space on underground and upper level. College and office sector share the main entrance which is located at the void part of the intersection of these two volumes, while public gathering space has its own lobby, which faces to roof pool and sunken garden, to connect underground exhibition hall and lecture hall on the upper level. The function of the college and offices is well kept in mind by using regular shapes for almost each unit. Yet applying zigzag corridor to connect these units creates abundant interests throughout inside and outside space. In the meanwhile, existing trees are incorporated into the design to add more charms to this complex.

Different materials and tectonics are applies to the different components of the complex. College sector is wrapped by Cor-Ten steel sheet panels. The unique texture and colour of the panels and the smoothness of the glass create delicate variation. Precoated cement panel is introduced into the office sector. Regular and irregular window bands provide sunlight to the office unites and corridors. Public gathering space is created by the combination of both Cor-Ten steel panel and precoated cement panel. The vivid colour and texture of Cor-Ten steel panel is contracted with plain grey cement panel. This treatment indicates the symbolic meaning in this project, the juxtaposition of two different cultures.

Landscape design plays a very important role in this design as wall. The retained existing metasequoias, surrounded by office sector, public gathering area and XuRi Building form an entry plaza of the complex. Connected with 12.9 Memorial Park, this space will become a very important outdoor space to serve the entire campus. The connection between two parts of the building formed a roof pool and a sunken garden, which becomes an intermedia between urban space and campus space. A semi private garden, created by college and office sector, gives a peaceful place for studying and relaxing.

Eventually, by applying different geometries, materials, colours and tectonics, the architects create a unique architectural piece that has a thorough and profound group of the meaning of cultural exchange between China and France.

同济大学中法中心位于校园东南角，西临校园内现存最老的建筑物一二·九大楼和一二·九纪念园，南侧为运动场，东侧紧靠四平路。

整个建筑分为既分又合的三个部分，分别用于教学、办公和公共交流。南北两条进深相同、由曲折连廊串联大小使用空间的教学、办公单元互相穿插后分别从空中和地下结合到最北端的公共交流单元。不规则的体量转折和穿插既最大限度地使9棵大树和水杉林得以保留，又创造了丰富多变的室内外空间，使巨大的体量消解于细腻的环境中，同时绝大部分使用空间仍保持规则形状。教学、办公单元的共用门厅位于它们上下穿插的虚空部分，通透高耸，强化了两者的穿插关系。公共交流单元另设一个独立门厅，并将地下的展厅、南侧的屋顶水池、下沉庭院和二层的报告厅联系起来。

三个不同单元采用不同的材质组合、色彩和构造做法来建构。教学单元用自然氧化的耐候钢板包裹网格状立面，均质的网格中开孔和玻璃微妙地变化；办公单元用轻质混凝土挂板覆盖立面，规则条窗和不规则条窗分别为办公室和走廊提供光线；公共交流单元是轻质混凝土挂板和耐候钢板的混合立面，外表皮为轻质混凝土挂板，大尺度开口部位为耐候钢板。这样的两种色彩和材质暗示了中法不同的文化传承的视觉表征。

相互耦合的空间体量与环境的互动形成了丰富多变的外部景观。保留的水杉林被办公单元、公共交流单元及旭日楼围合后成为建筑的入口庭园，并与一二·九纪念园一起，形成校园中一个重要的公共开放空间。建筑两个单元穿插处的屋顶水池和下沉庭院既丰富了景观层次，又使建筑本身成为纪念园空间和四平路城市空间的中介。南北单元在基地南部围合出另一个相对内向私密的绿化庭院，为师生提供一个安静的交流场所。这一庭院由透明的门厅与北侧水杉林园建立视觉联系。

West Shanghai Holiday Inn
竞衡西郊假日酒店

Location: Shanghai
Area: 5,926m²
Architect: Kris Yao / Artech Architects & Designers (Shanghai) Limited
Executive Architect: Shanghai Xian Dai Architectural Design (Group) Co., Ltd., Wei Dun Shan Design Studio
Photographer: Zheng Jinming
Completion Date: 2010
Client: Shanghai Blue Creek Real Estate Development Co., Ltd.

地点：上海
建筑面积：5,926平方米
建筑师：姚仁喜 / 会元设计咨询（上海）有限公司
执行建筑师：上海现代设计集团 魏敦山工作室
摄影：郑锦铭
建成时间：2010年
业主：上海蓝溪房地产开发有限公司

Shanghai, China
中国，上海

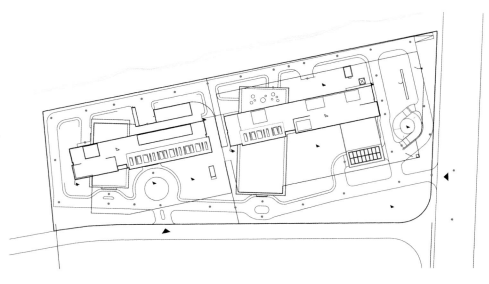

The main building of the project unfolds along east-west direction to overlook the green space at south side and the river view at north. The east side of the site is the 17-storey-high international hotel, and the west side is the 17-storey-high service apartment. The skirt buildings are weaved into the main building with a large angle to provide the required large spaces for various activities. The effect not only activates the overall spatial sense, but also organically combines the skirt buildings to the upper strip slab form of the main building. The interweaving and enclosures of the architectural layout also compose various interesting landscaping spaces. The building and landscape create surprising views along the circulation, forming addition, connection, fluidity and infiltration to the spatial layers.

The design of the entrance area takes on a central axial waterscape theme. The grand momentum is shown through the landscape central fountain in composition with the cantilevered entrance canopy. In order to present a relax and delightful atmosphere to the guests, large windows are used at the two-storey-high atrium to bring in the visions of the outdoor pools, platforms, and the green shades. Leisure and entertainment facilities such as restaurants, coffee shop, and dance hall are located at the skirt tower spaces on the first to third floors and are connected to outdoor areas and upper floor guestrooms through a rationally organised circulation arrangement.

酒店长条形的主楼耸立在三个相互穿插错落的裙房上，运用玻璃和金属等元素的结合塑造强烈的形体感。玻璃和金属结构的应用让整个建筑更加轻盈、透明，并通过外表皮上虚实、深浅的交错布置，巧妙地体现出生动活泼的外观，从另一方面展示了酒店富有活力的感觉，创造一个壮观、现代的地标式建筑。主楼立面上运用"虚"的处理手法，挖空局部体块，外露柱子。立面上以不规则的方式处理这些有趣的空间，同时形成空中花园，将绿色引入建筑。

底层裙房中布置的几个中庭，采用大面积的落地玻璃窗，充分引入周边的景观，客人置身其中可以欣赏南面的美景或北向的美丽河岸风光及景观园林。入口区的设计采用中轴水景、中央喷泉等景观处理，加上出挑的雨篷，气势不凡，两层挑空的入口大堂采用大面积的落地玻璃，将户外的水池、平台、绿树等映射入室内，给到来的宾客轻松愉悦的感觉。各种风格的餐厅、咖啡厅、舞厅等休闲娱乐设施布置在1~3层的裙楼空间，通过合理组织的交通流线与外界和楼上的客房相联系。

Plaza 353
353 广场

Location: Shanghai
Area: 40,000m²
Architect: Woods Bagot
Photographer: Woods Bagot
Completion Date: 2008

地点：上海
建筑面积：40,000 平方米
建筑师：伍兹·贝格
摄影：伍兹·贝格
建成时间：2008 年

Shanghai, China
中国，上海

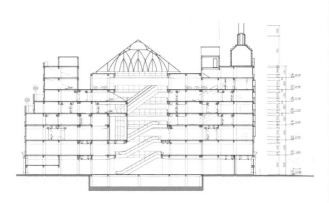

Plaza 353 is located on the pedestrianised Nanjing East Road – Shanghai's favourite shopping street. The original building was designed in a fashionable art deco style of the day, embellished with Bauhaus, art nouveau and renaissance influences by Architect Mr. Zhuang Jun in 1932. This eclectic design spirit has been transposed through a modern design language into the interiors and architecture of the retail tenancy design.
It has now been transformed into this through urban regeneration to retain its historic cultural value and is a successful example of multi-storey urban regeneration in Shanghai and China.
Through its planning, function, quality tenants and distinct mix of historic and modern design elements, it contributes positively to the re-establishment of the Nanjing East Road precinct as the premiere retail and entertainment district in China and an international destination for travellers to the City.
A secret roof garden being developed at the top allows you to look over Nanjing Road – Shanghai's favourite shopping street since 1920s. The redevelopment of the historic Plaza 353 restores the status of this landmark building with a lively and modern-retro design urban concept mall. Combining the best of new-age technology and cutting-edge retail design, Plaza 353 is the one-stop-centre for urban hipsters to shop, dine and hang out with friends.

353广场坐落在繁华的南京路上，原有建筑以"艺术装饰"风格为主，1932年曾被翻修过一次，包豪斯学派风格、新艺术风格以及文艺复兴风格被融合在一起。这种兼收并蓄的设计主旨将现代化设计语言运用到室内装饰中。
此次翻新旨在维护其原有的历史及文化价值，通过功能重划以及不同元素的融合，使其成为上海乃至中国翻新设计的典范。
私密的屋顶花园提供了一个欣赏城市风貌的绝佳场所。简言之，353广场的重新设计既保留了其作为城市地标建筑的地位，更诠释了独特的现代复古城市设计理念。现代技术及全新零售管理理念的运用，使得这里能够成为一站式服务的核心，集购物、餐饮及休闲功能于一身。

Chervon International Trading Company

泉峰国际贸易公司

Location: Nanjing, Jiangsu Province
Area: 30,700 m²
Architect: Perkins+Will Inc., Architectural Design & Research Institute of Southeast University
Photographer: James Steinkamp / Steinkamp Photography
Completion Date: 2007
Award: McGraw-Hill Construction 3rd Bi-Annual "Good Design Is Good Business" China Awards 2010, Best Commercial Project

地点：江苏，南京
建筑面积：30,700 平方米
建筑师：帕金斯威尔建筑师事务所、东南大学建筑设计研究院
摄影：詹姆斯·斯坦坎伯 / 斯坦坎伯摄影工作室
建成时间：2007 年
获奖：麦格劳－希尔公司《建筑实录》、《商业周刊》第三届"好设计创造好效益"中国奖项 2010 年最佳商业建筑

Jiangsu, China
中国，江苏

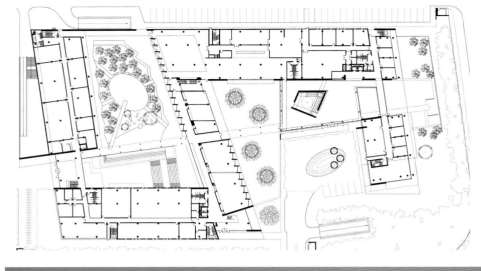

232 / EAST CHINA

Traditional Chinese garden design influenced the building configuration and landscaped spaces of the corporate headquarters for Chervon, a Chinese exporter of power tools. Located in the Nanjing Economic Development District on the outskirts of the city, the 30,700-square-metre building houses five major corporate departments in its five wings: management, sales, research and development, testing and training.

Taking its inspiration from the traditional zigzag garden path, the building bends to form two exterior spaces, the entrance court, which is open on its eastern edge and a more private west-facing garden. A narrow circulation spine cuts north/south across the site linking the five building wings and a series of courtyards. It alternately passes through the building wings and the courtyards, at times bridging over water like the traditional garden path. The two types of paths present in the building suggest the company's two cultural aspects, one represented by the non-direct contemplative zigzag and the other by the direct, practical line. The organisation of courtyards linked on an axis has precedent in Chinese monastery design.

The focal point of the entry sequence is the asymmetrical pyramid that pierces the two-storey lobby. Clad in aluminium of varying tooled finishes, it symbolizes the company's product and is intended as a space to display examples of Chervon's tools.

传统的中国园林式设计影响了泉峰集团——中国一家电动工具出口商——总部大楼的建筑规划及景观空间。此项目位于南京市郊的经济开发区，占地30,700平方米，共有五大分区，配备给集团的五大部门：管理、销售、研究与开发、检测及培训。

受到传统的"之"字型园林小径的启发，建筑采取了弯曲的造型，形成两个外部空间，有大厦的入口前庭，东面开放，另有一个朝西的较私密的花园。一条狭窄的中心隆起地点，横跨南北，连接了该建筑的五大主体部分和一些周围的庭院。它穿过大楼的各个部分和院落，有时也充当水面上的小桥，其作用与传统园林里的小径相似。两种不同风格的路径暗示出泉峰集团两种不同的文化战略，一种是非直接指引的"之"字型小路，另一种是直接实用的路径。由中轴统领院落组织的设计曾经在中国寺院的修建中有先例。

大楼入口的亮点是穿过二层大厅的不对称的金字塔，铝制外墙的变幻象征着集团的产品，并且意图提供展示泉峰集团工具的空间。

Huawei Research and Development Park
华为软件研发中心

Location: Nanjing, Jiangsu Province
Area: 332,000m²
Architect: RMJM
Photographer: Jason Findley
Completion Date: 2010

地点：江苏，南京
建筑面积：332,000 平方米
建筑师：英国 RMJM 建筑设计集团
摄影：贾森·芬德利
建成时间：2010 年

Jiangsu, China
中国，江苏

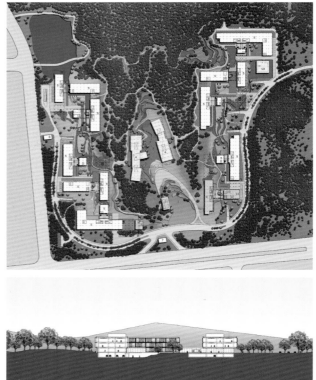

The masterplanning of the site has a synergy with the relationship between the history of Nanjing, the nearby Purple Mountain and the tranquility of the Yangtze River. Integral to the scheme is the integration of the natural typography of the surrounding hills and valleys, with the landscaping being brought through the scheme, blurring the edges between the built and soft environment.

This scheme is a new campus development for China's leading telecoms manufacturer, Huawei Technologies. The project incorporates research office and laboratory accommodation for 10,000 technical staff together with supporting canteen and data centre facilities.

The design proposes a low-rise, orthogonally arranged architectural composition of L-shaped structures to create a series of interlocking courtyards and prioritises harmony with its physical and climatic surroundings.

The building envelopes incorporate extensive solar shading to reduce heat gains and minimize the energy required for cooling purposes. Over 7,000 windows operate through an automated building control system that switches between air conditioning and full natural ventilation modes as external conditions alter. The extensive roof areas provide rainwater harvesting with the water naturally filtered on site through reed beds and providing 100% capacity for all landscape and irrigation requirements year round.

项目所在场地的总体规划设计考虑到南京这座城市的历史，以及附近的紫金山以及长江之间的联系。此外，周围山谷地形以及人造景观更是模糊了建筑与环境之间的界限。

建筑外观层安装有遮阳系统，减少太阳的灼晒，进而确保室内舒适的温度。7000多扇窗户全部采用自动控制系统调节，根据室外温度的变化实现在使用空调和自然通风条件之间的转换。屋顶可收集雨水，用于全年灌溉园区景观。

设计的主要任务即为中国电信巨头打造新的研发园区，包括研发办公区、实验室、食堂以及技术中心等，供大约10000名技术员工使用。

设计师提出打造一个低层的L造型建筑，以便于营造出相互联系的庭院，同时打造与周围自然环境以及气候环境相符合的空间结构。

POD – Jiangsu Software Park
建筑豆——江苏软件园南京徐庄基地

Location: Nanjing, Jiangsu Province
Area: 100,000m²
Architect: Wang Degang, Bu Yuanyuan / W2 Architects
Photographer: Li Gan
Completion Date: 2010

地点：江苏，南京
建筑面积：100,000 平方米
建筑师：王的刚、卜源远（南京佳的建筑设计事务所）
摄影：李敢
建成时间：2010 年

Jiangsu, China
中国，江苏

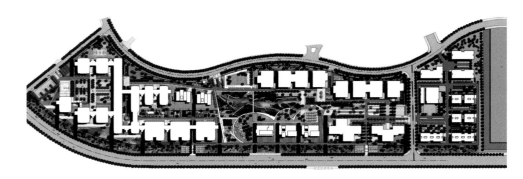

Located in the Xuzhuang Software Park to the east of The Mausoleum of Dr. Sun Yat-sen in Xuanwu District, the project was required to be no more than 15 metres in height. In addition, the building should not be too large in size according to the client's requirement. The building community thus was designed to serve companies of various scales.

The planning process quotes the POD (proper orthogonal decomposition) concept so that each building unit can be independent or combined according to requirements. All the structures are being arranged on the perimeter of the long, narrow site which is divided into eastern and western parts with the middle to be landscaped as public grassland. The individual volume is expressed through the shape of the building unit, most of the ground floor of which is designed as "grey zone". The objective of such design method is to make up for the site limit and interact with other buildings around. The landscape design focuses on the theme of "ecology and nature" with large public area to be planted with grass from prairie, maintaining the balance of natural ecology.

项目位于南京玄武区中山陵东侧，徐庄软件园内，地块狭长，建筑高度要求控制在 15 米的范围内。业主不希望建筑的体量过大，要求在一个相互有关联的建筑群体内各个建筑单元可以独立使用。

设计中借用了 POD (proper orthogonal decomposition) 组合概念来规划，形式分而不散。在狭长的地块内，建筑尽量沿着地块的边缘布置，分为东西两个区域，地块的中间作为园区内的公共绿地，服务周边的办公建筑。建筑形体以表现建筑的单元体量，各单元相互关联形成群体。整个建筑群体在地面层做了大量的灰度空间，意在适应狭长地形而导致的不太富裕的绿地空间与周边的建筑体有更多的互动。景观的设计是以生态和自然为主题。尽量减少硬质不透水泥铺地，植物大面积选用草原草以维护自然的生态平衡。

CIPEA No.4 House
CIPEA 四号住宅

Location: Nanjing, Jiangsu Province
Area: 500m²
Architect: Zhang Lei, Jeffrey Cheng, Wang Wang, Wang Yi
Photographer: AZL Architects
Completion Date: 2011

地点：江苏，南京
建筑面积：500 平方米
建筑师：张雷、杰佛瑞·成、王旺、王昳
摄影：张雷联合建筑事务所
建成时间：2011 年

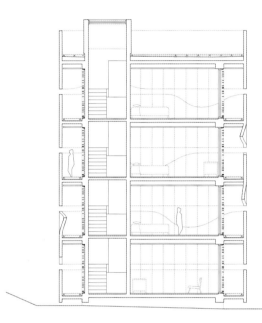

China International Practical Exhibition of Architecture is sited by the Buddha's Hand Lake near Pukou Ancient Mountain and Forest Park. A collective of 24 renowned architects from China and other countries all over the world were invited to design four public buildings and twenty detached residences. Furthermore, each residence was required to have at least five bedrooms with bathroom and the total area was not more than 500 square metres including living room, dining room as well as other public spaces. The main function of the residences was for family holiday and business reception.

The No.4 House designed by AZL Atelier Zhang Lei adopted the vertical layout which was not commonly seen in the construction of mountainous house. The volume of 500 square metres was divided into five cubes, stacking each other and forming a 4-storey structure. The design was not only corresponsive to the special typology of the site, but also made the original forest and the hillside land to be sustained to a large scale. The transverse fissures between every two connecting cubes ware enlarged at a special place on every level to frame the beautiful sceneries outside. What's more, the sceneries changed seen from different angels through the "fissures", a modern expression of the traditional Chinese landscape painting being unfolded horizontally but in a three-dimensional way.

中国国际建筑艺术实践展(CIPEA)坐落在南京浦口老山森林公园附近的佛手湖畔，项目邀请了来自中国和世界各地的 24 位建筑师参与设计，其内容包括四幢公共建筑和 20 幢小住宅。每幢小住宅设置不少于 5 个附带卫生间的卧室，再加上起居和餐厅等公共空间，建筑面积控制在 500 平方米以内，用于家庭度假和商务接待。

方案选择了山地小住宅不常见的垂直布局，将 500 平方米的体量分解成五个立方体叠为四层回应特定的地势，尽可能控制建筑基底的开挖范围，基本保全了北面的山林和坡地。四层建筑体量以贯穿横向裂缝的分离呈现垂直叠加的痕迹，裂缝在每层特定的位置扩大形成景框，通过观察方式的改变重新塑造了周围的景致，是传统中国山水绘画横轴展开的当代立体主义表现。

Concrete Slit House
混凝土缝之宅

Location: Nanjing, Jiangsu Province
Area: 270m²
Architect: Zhang Lei, Meng Fanhao, Cai Menglei, Lu Yuan, Tang Xiaoxin
Photographer: Iwan Baan
Completion Date: 2007

地点：江苏，南京
建筑面积：270 平方米
建筑师：张雷、孟凡浩、蔡梦雷、路媛、唐晓新
摄影：伊万·本
建成时间：2007 年

Jiangsu, China
中国，江苏

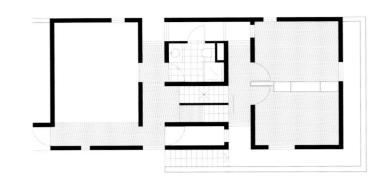

240 / EAST CHINA

The logical rationality of the Concrete Slit House can be expressed through the following points: the integrity of the interior elements, the harmony of the functional and spatial system, the transparency of the exterior façade and interior surface, the continuity between architecture and the surroundings, and the relationship between architecture and the paved site.

The staircase inside divides the interior into two parts and stands out from the entire structure, making the two parts vertically detached and forming a fissure. A zigzagging band of glass cleaves the 2,900-square-foot house in two and brings daylight inside. Slit is used to link two parts with half floor height difference, a 2-floor-high living area and a 1.5-floor-high dining area, both of which can overlook each other. Located in a historical neighbourhood in Nanjing and surrounded by grey-brick houses with a history of 100 years, the house differs in textile from the surroundings but shares the same quality in essence. The cast-in-place concrete wall and the precast façade and roofing emphasise the division of the two parts and the fissure is stressed at the same time to form an external contrast between the house itself and the surroundings.

混凝土缝之宅理性的逻辑结构主要表现在如下几个方面：内部空间要素的自洽；功能体系与空间体系的契合；内外面层的透明；建筑与环境外在肌理的连续；建筑与地表的关系。

设计将楼梯对应的两个朝向的外墙局部内缩并使其透明化，其结果是通过光线、附近空间的层次区分和尺度差异，将楼梯从整体中独立出来，并且导致左右两组垂直单元拉开距离并形成裂缝。裂缝把不同的房间加以分离，前后两部分也由此可以相互观望。因此，所有的空间序列都开始流动并变得立体化和视觉化了。身处敏感的民国文化保护区，面对周边近百年历史的青砖洋房，缝之宅小模板混凝土在肌理上有着神秘的差异和本质的呼应，现浇墙面和预制屋面极少主义式的交接突出了前后两个体量的分裂，于是裂缝被进一步强化并发挥了出人意料的作用，在主体和环境之间设置了外在的、最终的对立。

Poets Residences
诗人住宅

Location: Nanjing, Jiangsu Province
Area: 580m² (Ye Residence), 860m² (Wang Residence)
Architect: Zhang Lei, Shen Kaikang, Zhang Ang
Photographer: Iwan Baan, Nacasa & Partners
Completion Date: 2007

Jiangsu, China
中国，江苏

地点：江苏，南京
建筑面积：580平方米（叶宅）+860平方米（王宅）
建筑师：张雷、沈开康、张昂
摄影：伊万·班、仲佐写真事务所
建成时间：2007年

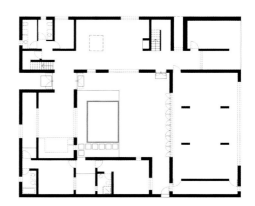

Located By Shijiu Lake in Gaochun County, Poets Residences are owned by two poets who are friends as well. The site enjoyed strong local flavour with a brick kiln to produce terracotta brick for nearby residences. The design adopts the theme of "courtyard" which is expressed in the site plan and interior spaces. The two houses are respectively designed with two different layouts: courtyard enclosed on three sides and courtyard enclosed on four sides. In addition, the functional spaces can be arranged in the shape of right angles as well as in a linear way for the sake of the expansive site. The various and continuous circulation lines for communication, movement, relaxation and exhibition composed the main elements of the interior which are as well the rational choice under the low technique condition.

The key feature of the two houses is the exterior wall: the main structure is closely wrapped up with red bricks to highlight the local style. Three different textures of brick made up the façade. An interlocking pattern leaves perforations between bricks and protruding bricks cast shadows along the wall. Windows punctuate the texture and the brickwork along the edges of these portals further develop abstract geometric relations while remaining loyal to the structural logic.

诗人住宅是两位诗人朋友的工作室兼私人居所，它的位置非常乡土——江苏省南京市高淳县凤山石臼湖边，周围还有砖窑在运转生产，供应远近农居修建所需的红砖。设计延续了"院落"的主题，"院落"体现在总平面和内部空间上，两幢房子各采用了三合院和四合院的布局，分别对应于西面和北面的湖面。基地的开阔使得建筑的功能能够以直角空间、线形序列等方式自由布置。大量且连续的交流、交通、休闲、展示动线成为建筑内部循环的主体成分，同时也是简陋的技术条件下合理的选择。

诗人住宅的外表是设计的重点：砖表皮将建筑严实的包裹起来，强化其材料的在地性。每一处墙面都是空洞、凹半砖和凸半砖两至三种砌法的混合。三种密度的砖肌理和无规律的窗洞共同进行着蒙德里安式的抽象编织，揭示了看似无序的窗洞隐藏在表皮形式背后立体主义的秩序。

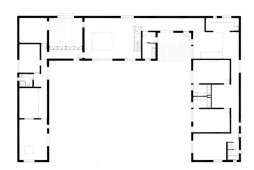

Three-Courtyard Community Centre
三间院

Location: Yangzhou, Jiangsu Province
Area: 1,900m²
Architect: Zhang Lei, Shen Kaikang, Yang Hefeng
Photographer: Iwan Baan
Completion Date: 2009

地点：江苏，扬州
建筑面积：1,900 平方米
建筑师：张雷、沈开康、杨鹤峰
摄影：伊万·班
建成时间：2009 年

Jiangsu, China
中国，江苏

Three-Courtyard Community Centre is located between the most famous waters converged by Beijing-Hangzhou Grand Canal and Liaojia River (the largest river around Yangzhou City). As the rapid development of urbanisation, the suburban area has become the rural-urban fringe district. In this way, several conflicts to the site came into birth. The project, composed of three independent parts, water courtyard, stone courtyard and bamboo courtyard, is designed with a strong local feature. Encircling, enclosing and concealing form the main characteristics of the Chinese traditional courtyard design.

The building is replete with playful conflicts, such as public function and private layout, the integrity of huge body and intimacy of small courtyard. Seen from afar, it seems like a micro village; looking close, it boasts interlaced brick façade with rich texture.

The architectural form of courtyard and the critical regeneration of traditional folk house have been two hot topics that architects explore and improve all the time. While, this project is inspired by both of these two forms: inn with continuous slope roof featuring simple structure with fewer materials in the district of Mongolian Prairie and the new dwellings with clear contour and traditional courtyard emphasising local material and technique, both of which are the eternal wealth.

项目坐落在京杭大运河和廖家沟这两条流经扬州城东的著名水系之间，都市化的快速发展已经使得这一近郊区域成为城乡结合部，包含了位置上的多重矛盾。三间院由水院、石院、竹院三个独立的院落组成，呈现出熟悉的本土造型。环抱、围合和隐藏这样的方式一直是中国传统庭院的突出特征。

这个项目结合了公共功能、私密形式、整体性大体量以及小庭院的亲切感，从形式到使用充满了有趣的矛盾。远观是乡野缩微村庄，近看是纹理丰富，凹凸交错的像素化砖墙表皮。院落形制和民居原形的批判性再现是建筑师一直以来探索和发展的类型，三间院的建构来自两方面的启示，一是蒙古草原连续坡顶的小客栈，它简单的建造和节省材料的做法；再就是江南新民居的轮廓和院落，地方材料和建造技能永远都是财富。

Dashawan Beach Facility
大沙湾海滨浴场

Jiangsu, China
中国，江苏

Location: Lianyungang, Jiangsu Province
Area: 7,761m²
Architect: Zhu Xiaofeng, Cai Jiangsi, Xu Lei, Xu Ye, Ding Xufen/ Scenic Architecture Office
Photographer: Shen Zhonghai
Completion Date: 2009

地点：江苏，连云港
建筑面积：7,761平方米
建筑师：祝晓峰、蔡江思、许磊、许曳、丁旭芬／山水秀建筑事务所
摄影：沈忠海
建成时间：2009年

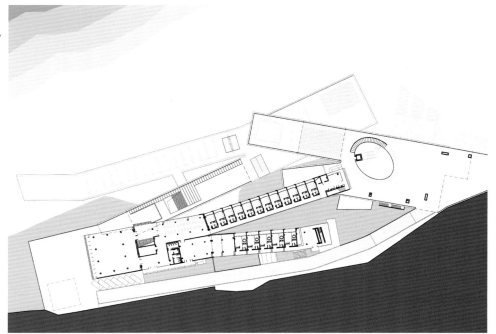

Dashawan Beach Facility is a newly completed project by Scenic Architecture Office. Located on the east seafront of Liandao Island of Lianyungang, an emerging harbour city booming in the middle section of the coastline of China, this slope building offers public beach facilities. Inspired by the morphological relationship of ocean waves, the architect evolved a simple set-back stacking section into a big seafront structure. When the programme experience is integrated with the scenic experience, the building then reconstructs the original scenery. These hundreds-meter-long slab buildings establish a scale dialogue with both the coastal mountain and the wave topography of the ocean, hence to present an ambition of the city to develop and grow.

The exterior of the main structure is constructed with concrete and without any furnishing process. Visually, the roughness of the grey and mottled exterior shares the same quality with the sandy beach. Most part of the roof is planted with grass to harmonise with the surrounding grassland. The roof of changing room adjacent to the beach is covered with sand to connect with the beach through a ramp. Paths and terraces on the roof are paved with anticorrosive floor board to create a relaxing atmosphere. Details in this project are of no great importance, and thus the original idea of installing glass curtain on the façade and handrails on the roof are finally ignored. The grand space, rolling hills and boundless stretch of sea can bring an unparallelled experience.

大沙湾海滨浴场位于连云港北部连岛度假区的东海岸，是一座建在山坡上的海滨公共设施，建筑师从海浪的形态关系中获取灵感，将一个简单的退台式剖面演化成自由奔放的海边巨构。人对建筑的使用体验融进了对风景的体验，建筑也反过来重构了原有的风景。这些超过百米长度的板状建筑与自然界的山坡和海浪同时建立了尺度上的关系，并表达了这座城市崛起的雄心。

设计中选择露明混凝土作为建筑主体的表面。出于对当地建造工艺的预判，并不追求高品质清水混凝土的完成度，而是放手让灰色斑驳的粗犷质感与粗粝的沙滩相匹配。屋顶大部分覆以草坪，与周围环境的草坡相融合，在紧临沙滩的更衣室屋顶铺上了沙子，通过端部的斜坡与沙滩连成一体。屋顶的步道和平台使用了防腐木地板，希望给散步的人提供轻松的感觉。虽然外墙上玻璃幕墙规格和屋顶栏板扶手的实施未能尽如人意，但在这样广阔的自然场景中，这些构造细节似乎已不再重要，决定人的体验的，是宏观的建筑空间、起伏的山峦和一望无际的大海。

Pavilions at Lake Yangcheng Park, Kunshan

昆山阳澄湖公园景观建筑

Location: Kunshan, Jiangsu Province
Area: 160m²
Architect: Miao Pu, Jiang Ningqing / Shanghai Yuangui Structural Design Inc.
Photographer: Miao Pu
Completion Date: 2010

地点：江苏，昆山
建筑面积：160平方米
建筑师：缪朴、蒋宁清 / 上海源规建筑结构设计事务所
摄影：缪朴
建成时间：2010年

Jiangsu, China
中国，江苏

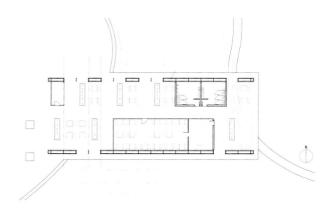

Located at a new suburban park, this is the constructed one of three proposed pavilions that share one prototypal form and structural system. The design explored the theme "multi-valency" in two aspects. First, Chinese traditional garden buildings took the form of courtyard which combines both interior and exterior spaces, providing users with two kinds of enjoyment within a short distance. This 36X14X3.34m building volume contains indoor, outdoor and semi-outdoor spaces. The 14m span of the structure allows the architects to juxtapose these spaces freely, rather than grouping them into two chucks.

Next, double-skin glass façades line up the south and north sides of the building. The use of such a façade had been limited to high-rise office buildings and mainly for energy saving. The design tried to add two new meanings into it. The architects added two tiers of shelves in the cavity to allow potted plants to grow there, making the façade a "green" wall (that can also be used for displays). They also partially sand blasted the glass skins to create small transparent "windows" that overlap each other in the two skins. People will see three kinds of depths, with varied overlapping effects when the viewer moves. Both skins of the wall are openable. In mild climates windows on both skins can be opened to afford natural ventilation. While all windows are closed in winter, only the windows in the outer skin are opened in the hot season. The shelves only cover two thirds of the cavity space to make room for the vertical air flow.

本工程是昆山市郊一个新建大型公园中的三个服务设施中先建成的一个。三个建筑将共用一个形式母题及结构体系。本设计在两个层面上探讨了"复合"这一主题。首先，我国的传统园林建筑大多为包含室内外空间的庭院建筑，能在一个小范围内让人同时享受到不同环境各自的优点。为此，设计师在这个36X14X3.34米高的建筑形体中复合了室内、室外及半室外空间。跨度14米的结构使设计师得以在三个单体中对室内外空间按使用做交错的布置，而不再捏合成两组。

其次，建筑的南、北两面各是一道双重玻璃外墙。该类外墙到目前为止基本上局限在高层办公楼中，只为节约能源而用。设计师把它与两种新的意义复合起来。在双重外墙中设计了二层搁板，使其成为可生长盆栽植物的"绿色"外墙（也可陈列商品）。设计师还在内外玻璃面上用局部磨砂创造出内外错开的透明"窗口"。使外墙呈现出三种层次的通透感。当人移动时，这些窗口会彼此掩映。复合外墙的内外层均可开启。在气候宜人时可打开内外窗利用自然通风。在冬季可同时关闭内外层。在炎夏可只开启外窗。为了保证空气的垂直流动，搁板只覆盖了三分之二的空腔平面。

Kangju Community Centre
昆山康居社区活动中心

Location: Kunshan, Jiangsu Province
Area: 420m²
Architect: Miao Pu / Shanghai Landscape Architecture Design Institute
Photographer: Miao Pu
Completion Date: 2006

Jiangsu, China
中国，江苏

地点：江苏，昆山
建筑面积：420平方米
建筑师：缪朴 / 上海市园林设计院
摄影：缪朴
建成时间：2006年

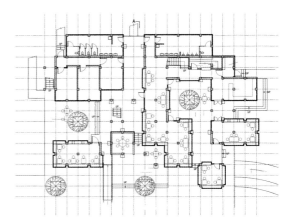

Located at the park of a new suburban residential area, the building is intended to be used by nearby residents to socialise and to play cards and chess.

Reflecting people's desire for a "refuge" with a "prospect" in a public space, the Centre has a porous exterior which resembles a forest or honeycomb that attracts people to enter. The building plan has many boundaries and corners to create numerous territories for small groups to claim. Columns are moved away from the corners to highlight the corner as a territory.

The overall spatial structure juxtaposes rooms and courtyards/roof decks, which not only produces rich and "suspenseful" views within the building and towards the surrounding park, but also generates an intimate relationship between the indoor and the outdoor spaces, a characteristic of traditional Chinese architecture.

中心坐落在昆山市西郊一个新建小区中的社区公园内，目的是为周围居民提供喝茶聊天，下棋打牌的场所。

人在休闲时最喜欢有庇护感同时又可以观察其他空间的地方。本建筑因此被设计成一个由许多小空间组成的"丛林"或"蜂窝"状建筑。这不仅增加了能创造庇护感的边界，同时也方便了各种小型聚会的形成。设计师还试验了转角无柱的独特结构体系来强调每个转角都可以是一个领域。为了让人可以观察探索，茶室的所有房间都开向多种室外景观，包括公园中的湿地和屋顶上种植的花木。设计师在建筑形体组合上还有意创造了一些半隐半显的"悬念"空间。使人们无论是在建筑内外，都可以不时透过一个尺度不大的开口，看到远处某个院落或房间的一角。

将室内外空间配对服务于一个主要建筑功能的做法，是我国传统建筑的宝贵遗产之一。设计师在本工程中特别强调了这一点。无论是在地面还是二层，人们都可以从室内移座到附近的一个庭院中或带遮阳花架的屋顶平台上。

Sichang-Road Teahouse, Kunshan

昆山思常路茶室

Location: Kunshan, Jiangsu Province
Area: 300m²
Architect: Miao Pu / Shanghai Landscape Architecture Design Institute
Photographer: Pu Miao
Completion Date: 2007

地点：江苏，昆山
建筑面积：300 平方米
建筑师：缪朴 / 上海市园林设计院
摄影：缪朴
建成时间：2007 年

Jiangsu, China
中国，江苏

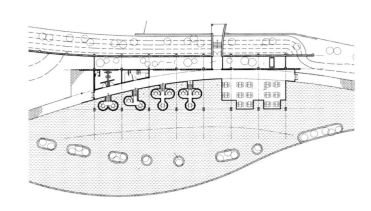

The urban renewals of the past three decades have made Chinese urban residents more and more detached from nature. Located next to a river preserved in a new residential area, the project provides an opportunity to reconnect residents with water. The river level fluctuates greatly. Therefore the architects designed an intermediating pool that draws its water from the river. Viewed from the teahouse, the pool appears to merge with the river. A row of metasequoia trees along the river bank is continued into the pool. Ten private tea rooms take the form of glass pods surrounded by water and half-sunken into the pool. Through the operable windows users can touch the water under their elbows, just like in a boat. Tiny fountains are bubbling in the gaps between the pods. A layer of wood trellises with vines above the glass roof affords people the feeling of peeking into the bright river from under the dark shades.

The design also tries to experiment with the "flat curve" observed in traditional Chinese architecture. As a gradual transformation of the orthogonal composition rather than a sharp contrast to the latter, the flat curve creates a subtle variation instead of noisy drama. The trellis roof adopts such a form to better relate the building with the river. Local construction materials and methods were adopted to build this double-curvature form inexpensively.

在近30年的城市更新中，城市居民离自然环境越来越远。本建筑位于昆山西郊一个新建住宅区中保留下来天然河道边，设计师因此着重探讨了如何更好地让居民接触自然水体。天然河道水位落差很大，设计师为此在建筑与河道之间设计了一条中介水池，池水取自河中。从茶室内望去，池水与河水融为一体。现有沿河的水杉林带被延续到水池中。每个包厢茶座均设置在一个被水环绕的玻璃圆柱体内。圆柱内的地面是半沉在水中，并有大窗可让人触及水面。就在胳膊肘下的水面让人有舟中之感。下沉茶座之间的水面中安置了多个涌泉。圆柱体的玻璃屋顶有一层将爬满藤蔓的木花架，给人以从被庇护的暗处眺望明亮风景的感觉。

设计师同时在建筑实体的几何形式上探索了新的语言。我国传统建筑形式中通常总含有一种"平缓曲线"，它们是正交直线体系的延伸而不是对立。在总体效果上形成一种含蓄的变化，而非某些时髦风格的碰撞叫嚣。为了使建筑与小河在几何形式上更好的联系起来，设计师在河边茶室的花架屋顶中实验了上述概念。设计师还特地探索了如何用当地施工单位能理解的二维图纸表述及施工技术来实现一个三维曲面。

Chinese Fourth Army Jiangnan Headquarters Exhibition Hall
新四军江南指挥部展览馆

Location: Liyang, Jiangsu Province
Area: 4,200m²
Architect: Zhang Lei, Shen Kaikang
Photographer: Lv Hengzhong, Jia Fang
Completion Date: 2007

地点：江苏，溧阳
建筑面积：4,200平方米
建筑师：张雷、沈开康
摄影：吕衡中、贾方
建成时间：2007年

Jiangsu, China
中国，江苏

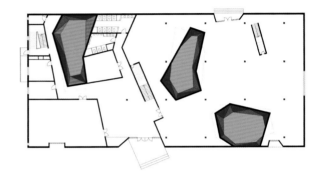

254 / EAST CHINA

Chinese Fourth Army Jiangnan Headquarters Exhibition Hall is located in Shuixi village, Liyang, 70 kilometres from southeast of the Nanjing. The memorial hall has a special and significant historical meaning for the 300,000 people in Liyang.

The letter and yard are all atoms of architecture, representing some basic materials that cannot be in further decomposition. The visual elements in Chinese Fourth Army Jiangnan Headquarters Exhibition Hall design are almost signifying. Geometric form shows that the building itself is memorial; flagstone attaching veneer is a symbol of classical meaning of the memorial. The sculpture garden skin which is made of red aluminium stands for revolution and blood. Several grooves in the east corresponding with the yard make graphics on the front elevation in an abstract way with three letters "N4A". In the exhibition hall, we also see the connection between symbol activities and building content, the connection between crushing modernism and China's current reality. More importantly, we also see the independent revolution which has the absolute value; pure connotation and one-dimensional direction began to gradually come to the wider human history. No longer like a lot of old-fashioned memorial buildings, it is not just about primary trauma memorial or the consolidation of the reality order. It shoulders the common whole that participates in human activities.

新四军江南指挥部展览馆坐落于距南京城东南70公里的溧阳水西村，对于拥有30万人口的溧阳而言，纪念馆有着特殊而重要的历史意义。

字母和院落都是建筑的原子，是代表着某种无法深入分解的基础物质。新四军江南指挥部展览馆设计中的视觉要素几乎完全符号化了。几何母体表示建筑本身就是纪念物，石板贴面则象征着石碑的纪念性经典含义。用红色铝板做成雕塑庭院的表皮象征着革命和鲜血，也有铭刻历史的意象。东面几个与院落对应的凹槽把正立面用"N4A"三个字母抽象性地图解出来。在展览馆中，我们同时也看到了符号活动和建筑内容的连接、破碎的现代主义和中国当下现实的连接。更重要的是，我们还看到了拥有绝对价值、纯粹内涵、一维方向的独立革命史开始逐渐走向更为广阔的人类史。已经不再像很多老式的纪念馆建筑仅仅是关于原始创伤的纪念，或者是关于现实秩序的巩固。它肩负起了参与进人类的普遍的整体。

Suzhou Museum
苏州博物馆

Location: Suzhou, Jiangsu Province
Area: 17,000m²
Architect: I. M. Pei Architect with Pei Partnership Architects (New York, NY), Suzhou Institute of Architectural Design Co., Ltd.
Completion Date: 2006

地点：江苏，苏州
建筑面积：17,000平方米
建筑师：贝聿铭建筑师（美国）暨贝氏建筑事务所（美国）、苏州市建筑设计有限公司（中国）
建成时间：2006年

Jiangsu, China
中国，江苏

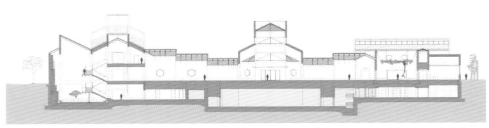

The new Suzhou Museum is located at an important historic and cultural block, neighbouring with famous gardens such as the Humble Administrator's Garden and Zhong Wang Fu, which makes it a big challenge to the designer. On the basis of solid research and full comprehension on the local culture, I. M. Pei interprets the connotation of Suzhou traditional gardens with modern architectural language.

The new museum is in harmony with the original environment by its scattered layout. It is divided into three sections. As the main circulation space, the Centre part includes the entrance, the atrium, the central hall and the main garden; the west wing is the main exhibition area; the east wing includes the secondary exhibition area and the administration offices. The main garden is neighbouring with the Humble Administrator's Garden that is to the north of the new museum. To the east, the Zhong Wang Fu (the former Suzhou Museum) has been rebuilt to be part of the new exhibition hall. Enjoying a harmonious coexistence, the new and the old architectures bring out the best in each other. The design follows principles regarding the volume that are "not too high, not too large and too abrupt". The architecture mainly consists of one-floor buildings and underground buildings. The cornice height of the main building is less than 6 metres; the two-floor buildings, arranged at the Centre and the west that are far from the conserved architectures, do not exceed the height limit for the adjacent historic architectures. By such ingenious layout and rational scale control, the museum homogeneously blends into the original environment. In addition, streets near the museum have been repaired to revive their former appearance, which creates an atmosphere of traditional culture with an ancient flavour.

A lot of new technologies, new materials and new design techniques are applied in the design and construction of the new museum, which not only keeps the traditional style of Suzhou gardens, but also presents the features of modern architecture as a cutting-edge museum.

苏州博物馆新馆的选址在苏州重要的历史文化街区，紧邻拙政园、忠王府等名园而建，对于设计者是极大的挑战。贝聿铭在充分研究和理解当地文化的基础上，用现代的建筑语言诠释了苏州传统园林建筑的内涵。

新馆建筑采用分散布局的方法有机地融入原有的环境中。建筑群分为中、东、西三块，中部为主要交通空间，由南到北依次为入口、前庭、中央大厅和主庭园；西部为主展区，东部为次展区和办公区。主庭园和新馆北边的拙政园隔墙相连，从新馆望去，拙政园的高大古树随风摇曳，新旧园景笔断意连。紧邻新馆东侧的"忠王府"（原苏州博物馆）也整修一新成为新馆展厅的一部分。新老建筑和谐相处，相得益彰。博物馆设计在体量上遵循的原则是"不高、不大、不突出"。建筑以一层和地下一层为主，主体建筑檐口高度控制在6米内，局部二层主要位于远离控保建筑的中西部，也未超出周边古建筑的限高点。通过巧妙布局和尺度控制，博物馆和周围原有建筑环境有机融为一体。此外，新馆外围街区按"修旧如旧"原则进行了整体修缮，形成了古香古色的传统文化氛围，新馆不仅成功融入了历史街区，并为老区注入苏州博物馆了新的活力。

在新馆建筑设计中，大量新技术、新材料和设计手法的运用，使这组新建筑既有传统苏州园林建筑的特色，又处处散发着时代的气息，而不失为一座十分现代的、高标准的博物馆建筑。

China Museum of Sea Salts
中国海盐博物馆

Jiangsu, China
中国，江苏

Location: Yancheng, Jiangsu Province
Area: 17,800m²
Architect: Cheng Taining, Cheng Yuewen, Wu Nina, Yang Tao, Li Shutian, Wu Wenzhu
Completion Date: 2009

地点：江苏，盐城
建筑面积：17,800 平方米
建筑师：程泰宁、程跃文、吴妮娜、杨涛、李澍田、吴文竹
建成时间：2009 年

The China Museum of Sea Salts is located in Yancheng City, Jiangsu Province. Sitting on the east bank of Chuanchang River, an important river in the city, the museum totals 17,800 square metres in building area.

The building draws inspiration from salt crystal. The extensive shoal along the sea provides the unique setting for sea salt production. How to integrate these elements into the architectural design is the primary issue the architects confronted with.

The rotating "crystals" on the top, combined with different base layers, seem like shining crystals scattered on the shoal along the Chuanchang River. The unique shape makes the building become a landmark of the city.

中国海盐博物馆位于江苏省盐城市，基地位于贯穿盐城市的重要河流"串场河"以东，总建筑面积17800平方米，是全国唯一一座反映悠久的中国海盐历史文明的大型专题博物馆。

中国海盐博物馆设计理念为：全面系统地反映和研究中国海盐发展史，介绍、研究中国海盐文化的形成、发展和最新成果，收藏保护和陈列展示中国海盐历史的文物和资料，科学地表现与再现中国海盐文化的丰富内涵，高标准地构建盐城历史文化名城形象。

博物馆分为主体楼、海盐文化广场和海盐产业区，包括搭建反映海盐生产和盐民生活的多层场景和雕像。建筑造型试图演绎海盐的"结晶之美"，广阔的海边滩涂为海盐的生产提供了独特的环境，如何把这些元素融入到建筑设计之中，是设计师首要研究的课题。

海盐博物馆体量通过晶体的组合叠加，结合层层跌落的台基，就像海盐的结晶体随意地散落在串场河沿岸的滩涂上，意境开阔。

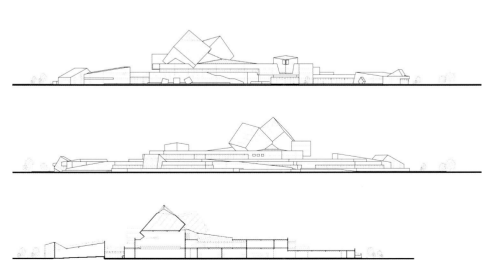

Jiangsu Provincial Art Museum
江苏省美术馆新馆

Jiangsu, China
中国，江苏

Location: Nanjing, Jiangsu Province
Area: 27,449m² (above ground: 18,210m²)
Architect: KSP Jürgen Engel Architekten
Photographer: Shu He Photography
Completion Date: 2009
Award: Competition June, 2006, 1st prize

地点：江苏，南京
建筑面积：27,449 平方米（地上：18,210 平方米）
建筑师：德国 KSP 尤根·恩格尔建筑师事务所
摄影：舒赫摄影工作室
建成时间：2009 年
获奖：2006 年 6 月竞赛一等奖

The two structures of the Museum that stand at slight angles to one another follow the two flanking thoroughfares: Zhongshan (or Revolution) Road, and Changjiang (or Culture) Road.

The two interlocking U-shaped buildings also create a space that is modelled on a canyon to function as the access zone between two soaring walls of natural stone. It is lit from above and covered with a light glass roof. This 17-metre-high access area, which narrows at its two main entrances, links the two stone halves of the building and guides visitors into the Museum. In the northern building, clear exhibition rooms of varying sizes offer ideal conditions for presenting the works of art. Two bridges spanning the glass-covered intermediate space connect the exhibition area with the southern element. In addition to training, conference and office space, this building also contains the VIP area and the auditorium with seating for around 400 people.

The travertine natural stone facing with its narrow window indentations obscures the sheer number of storeys and as such reinforces the overall monolithic impression of the museum building. Simultaneously, the alternation between vertical stone panels and window slits with sheet metal jutting out at the sides creates rhythm in the façade. The structural frame and the delicate construction of the glass roof were developed in collaboration with Stuttgart-based German engineers Breuninger.

美术馆的主入口朝向城市广场"大行宫市民广场",相互交错的两个建筑毗邻两条大街——被称为革命大道的中山路和被称为文化大道的长江路。

相互咬合、限定的两个U形石材建筑形成了一个由轻盈的玻璃屋顶覆盖的中庭过渡空间,中庭17米高,两侧的主要入口处空间渐渐变得狭长,将参观的人流吸引到美术馆内。北侧的展览楼内设有不同规模的展览空间,为艺术作品的展示创造了理想的环境。中庭玻璃屋顶下的两个空中连桥,将展览厅的前厅和南侧的办公楼连接在一起。办公楼内设有培训、会议、办公、贵宾区和一个约400座的小剧场报告厅。

以罗马洞石饰面的建筑立面搭配细长的竖向窗带,消除了一般建筑楼层叠加的视觉感受,塑造了美术馆整体上的雕塑感。深浅交错的竖向洞石立面和窗带,以及窗带侧面的金属挑板,创造了建筑丰富的韵律感。简洁的玻璃屋顶钢结构,是与来自德国斯图加特的Breuninger结构工程师事务所一起合作设计的。

Nanjing Sifang Art Museum
南京四方美术馆

Location: Nanjing, Jiangsu Province
Area: 2,787m²
Architect: Steven Holl Architects
Photographer: Iwan Baan, Shu He, Steven Holl Architects
Completion Date: 2010

地点：江苏，南京
建筑面积：2,787 平方米
建筑师：美国斯蒂文·霍尔建筑事务所
摄影：伊万·班、舒赫、美国斯蒂文·霍尔建筑事务所
建成时间：2010 年

Jiangsu, China
中国，江苏

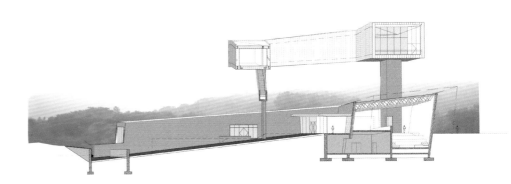

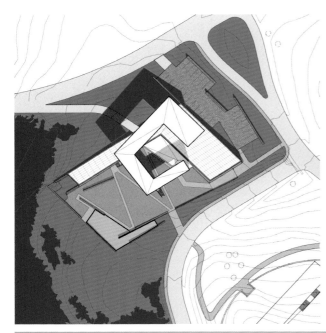

The new museum is sited at the gateway to the Contemporary International Practical Exhibition of Architecture in the lush green landscape of the Pearl Spring near Nanjing. The museum explores the shifting viewpoints, layers of space, and expanses of mist and water, which characterise the deep alternating spatial mysteries of early Chinese painting. The museum is formed by a "field" of parallel perspective spaces and garden walls in black bamboo-formed concrete over which a light "figure" hovers. The straight passages on the ground level gradually turn into the winding passage of the figure above. The upper gallery, suspended high in the air, unwraps in a clockwise turning sequence and culminates at "in-position" viewing of the city of Nanjing in the distance. The meaning of this rural site becomes urban through this visual axis to the great Ming Dynasty capital city, Nanjing.
The courtyard is paved in recycled Old Hutong bricks from the destroyed courtyards in the centre of Nanjing. Limiting the colours of the museum to black and white connects it to the ancient paintings, but also gives a background to feature the colours and textures of the artwork and architecture to be exhibited within. Bamboo, previously growing on the site, has been used in bamboo-formed concrete, with a black penetrating stain. The museum has geothermal cooling and heating, and recycled storm water.

四方美术馆坐落在中国国际建筑实践展大门口——珍珠泉绿丛掩映的自然景观内，其设计以探索视角变化、空间层次以及烟波浩渺这一在中国早期画作中呈现的空间奥秘为主要理念。
美术馆建筑由一系列平行的透视空间及黑色竹子包裹的水泥花园墙壁打造而成。一层笔直的通道逐渐蜿蜒，一直通往透视空间下方的曲折小径。上方的画室好似"飘浮"在空中，沿着顺时针方向展开，最终"定格"在朝向南京市区的方向，通过视觉的变换使得这一偏远的地区与城市连通。
庭院地面采用南京市区老胡同住宅区内回收的砖石铺设，仅有黑白两种色彩的运用让人不禁联想到古老的画作，同时也为展出的作品提供了一个简约的背景。原来生长在美术馆场址上的竹子被用作建筑材料，突出环保特色。此外，美术馆采用地热供暖、冷却设备，并具有雨水收集利用功能。

Fei Xiaotong Memorial Museum
费孝通江村纪念馆

Location: Wujiang, Jiangsu Province
Area: 10,000m²
Architect: 9-Town Architects, College of Architecture Design and Urban Planning at Tongji University, Li Li
Photographer: Yao Li
Completion Date: 2010

地点：江苏，吴江
建筑面积：10,000 平方米
建筑师：苏州九城都市建筑设计有限公司、同济大学建筑与城市规划学院、李立
摄影：姚力
建成时间：2010 年

Jiangsu, China
中国，江苏

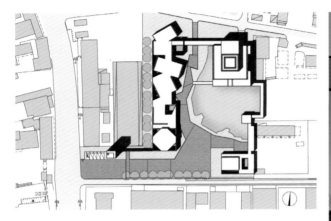

The project composed of Fei Xiaotong Museum, Fei Dasheng Museum and History and Cultural Museum of Jiang Village, is located in an abandoned plot at the edge of Jiang Village. As the improvement of the traffic condition surrounding the village, the plot has the real potential to become the centre of new village. Taking this into consideration, the architect defines the design scheme based on the concept of "improving people's livelihood and serving the village". After researching the site carefully, the camphor trees (the key landscape in the village and the axis to link the bridge across the village) growing on the northern side become the crucial elements for architectural organisation. Finally, the main layout is formed by arranging all the buildings along the perimeter of the site with the central part left unbuilt.

In addition, the buildings are connected to the surrounding market, village committee and elementary school in an appropriate way, rendering it the public centre of the village. The architectural structures of hall, corridor, pavilion, alley and courtyard are fully employed to show the characteristics of Jiangnan Garden. The exterior and interior compliment each other and integrate together in a flowing and perfect way.

建筑功能由费孝通纪念馆、费达生纪念馆以及江村历史文化馆组成。建筑选址位于村落边缘的一处废弃用地，周边环境混杂。设计者意识到随着村落周边交通条件的改善，这块废弃用地存在着转化为新的村落公共中心的潜力，所以设计策略立足于改善民生、服务乡村这个根本目标。设计充分研究了场地特点，基地北侧的香樟树成为决定建筑布局的最关键因素。这几株香樟树是村落的重要景观标识节点，也是连接村内跨河桥梁的枢纽，于是建筑化整为零、让出中间视廊的布局形成。建筑群体沿基地周边布置，空出了中心场地。

建筑没有与周边的农贸市场、村委会和小学校隔离，而是适度连通，使得纪念馆作为必经之路真正成为村落的公共场所。在建筑类型构成上以堂、廊、亭、弄、院等元素回应了江南园林特点，并通过形体扭转与扭转间隙精确的对景处理给分散的建筑群体增强了视觉张力，丰富了行进中的空间体验。建筑室内外设计一气呵成，创造出完整连贯的空间意象。

Nanjing Massacre Memorial Hall Extension Project
侵华日军南京大屠杀遇难同胞纪念馆扩建工程

Location: Nanjing, Jiangsu Province
Area: 74,000m²
Architect: He Jingtang / Institute of Architectural Design at South China University of Technology
Completion Date: 2007

地点：江苏，南京
建筑面积：74,000 平方米
建筑师：何镜堂 / 华南理工大学建筑设计研究院
建成时间：2007 年

Jiangsu, China
中国，江苏

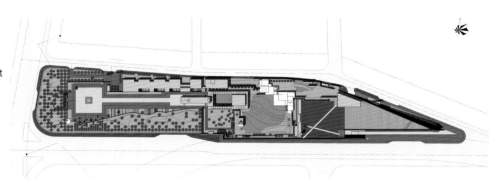

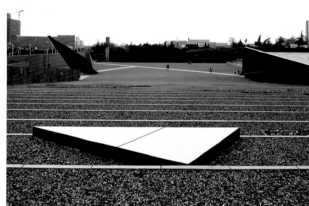

266 / EAST CHINA

Located on the east and west sides of the existing Memorial Hall, the extension includes new addition of Memorial Hall, renovation of Skeletal Remains and Peace Park. Highlighting the historic site theme and respecting the existing architecture, the architects create an overall atmosphere, with wall, scars, courtyard of death, memorial ceremony courtyard and path of candles as architectural elements to express a unique sense of place. The concept combines war, massacre and peace from east to west, with "Broken Sword", "Courtyard of Death" and "Casting Sword into Plough" as counterparts. The three spaces form a complete spatial sequence of "Prelude - Bedding – Climax - Ending". From the enclosure and isolation in the east to the opening in the west, the spaces are integrated with the urban and natural environments.

Respecting the Existing Building: Since it is an extension project, the architects always focus on keeping harmony with the existing building. The architects adopted a "volume elimination" design method, and buried the main body of new building underground when they decided the building shape, reducing influence of this new building towards the existing one. The new building gradually rises to the east, and the roof is treated as memorial plaza, which emphasizes the special features of the new memorial and reduces pressure that the new building brings to the existing one. In addition, the architects take the existing memorial building scale as model, and arrange a series of small volumes with similar size, including Skeletal Remains Protection building and Meditation Hall, which are connected by a central axis, to achieve unification of space scale and order. The memorial collection communication area in the west was also divided into smaller volumes. Meanwhile, the new and existing buildings have the same architectural language and design methods, which bring the impression of whole building to visitors.

Creating a Memorial Place: The architects thought that this extension project was not only an architectural design, but also a series creation of memorial places, in order to emphasize the spirit of the place. In the entrance plaza, the architects use lifeless-featured detritus to pave the plaza, in order to express the spirit theme of "life and death". The interior space of the new building is unified with the building shape, and uses gradient wall and ramp floor to composite a kind of unusual inner spaces with disorder and conflict, to express the space spirit, which meets the exhibition theme. In the Peace Park, the architects use a huge strip-like pool to directly lead people's eyes to the statue of Irene at the end of the pool, expressing the space spirit of "yearning towards peace".

The architects re-adjusted design connecting with the skeletal remains, which was found while constructing, and added transparent flooring and skylight to show the original remains. The length of south elevation of the new building is more than 180 metres. The architects specially think about the stone skin and unit size change to meet the required feeling of strength and vicissitudes. In this skeletal remains protection building, sand-stone flooring, aged steel bridge and dim lighting were used to show the historic feeling coming from the spot of the victims.

扩建范围位于现有纪念馆的东西两侧，主要包括新扩建纪念馆、万人坑遗址改造以及和平公园三部分。本设计主要希望突出遗址主题，尊重原有建筑，塑造整体氛围，以墙、伤痕、死亡之庭、祭奠庭院、烛之路等为建筑元素表现特定的场所精神；总体构思以战争、杀戮、和平三个概念组合，由东到西顺序而成，与此相对应的是"断刀"、"死亡之庭"、"铸剑为犁"三个空间意境的塑造，形成序曲—铺垫—高潮—尾声的完整空间序列；建筑空间从东侧的封闭、与世隔绝过渡到西侧的开敞，与城市、自然融为一体。

尊重原有建筑：作为扩建工程，建筑师始终注意与原有纪念馆保持协调。在建筑尺度上，为避免新建部分对原有纪念馆的影响，建筑师在确定建筑形体效果时采用了"体量消隐"的设计手法，将新建的纪念馆主体部分埋在地下，向东侧逐渐升高，屋顶作为倾斜的纪念广场。既突出了新馆的特殊风格，又减少了对原有纪念馆的压迫感。再次，建筑师以原有纪念馆的体量为模数，在扩建工程的设计中安排了"万人坑"遗址保护建筑、冥思厅等一系列尺度相近的小型建筑体量，并在设计中用一条中轴线将这些主要体量统一起来，形成空间尺度和秩序的统一。此外，园区西侧的馆藏交流区也采用了化整为零的手法。新老纪念馆建筑语言和手法统一，令参观者感觉浑然一体。

营造纪念性场所：建筑师认为，这项扩建工程重要的不只是设计建筑，而是要营造一系列纪念的场所，形成突出的场所精神。在入口纪念广场部分，建筑师以无生命特质的级配碎石铺装广场，通过这一特殊的铺装材料来反映"生与死"的场所精神主题。新馆建筑内部空间结合形体特点，运用倾斜的墙体和缓坡的地面，组合成一种错乱、冲突的非常态空间，表达同展览主题相适应的场所精神。在和平公园，建筑师用巨大的长条形水池将人们的视线直接引向水池终点的和平女神塑像，映衬出向往和平的场所精神。

精心处理细节：结合施工过程中现场发现的骸骨遗址，建筑师重新调整设计，加做了透明地面和天光，使这一遗迹原貌展示。新馆南立面长度达到180米以上，建筑师就重点考虑石材文理和单元尺寸的变化组合使其达到力度感和沧桑感的要求。重新设计的万人坑遗址保护建筑内，采用了沙石地面、锈蚀钢桥与昏暗的灯光的组合，充分表达了遇难现场的历史感。

Nanjing Changfa Centre
南京长发中心

Location: Nanjing, Jiangsu Province
Area: 140,000m²
Architect: WSP Architects
Photographer: Shu He, Yao Li
Completion Date: 2006
Award: The International Competition, 1st prize

地点：江苏，南京
建筑面积：140,000 平方米
建筑师：维思平建筑设计
摄影：舒赫、姚力
建成时间：2006 年
获奖：国际竞赛一等奖

Jiangsu, China
中国，江苏

268 / EAST CHINA

Nanjing Changfa Centre is located at a bustling area in Nanjing City – Daxinggong area, and it is the first class land used for business and office, with convenient communication and complete auxiliary facilities. The land for the project is located at CBD of Nanjing, on the east side of the high-rise buildings. Xuanwu Lake – President Building – Confucius Temple form a row of sequence, setting off each other. Nanjing Changfa Centre joins in the sequence by letting its axis pass through the middle of the twin office tower. Designer hopes that Nanjing Changfa Centre can integrate with the city authentically and become an indispensable part of the city life.

Nanjing Changfa Centre is composed of two 150m high office twin towers, and two 135m high apartment towers in the south. The business is concentrated around the sinking square at the foot of the office twin towers in the south and under the huge grass slope which is connected with the city at the foot of the apartment twin towers in the south. Nanjing Changfa Centre adopts "low-tech high-efficient" design strategies, i.e. to create highly comfortable and highly efficient environment with simple, energy-saving and low-cost techniques.

The "core in core" structure system is adopted in the design. The rectangular grid composed by close columns and beams of the outer core can be detected clearly from the façade. The "dual surface" structure is adopted by the design for façade: the inner layer is French Windows that can be opened and plain frame columns and beams, and the outer layer is the large area of perforated aluminium panel curtain wall. The inner layer and the outer layer are connected by steel frame. The glass windows that can be opened make the ventilation be realised by opening the windows in high-rise buildings. The perforated aluminium panel can filter the "high building wind" that impacts the building in horizontal direction, and shield 40% excessive sunlight.

The double-height concept is introduced into the interior space of the office and residential towers, with the height of 5.4m and 4.95m individually, to effectively meet the requirements that the expanding enterprises and families can re-divide the indoor space vertically. At the same time, the rectangular office and residence have high-efficient plane and variable and flexible usable space.

南京长发中心地处南京市繁华地段——大行宫地段，是南京市一类商业办公用地，交通极为便利，周边公用配套设施齐全，项目用地地处CBD超高层区的东端，玄武湖—总统府—夫子庙形成遥相呼应的一行序列，南京长发中心以让轴线从办公双塔之间居中穿过的方式加入到这个序列当中。设计师希望南京长发中心能够真实地与城市融为一体，成为这个城市生活中不可或缺的一部分。

南京长发中心由两栋高150米的办公姊妹双塔，以及南侧两栋135米高的塔式公寓所组成。集中商业分别设置于北部办公双塔下的下沉式广场周边以及南部公寓双塔下与城市相衔接的巨大草坡之下。南京长发中心所采用的设计策略之一是"低技高效"，即通过简单的节能材料和低成本的技术营造了高舒适度、高效能的环境。

南京长发中心的结构选型采用钢筋混凝土筒中筒结构体系，外筒的横梁和密柱构成的矩形网格在立面上被清晰地表达出来。外立面设计采用了独特的"双层表皮"构造，内层是可开启的落地玻璃窗和朴素的框架梁柱，外层是大面积的穿孔铝板幕墙。内外表皮之间以钢构架相连接。可开启的玻璃窗使超高层也可以通过开窗实现空气流通；穿孔铝板可以帮助过滤横向冲击大厦的"高楼风"，并屏蔽40%的多余阳光。

写字楼和公寓的室内空间均采用了双层高度设计，层高分别为5.4米和4.95米，以便成长的企业和家庭在必要时从竖向上重新划分室内空间。同时，矩形的办公和住宅具有高效率的平面和灵活可变的使用空间。

Nanjing International Conference & Exhibition Centre

南京国际会展中心

Location: Nanjing, Jiangsu Province
Area: 108,000m²
Architect: tvsdesign
Photographer: Dingman Photography, Blain Crellin Photography
Completion Date: 2008

Jiangsu, China
中国，江苏

地点：江苏，南京
建筑面积：108,000 平方米
建筑师：美国 tvs 建筑设计公司
摄影：丁曼摄影公司、布莱恩·克莱林摄影公司
建成时间：2008 年

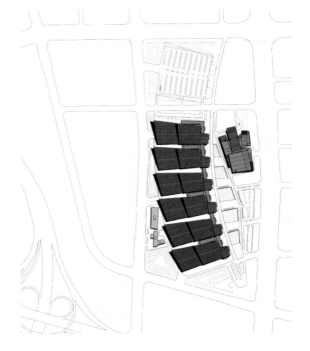

The design, dubbed "Curling Dragon, Crouching Tiger", is based on Nanjing's local history and natural environment. The complex, spanning more than 50 acres, is the largest conference and exhibition centre in Jiangsu Province. In the first phase it features six identical halls, with more than 80,000 square metres of world-class exhibition space arranged along a gracefully curving grand concourse. The new conference centre features a unique multi-purpose hall accommodating banquet, assembly or exhibition events. A future second phase will include an additional 20,000 square metres of exhibition space, a 400-room hotel, underground parking and a below-grade retail concourse linked to the city subway system.

The scheme for the Nanjing Conference and Exhibition Centre was achieved through the careful creation and overlap of three diagrams. First, a functional diagram was generated for a 100,000-square-metre multi-hall exposition centre, comprised of light-filled pavilions, coupled to a free-standing 30,000-square metre conference centre. The goal of this diagram was to explore multi-use flexibility, rigorously separate traffic types and minimise walking distances for the visitor. The second diagram, an urban design diagram, idealised how the exposition and convention components, along with the non-core programme of hotel, retail and commercial developments, could support and engender the long range urban planning goals for the HEXI district as envisioned by the Nanjing Planning Bureau. A third diagram abstracted the dueling tiger/dragon figural landscape along with the Yangtze River onto the site as a handprint of the Nanjing experience.

这一设计以南京历史及自然环境为基础，被命名为"卧虎藏龙"。整幢建筑占地超过50公顷，是江苏省最大的会展中心，一期工程包括6个相同的展示大厅，建筑面积超8万平方米，沿着一个蜿蜒连绵的广场一次排列。会议中心以独特的多功能厅为特色，用于举办宴会及展览等。二期工程包括2万平方米的展示空间、一个400间客房的酒店、地下停车场以及与城市地铁相连接的地下商业广场。

南京会展中心建筑通过将三个不同规划目标而实现。首先，打造10万平方米的多功能展示厅，包括光线充裕的展厅以及独立的3万平方米会议中心。这一功能区的主要涉及目标即为营造多功能的灵活性，将不同类型的通道分离，减少参观者行走的距离。其次，如何使得酒店、商业广场等辅助空间与会展功能结合，使其发挥最佳作用。最后，营造独特的景观环境，带来独特的体验。

Suquan Yuan, Suzhou
苏泉苑

Location: Suzhou, Jiangsu Province
Area: 254m²
Architect: Tong Ming
Photographer: Tong Ming
Completion Date: 2007

地点：江苏，苏州
建筑面积：254平方米
建筑师：童明
摄影：童明
建成时间：2007 年

Jiangsu, China
中国，江苏

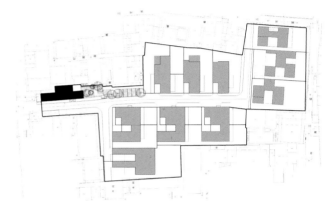

Suquan Yuan is the redevelopment project of an area which used to be the large coach parking lot of Suzhou Travel Vehicle Company. There are two major challenges: one is that the only access to this spacious area from Shiquan Street is less than 6 metres wide. Once the area is put into commercial use, with the only access to this site being the narrow street line, the bottleneck problem will arise; the other is the functional uncertainty.

Based on the concept and memory of traditional architecture, this building at the entrance is designed as a grey brick box with strip wooden windows. The façades can change according to use. It is a box full of light and shadows, unfolding the texture of materials and construction. Its structure maintains the domino system, with the internal space divided into service area and subsidiary area. This is a structure of feasibility that is adaptable for more future usage, and ready to stage more events.

苏泉苑是一个再开发项目，基地原先是苏州外事旅游车船公司的一片大型客车停车场地。该项目具有两个挑战：第一个是内部巨大的停车空间仅由一个不到6米宽的出入口与十全街相连接。一旦该地形被转变为商业使用，并且只能采用狭小的沿街面作为基地的出入口，这就产生了一种瓶颈性的不利因素。第二个是使用功能的不确定性。

基于对传统建筑的印象和记忆，这座入口处的建筑被设计成了一个镶有长条木窗的青砖构成的匣子。建筑的立面可以根据使用情况发生改变。它是一个纯净的匣子，充满光和影，展现材料和构造的质感。在结构上，它保持了多米诺体系，内部的空间分成服务区和辅助空间。这样的一个匣子能够适应将来更多未知的使用者，它将会是一个为未来上演的事件所准备的舞台。

Nanjing University Performing Arts Centre

南京大学表演艺术中心

Location: Nanjing, Jiangsu Province
Area: 16,000m²
Architect: Preston Scott Cohen, Inc. / Zhang Lei
Photographer: Preston Scott Cohen, Inc
Completion Date: 2009

地点：江苏，南京
建筑面积：16,000 平方米
建筑师：雷斯顿·斯科特·科恩设计事务所 / 张雷
摄影：雷斯顿·斯科特·科恩设计事务所
建成时间：2009 年

Jiangsu, China
中国，江苏

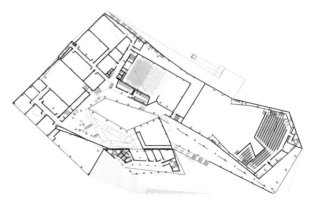

274 / EAST CHINA

The Nanjing University Performing Arts Centre, located centrally in the master plan for the new Nanjing University campus in Xianlin, offers a singular expression of the dialogue between two opposed paradigmatic forms of symbolic significance: a curving roof related to the landscape of the larger campus context, and a tower which acts as a beacon and observation point.

The design exploits the techniques and economy of local construction practices as a means to develop an exceptional form. Poured-in-place concrete construction, using adjustable and recyclable frame, gives shape to a landscape-like roof that acts as a unifying "umbrella". The roof form is derived from a series of hyperbolic paraboloids, the ruling lines of which become reinforcing beams, all based on the same cross section, and distributed at regular intervals. As such, the roof creates the effect of a remarkably variable form, despite its underlying logic of regularity and economy of means. The roof landscape surrounds the tower in such a way that it appears as if the tower is an anchored point of resistance or a buoy atop the surface of a roiling seascape.

The design of the building was driven by an economical and efficient passive energy strategy. By strategically dividing the building into several functionally independent zones, parts of the building that are not in use can be closed off thermally from those parts that are, dramatically reducing heating and cooling loads. Working with fluid dynamics modelling software, the tower's interior organisation and exterior form (a narrow floorplate oriented towards prevailing winds) allow for cross-ventilation satisfying the building's summertime cooling demands.

南京大学表演艺术中心位于仙林新校区中心地带，其建筑设计成功实现了两个造型完全不同的结构的融合，并体现出各自的象征意义：弯曲的屋顶与校区的整体景观环境统一；尖塔则用作灯塔及观景台。

这一设计利用本地建筑公司的技术及经济条件，以创造出一种超乎寻常的建筑样式。现场浇制的水泥结构以及可调节的回收框架奠定了屋顶的造型，犹如一把大伞。此外，屋顶造型灵感主要来自于一系列曲面抛物线，标尺线的位置布置钢性梁，并以固定的间距排列。因此，屋顶造型呈现出多变性。独特的屋顶设计使得尖塔结构犹如定位点一般，又好似矗立在大海上的浮标。此外，建筑设计采用经济的被动节能技术。整体结构根据各自的功能被分割成独立的小空间，部分不使用的区域停止供热及制冷，极大地降低了能源需求。室内外空间的合理安排营造了交叉通风系统，满足建筑夏季的制冷需求。

Nam Gallery
南画廊

Location: Nanjing, Jiangsu Province
Area: 230m²
Architect: Zhang Lei, You Shaoping, Yuan Zhongwei
Photographer: Nacasa & Partners
Completion Date: 2007

地点：江苏，南京
建筑面积：230平方米
建筑师：张雷、游少萍、袁中伟
摄影：仲佐写真事务所
建成时间：2007年

Jiangsu, China
中国，江苏

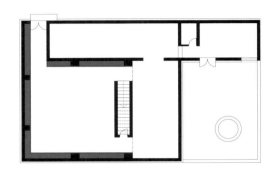

Nam Gallery was renovated from a used boiler house in the local chemical engineering institute. The corrugated panel was employed to cover the façade and the Corten steel for the main entrance gate. The building's weathering pattern enmeshed it into the context and fully expressed the post-industrial character.

The refurnishing of the inside space was based on the protruding brick columns and all the walls were painted pure white. The seven-metre-high exhibition space was very conducive to exhibitions with nuanced architectural expression in clear difference to artwork on display. White walls of different volume and white ceiling made the exhibition space itself more like a contemporary art installation under the illumination of ambiguous light. Lack of furnishing elements rendered the empty space more solid. The transition from outside to inside couldn't be more stark. With clean white, the gallery space was made all the more pure in contrast with the rough weathered exterior. The unity of contrast, realistic materials and visual games foster a sense of balance and self-sufficiency.

南画廊由原化工研究所锅炉房改造而成，其覆面采用的是普通的石棉瓦，画廊主入口采用了更为触目的耐候钢板门。通过石棉瓦和耐候钢板这两种质朴的材料，南画廊的表皮将场地的前符号现实的后工业气息挖掘出来，不动声色地保持了建筑底色的连续性和场景的历史感。

和外层相反，内部处理以凸出的砖柱为基准包裹成连贯的白墙面。一墙之隔，设置了巨大的落差。7米高的展厅里几面宽窄不一的白墙、白屋顶在暧昧的灯光烘托下，更似当代抽象艺术装置。这一内在视觉符号的空乏却起到一个特殊的作用——使空间趋向实体化，虚无的空间质变为一种实在的空无。南画廊内与外的强烈反差，使得彼此中和，对立统一的辩证关系得到体现，建筑自身也由此达到高度的均衡感和自足性，成为一个独立的非历史性场所。

Spgland Xi Shui Dong Sales House

无锡西水东售楼处

Location: Wuxi, Jiangsu Province
Area: 3,437m²
Architect: Kokai Studio
Project team: Filippo Gabbiani, Andrea Destefanis, Li Wei, Trill Zhang, Song Qing, Yu Feng
Photographer: Kokai Studio
Completion Date: 2009

地点：江苏，无锡
建筑面积：3,437 平方米
建筑师：柯凯建筑公司
设计团队：菲立波·加百尼、安德烈亚·黛安特法尼斯、李伟、张怡、宋庆、俞峰
摄影：柯凯建筑公司
建成时间：2009 年

Jiangsu, China
中国，江苏

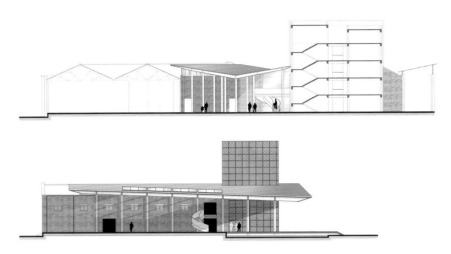

Encapsulating the spirit of the project in a sales centre, the choice was made to restore three existing warehouses of the previous factory and integrate them with a new, modern architectural building topped by a light roof composed by two flying wings that embrace and complete them. The effect is an utterly compelling contemporary building that serves as the symbol of the entire project. The effect of this light roof is amplified by the tri-dimensional design conceived by setting the sloping to the front and creating a strong perspective effect to the incoming visitors. The entire system determined by this new light roof, is supported on the front part of the building by the creation of a water pond that collects the rain water, light up the building and create an experience for the public.

The light roof has been conceived to connect the existing warehouses and create a new indoor space with high ceiling height, impressive reception lobby of the sales centre. The lightness of the roof and the comfort for the people is amplified by the design of skylights that together with a special light wooden ceiling create different pleasant lighting effects during day and night.

为了将整体项目的总体理念融入到这个售楼处的设计中，设计师选择对三个基地现存的历史仓库进行改造，同时新建一栋顶部为一对飞翔的翅膀造型的透光屋顶构成的现代建筑，翱翔展翅的羽翼将这组建筑群围合起来，形成一个全新的整体群落。售楼处整体的效果给人留下强烈的现代感，成为了整个项目的标志。透光屋顶通过三维设计在视觉上得以扩张，为来访的客人营造强烈的视觉透视效果。整个建筑体系透过前部的水池得以强调和完整，水池既可以收集天然雨水，又点亮建筑本身，为公众提供了有趣的体验。

透光屋顶的设计不仅用以衔接现存的几个仓库，同时也为售楼处室内提供了极高的层高空间，也强化了接待大厅的视觉效果。天窗的设计辅以轻盈的木质天花装饰为室内营造了日夜各有变化的光线效果，也使得屋顶的整体感觉更轻盈，给人更大的愉悦感。

Ningbo Southern Business District
宁波南部商务区

Location: Ningbo, Zhejiang Province
Area: 175,000m²
Architect: MADA s.p.a.m
Photographer: MADA s.p.a.m
Completion Date: 2010

地点：浙江，宁波
建筑面积：175,000平方米
建筑师：马达思班建筑设计事务所
摄影：马达思班建筑设计事务所
建成时间：2010年

Zhejiang, China
中国，浙江

The proposal attached importance to the mix and match of various architectural functions and combined the individual conditions of different plots as well as varied requirements of different commercial activities. Multi-functional architectural patterns and various spatial layouts were achieved to keep the additional functions of commercial, financial, cultural and living in line with the main theme of official business. It is a lively district everyday in one week and every hour in one day.

On the east side of the plot, a public passage links the commercial activities, sharing information centre and public communication platform together, while the individual buildings are connected with sky-bridge. Buildings on the south and north side are a little higher than the ones in the middle and serve as the headquarters of large enterprises. On the western side of the plot, river is the "theme" with all the cultural facilities and catering spaces along the two sides. Water landscape forms the main element both along the pedestrian street and in the individual buildings' interior and exterior.

规划方案重视相关建筑功能的混合匹配，结合不同地块自身的既定要素以及不同商务活动的配套需求，形成多元化建筑类型和空间布局形态。从而避免了商务办公与商业、金融、文化、居住等配套功能相脱节的弊病，形成7×24精彩活力区（每周7天，每天24小时全天候）。

在地块的东半部，通过一条公共服务走廊将商业活动、共享信息中心以及公共交流平台等元素进行总体整合，而建筑单体可以通过空中连廊与之相连。地块东半部的建筑高度呈南北两侧较高，中间较低的分布格局，并在南北两侧布置全区的制高点建筑，借助富有标志性的建筑形象，定位为大型企业的总部。在地块的西半部，则以河流为主脉，组织滨河活动空间。沿河两侧组织文化休闲设施和商业餐饮设施，打造沿河商业文化步行街，并结合建筑单体的室内外空间，经营多样化水体景观。

Congress Centre Hangzhou
杭州会议中心

Location: Hangzhou, Zhejiang Province
Area: 29,000m²
Architect: Pysall Ruge Architekten
Photographer: Jan Siefke
Completion Date: 2011

地点：浙江，杭州
建筑面积：29,000 平方米
建筑师：皮赛尔·鲁格建筑事务所
摄影：简·赛风
建成时间：2011 年

Zhejiang, China
中国，浙江

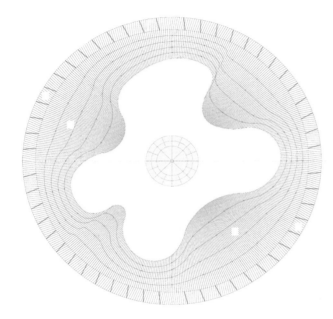

282 / EAST CHINA

The new building ensemble is situated close to Qiantang River not far from the city centre. It will be a focus building in the new business and administration district in the city. The new fascinating complex consists of six high-rise office buildings arranged in a circle and connected on the upper floors through a circular bridge. The high-rise buildings are flanked with flat multi-functional buildings including four main entrances from all directions.

The façade design should support on one hand the unique modern architecture of the building ensemble, but on the other hand it should take up typical local or traditional aspects of the region also. Zhejiang Province is known for its tea-producing region. To express the building's regional characteristics, design of the façade is based on the superimposed configurations of the tea cultivation pathways and the planting nets. As a result, the building is enveloped by a multi-layered fabric, giving it a true architectural plasticity. Seen from a distance, the façade appears like a rigid volume, but dissolves into a network of structures and levels as you come closer. The main idea for the design of the roof was to use it as the fifth façade of the building to set up a strong and typical local image in the shape of a lotus blossom, which you can see from all upper floors of the surrounding high-rise buildings. The façade structure would be extended unto the roof of the congress centre to cover up it partly. Through the different lengths and fixed height of the steel beams the structure is waved and form the abstract blossom of lotus in the centre of the roof. This part isn't covered and is designed and planted as a green landscape.

The aim is to combine and express all the regional natural features within the Centre, so that the local people will be able to identify themselves with the City of Hangzhou.

会议中心离市区不远，靠近钱塘江，属于城市大型商务综合建筑群，6个建筑单体围绕中心一个圆状单体，彼此之间通过圆形桥梁在高空中进行联系。有四个主要入口到达这个宛如巨大宝石的中心建筑体。
其立面设计兼顾现代与传统地域性。浙江省是著名的产茶地，为了表达这一特色，建筑外立面的设计采用了很像种茶使用的栽培网，形成多层幕墙系统。从远处看，建筑清晰明确，走到近处，建筑演变成网状的结构与层面。屋顶的设计则是想利用建筑这第五立面塑造一个强大和典型的地域形象——盛放之莲。周围建筑像花瓣一样围绕着中心花蕊。
建筑师力求将当地自然特征与会议中心完美融合，让其能被当地人民认同为杭州的地标。

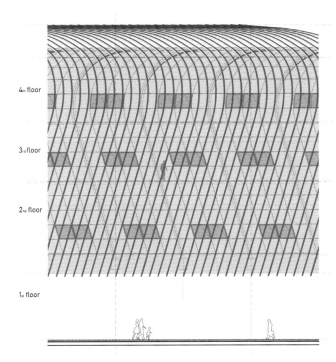

China Textile City International Convention and Exhibition Centre

中国轻纺城国际会展中心

Location: Shaoxing, Zhejiang Province
Area: 140,000m²
Architect: Xu Lei / Atelier 11, China Architecture Design & Research Group
Photographer: Zhang Guangyuan
Completion Date: 2009

地点:浙江,绍兴
建筑面积:140,000 平方米
建筑师:徐磊 / 中国建筑设计研究院拾壹建筑工作室
摄影:张广源
建成时间:2009 年

Zhejiang, China
中国,浙江

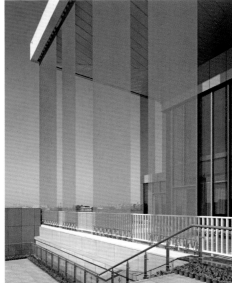
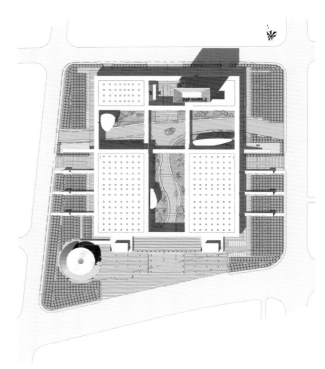

The design of China Textile City International Convention and Exhibition Centre was inspired by the form of silk box, a traditional Chinese handcraft. By adopting a clean box shape for the exterior, the Centre presents a complete volume and appropriate scale to react with the surroundings so that the space continuity is well preserved around the site. The interior space of the Centre is carefully planned based on the different needs for possible exhibition activities. Except the grand exhibition hall, the entrance square for opening ceremonies, parking lot, and terrace on the first floor can all be transformed into exterior or partially exterior exhibition area, so that the flexibility of the space use could be maximised.

As for the function of the building, the design of the Centre takes a concise and practical approach. The use of the steel mesh on the exterior of the building is an abstract embodiment of the partition door design in the traditional architecture in South China. The building's colour scheme also displays a reflection on the local architectural tradition by using an understated grey and white combination. As a modest contrast, the billboard section inside the exhibition hall adopts a more dynamic colour scheme to give the space a more contemporary touch.

在功能设计上引入了锦盒中的展示城这样一个概念，根据可能发生的展示活动来设定不同的位置和使用空间高度。除了室内的正式展位，入口开幕式广场、停车场、二层外廊都可根据实际需要转换为室外或半室外展位，以适应不同的展览要求。场地中有T形的河道，设计中利用河道形成T形的中央景观院落。建筑尽量靠近内部红线，在外围形成完整的绿化和场地，在土地的长远利用上也留下了余地。为增加未来的土地利用强度提供了可能。

展览功能本身决定了形象的简洁实用，本方案建筑内在的空间结构形成了外部形态的美学韵律和视觉冲击力。建筑设置的外廊也符合江南通透明快的审美取向。立面上的丝网表现了对江南建筑上格扇的一种抽象再现。建筑主体色调以灰白为主，反映了江南传统建筑的色彩体系，而在内置广告的设置上，又采用了鲜明的色彩以体现现代的审美观。

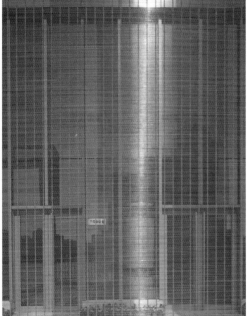

Liangzhu Museum
良渚文化博物馆

Location: Hangzhou, Zhejiang Province
Area: 9,500m²
Architect: David Chipperfield Architects
Photographer: Christian Richters
Completion Date: 2007

地点：浙江，杭州
建筑面积：9,500 平方米
建筑师：大卫·奇普菲尔德建筑事务所
摄影：克里斯琴·里希特
建成时间：2007 年

Zhejiang, China
中国，浙江

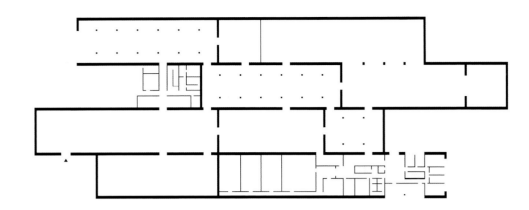

The museum houses a collection of archaeological findings related to the Liangzhu culture, also known as the Jade culture (3000 B.C.). It forms the northern point of the Liangzhu Cultural Village, a newly created park town near Hangzhou.

The building is set on a lake and connected via bridges to the park. The sculptural quality of the building ensemble reveals itself gradually as the visitor approaches the museum through the park landscape. The museum is composed of four bar-formed volumes made of Iranian travertine stone, equal in 18m width but differing in height. Each volume contains an interior courtyard. These landscaped spaces serve as a link between the exhibition halls and invite the visitor to linger and relax. Despite the linearity of the exhibition halls, they enable a variety of individual tour routes through the museum. To the south of the museum is an island with an exhibition area, linked to the main museum building via a bridge. The edge areas of the surrounding landscape, planted with dense woods, allow only a few directed views into the park.

The entrance hall can be reached via a courtyard, the centrepiece of which is a reception desk of Ipe wood, lit from above. The material concept consists of solid materials that age well, Ipe wood and travertine stone, and extends to all public areas of the museum.

博物馆是杭州附近的良渚文化村的一部分，用来展览良渚文化的考古学发现，这个时期也被称为"玉文化"时期（公元前3000年）。

博物馆位于湖上，以桥和公园连接，游客在公园中会逐步感受到建筑雕塑般的体量。博物馆由4个平行的体量构成，建筑表皮是灰黄色石灰石，每一个体量内都设有一个内院。内院作为展厅之间的连接，更起到供游人休息的作用。每个馆之间是连通的，不过，游客可以有不同的参观流线。博物馆的南面有一个岛，通过桥与主馆连接。建筑周围植物繁茂，景观优美，遮挡了视线。通过一个院子到达入口大厅，中心是接待桌，天井采光。建筑的所有公共区域采用坚实和古老概念的材料——木和石灰石，打造而成。

Ningbo Fellowship Museum
宁波帮博物馆

Location: Ningbo, Zhejiang Province
Area: 24,107m²
Architect: He Jingtang / Institute of Architectural Design at South China University of Technology
Photography: Institute of Architectural Design at South China University of Technology
Completion Date: 2008

地点：浙江，宁波
建筑面积：24,107平方米
建筑师：何镜堂 / 华南理工大学建筑设计研究院
摄影：华南理工大学建筑设计研究院
建成时间：2008年

Zhejiang, China
中国，浙江

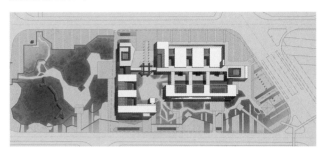

288 / EAST CHINA

The Ningbo Fellowship is a special business group that blends wisdom in business and profound patriotism. The Museum, with a total floor area of about 20,000 square metres, is located on a 4.7-hectare site in the south of Zhenhai New Town in Ningbo, just at the central north part of the "Central Park – Cultural Park – Yong River" urban axis landscape. The plain plot is surrounded by water on all sides. It is a theme museum dedicated to the profound history. It narrates the great legend of the Fellowship and demonstrates the characteristics of the city in terms of people, geography and culture. How to narrate the legend with the architecture has always been the topic under discussion for the architects.

Ningbo Fellowship Museum will be a landmark for Ningbo City and its citizens. The architects took advantages of the site as an urban axis, establishing the friendly interconnection relationship between the architecture and the city, successfully spreading the power of existence of the Museum into the city. If the Museum is to be deeply rooted in Ningbo, it has to set up proper relationship with the city, and to continue the feature of local architecture. With modern theories of typology, structuralism, etc., the architects created courtyards, landscapes and detail patterns with a local feel.

宁波帮是一个独特的商帮，集杰出的商道才智与深厚的家国情怀于一体。宁波帮博物馆总建筑面积约2万平方米，基地位于宁波市镇海新城南区"中央公园——文化公园——甬江"城市轴线景观轴带中北部，四面均有水系贯通，用地平坦，总用地面积约4.7公顷。这是一个有着浓厚人文色彩的主题博物馆，它所展示、所叙述的主题是宁波帮这个特殊人群的传奇历史以及宁波城市的人文地理特色。如何通过建筑的本体内容对展览主题展开一系列的叙事，成为设计师在设计中始终关注的议题。

宁波帮博物馆将会是宁波城市与宁波人心目中的一个具有标志性的建筑。设计师认为在设计上应该把基地"城市之轴"的特性作为一项主要的设计资源加以发挥，通过建立建筑与城市之间互相支持、相互加强的"共振"关系，使博物馆存在的"力场"在基地之中向外弥漫开去。博物馆要把根扎在宁波，除了处理与城市的关系，还需接续地域建筑的根脉，运用类型学、结构主义等现代思维方式，通过铺排庭院布局、营造场所情景、转化细部纹样等不同层面的设计手段把地方性的情感注入新建筑当中。

Yu'niaoliusu Street in Liangzhu Culture Village

杭州万科良渚文化村商业街区"玉鸟流苏"

Location: Hangzhou, Zhejiang Province
Area: 8,000m²
Architect: AZL Architects
Photography: Jia Fang
Completion Date: 2008

地点：浙江，杭州
建筑面积：140,000 平方米
建筑师：张雷联合建筑事务所
摄影：贾方
建成时间：2008 年

Zhejiang, China
中国，浙江

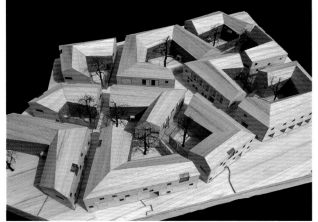

The concept comes from the analysis of the organic organisation of the original village. The basic unit of the village is family and each family residence exists in the same form. The separation of the exterior and interior of each residence is expressed through the orientation form as well as courtyard and light well. All the residences "grow" in a natural way like cells and then are organised together to form a village.

The A and F plots of Yu'niaoliusu share the same prototype in village formation. The critical relationship of the exterior and interior of the existing residence is reserved as the starting point of this project. At the same time, the typical appearance of traditional residence with white walls and grey tiles is not adopted. Courtyard plays an important role as both the extensive space and the backdrop for the inner activities. Moreover, it can be used as a public square or a private space through the opening and closing of walls.

方案的构想来自基地上自然村落有机形态的分析。村落的基本构成单位是家庭，每个家庭宅院以相似的方式构成。内与外的区分一方面表达在建筑南向与北向迥异的形态上，另外还有院落和天井以实墙宣示里外。很多类似的细胞以自然生长的方式逐渐聚集起来，形成聚落，他们以个体内敛的姿态界定街巷，形成公共场所。玉鸟流苏A、F地块的组成单元在类型学上有着相同的原型。现存村落普通的家庭内与外辨证的关系在这里被传承下来，成为重构的出发点，而不是模仿白墙灰瓦一目了然传统的外形。庭院在方案里承担的角色是重要的，它是建筑内部活动拓展的容器，也是充满激情的内部活动的背景，通过今后院墙的开洞及不同的围合材料与方式的选择，它可以变成公共小广场，也可以成为完全封闭的辅助性后院，以满足今后可能改变的实用需要。

Ninetree Village

九树公寓

Location: Hangzhou, Zhejiang Province
Area: 23,500m²
Architect: David Chipperfield Architects
Photographer: Christian Richters, Shu He
Completion Date: 2008

Zhejiang, China
中国，浙江

地点：浙江，杭州
建筑面积：23,500 平方米
建筑师：大卫·奇普菲尔德建筑事务所
摄影：克里斯琴·里希特、舒赫
建成时间：2008 年

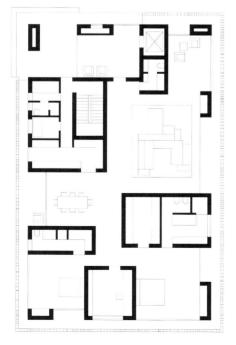

A small valley, bordered by a dense bamboo forest, forms the site for this luxury housing development, situated near the Qiang Tang River in Hangzhou, southeastern China. The particular charm and beauty of the place are the determining factors. Twelve individual volumes are arranged in a chessboard pattern to create the maximum amount of open space for each building. The individual apartment buildings contain five generously proportioned apartments, each accommodating a full floor of approximately 450m². The floor plan concept creates a flowing interior space defined by solid elements which accommodate auxiliary functions. Through planting new vegetation, each apartment building is set in its own clearing in the forest. The buildings adapt to the topography, creating a flowing landscape through a slight turning of the blocks.

The buildings adopt traditional wooden materials, enhancing harmony of nature with architecture. Based on a traditional principle of Chinese housing, an exterior skin using wooden elements protects the privacy of the residents. This skin differs in density, depending on the interior functions, sunlight and the conditions of the site. Skylights let natural light deep into the rooms.

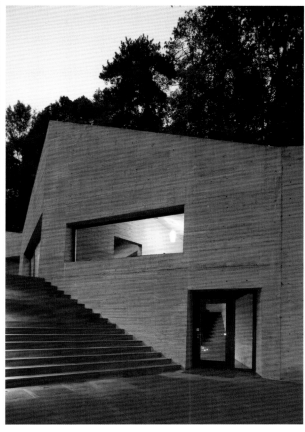

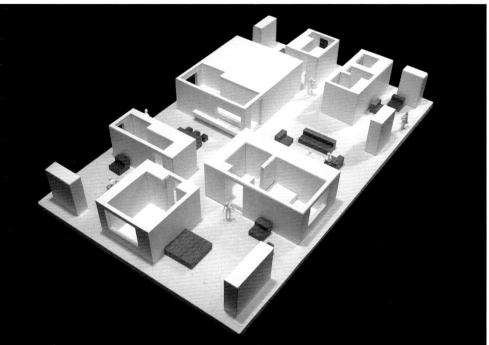

九树公寓位于杭州钱塘江附近一片茂密竹林怀抱的小山谷中，周边环境异常幽静美丽。12栋单体低层住宅楼错落分布，每栋仅有5间公寓，每间公寓建筑面积约450平方米，堪比豪华别墅的气派。室内格局设计以营造流畅空间为主。每一栋建筑都在树林中拥有属于自己的一片空地，用于种植新的植被。建筑在造型上顺应地势特征，进而打造了连续流畅的景观环境。

设计中利用传统的木制材料，与周围自然树木环境结合，在建筑与自然中创造出趋于极致的和谐。九树公寓里每栋建筑均穿着一套"木制的衣裳"——百叶窗，根据方位的不同和建筑密度的大小调整光线的导入和日晒的阻隔。更重要的是，根据居住者的爱好需要，设置不同的隐私空间。天窗设计让室内房间的采光更加通达明亮。

Tianhe Residence

天河家园

Location: Ningbo, Zhejiang Province
Area: 241,000m²
Architect: Atelier 11, China Architecture Design & Research Group
Photographer: Zhang Guangyuan
Completion Date: 2010

地点：浙江，宁波
建筑面积：241,000 平方米
建筑师：中国建筑设计研究院拾壹建筑工作室
摄影：张广源
建成时间：2010 年

Zhejiang, China

中国，浙江

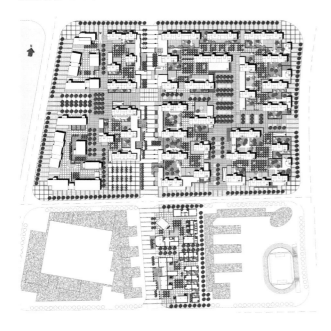

The design of Tianhe Residence focuses on the exploration of combining traditional and contemporary ways of life. The architects identified two non-material characteristics from the traditional architecture that can be transcended into modern life: courtyard-ism and naturalism. The courtyard is the soul for traditional space planning, which is evident in many traditional Chinese dwellings. Tianhe Residence's public space is divided into four different categories: public spaces that are open to the city, semi-public space formed by five to eight residential units, semi-private courtyards in front of each tower and finally relatively private roof-terrace for each individual unit. The boundaries of these four categories are reinforced by building, green walls and landscape elements. Moreover, different atmospheres are created in the different courtyards, so the progressive relationship among the spaces can be felt clearly as it does in traditional architecture. Then the traditional courtyard spirit is closely connected to the contemporary community life. Naturalism, the second aspect the architects have identified, is to emphasise the total integration between natural environments and built structures, because traditional Chinese architects always believed that man and nature are one. In Tianhe Residence, some of the design concepts are repeated throughout the project, such as raised ground floor, public meeting place, underground garage, sight-seeing lift, gardens on roof terrace, and simple material and colour choice. All of these elements have the same purpose, which is to provide a healthy and nature-oriented lifestyle for Tianhe residents.

天河家园的设计着重探讨传统与现代的结合。找到了传统建筑中能够被现代生活所继承的两个最突出的非物质特征：第一，院落精神。院落是传统建筑中空间组织的灵魂。天河家园的社区空间分为四级——由泛会所围合而成的对城市开放的公共空间；由五到八栋住宅组合成的半公共院落；每栋住宅楼前的半私密庭院以及住宅内部相对私密的花园平台，这四级院落的边界被建筑、景观墙体和绿化所强化，不同的院落中还进一步营造出不同的环境氛围，使小区空间的层次递进关系就如同在传统建筑中一样清晰，传统的院落精神与现代的小区生活紧密地结合到一起。第二，自然精神。建筑与自然相结合是中国传统的"天人合一"观念在建筑中的体现。中国传统的居住建筑追求人与自然的充分结合。在天河家园中，底层架空、泛会所、地下车库、观光电梯、大进深的空中双层花园平台以及朴素的材料与色彩设计等，这些设计元素增加了社区居民与自然的接触，使社区生活与自然紧密结合。

Banyan Tree Hangzhou
西溪悦榕庄

Location: Hangzhou, Zhejiang Province
Area: 300,000m²
Architect: Architrave Design and Planning Pte Ltd.
Photographer: Banyan Tree
Completion Date: 2009

Zhejiang, China
中国，浙江

地点：浙江，杭州
建筑面积：300,000 平方米
建筑师：悦榕酒店集团设计部
摄影：悦榕酒店集团
建成时间：2009 年

Banyan Tree Hangzhou is less than two hours' drive from Shanghai and is located 15 minutes away from downtown Hangzhou and the famed West Lake. The development is within a 50-minute drive from Hangzhou Xiaoshan International Airport. Enveloped in lush greeneries and calm waters, Banyan Tree Hangzhou sits within China's first wetland reserve, the Xixi National Wetland Park, home to an amazing wealth of thriving plant and animal life.

From its tiled roofs and ivory walls, to the elaborate décor of each suite and villa, Banyan Tree Hangzhou is an exemplary depiction of traditional Chinese architecture and modern comforts, set within tranquil natural surroundings.

Exuding the essence of modern chinois chic, all 72 suites and villas of Banyan Tree Hangzhou combines the hallmarks of oriental and contemporary design. Reflecting the four glorious seasons of the year, all suites and villas are available in four colour themes of autumn, winter, spring and summer. The high ceilings are accented by hand-painted silk brocades hanging over the beds.

Coupled with the serenity of this intimate sanctuary are the essential contemporary comforts such as an LCD TV, stereo system, air-conditioning, heated floors, and mini safe. Other features include a mini bar, hairdryer, bathrobes and turndown service in the evenings.

西溪悦榕庄距上海不到两小时的车程，离杭州萧山机场不到50分钟车程，而与杭州市中心及西湖仅有15分钟车程。

酒店四周被郁郁葱葱的绿色植被及静静的小河环绕，其所处地块是中国第一个湿地保护区（西溪湿地公园），这里有丰富的植被资源以及珍稀动物。

青瓦屋顶、白色墙壁以及精美的装饰是传统中式建筑风格的真实刻画，宁静悠远的自然环境更衬托出现代化的舒适感。

72间套房及别墅内集东方特色及现代装饰于一身，散发出独特的魅力。此外，春、夏、秋、冬四个季节分别用不同的色彩象征，并被用作四种主题。高高的天花在悬垂在床上方的手工织锦的衬托下，更加突出。

当然，现代化设备如LCD电视、音响、空调、地热等的运用营造了一定程度的舒适感，为客人提供了完美的休憩环境。

Ningbo Wulongtan Reception Centre
宁波五龙潭接待中心

Location: Ningbo, Zhejiang Province
Area: 5,000m²
Architect: Bu Bing / Ningbo Civil Architectural Design Institute
Photographer: Bu Bing
Completion Date: 2006

地点：浙江，宁波
建筑面积：5,000 平方米
建筑师：卜冰／宁波市民用建筑设计院
摄影：卜冰
建成时间：2006 年

Zhejiang, China
中国，浙江

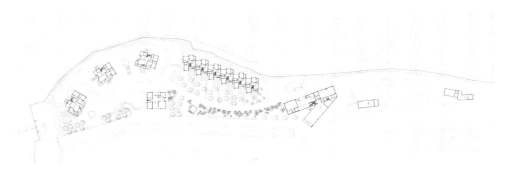

Wulongtan Resort Hotel scatters itself in the narrow valley about 400 metres long and 40 metres wide. It is situated on the only path to the Cloud Ladder at the depths of the scenic area. A clear creek runs through the whole valley, while at the distance several little waterfalls pouring in an absent-minded way on the green steep hillsides are faintly visible. The sound of water can be heard anywhere, but it is difficult to tell whether it is from a stream nearby, or from a distant waterfall, or from the hydropower station at the foot of the mountains.

In this space of compactness and ambiguity, the nearly ten construction monomers can be roughly divided into three categories: guest houses, clubs and other buildings. The guest houses feature slope roofs: some are upward offering a view of the sky beyond the narrow valleys, some are downward leading the eye to enjoy the creek. Since space is limited, the buildings are drawn close to each other, yet reserve is kept while they look up and down, drawing the attention into the sky and the water.

五龙潭山庄度假酒店稀稀落落的散布在一个长约 400 米、宽约 40 米的狭长山谷之中。这里是通向风景区深处青云梯的必经之路，一条清澈见底的小溪穿流而过整个山谷，远处隐约可见若干小小的瀑布从绿绿而陡峭的山坡上随意倾泻出来。时刻能听得到水的声音，但却难以分辨是近处的溪水，还是远处的瀑布，亦或是山脚水电站喷涌而出的水流。

在这样一个紧凑而暧昧的空间环境里，近 10 个建筑单体大致可以分为三类：客房、会所、其他建筑。客房建筑是坡屋顶的，有向上的坡也有向下的坡，向上的是为了能够越过狭窄山谷的视线阻隔看见天空，向下的则是为了引导视线去亲近溪水。因为环境的紧凑，各个建筑不得不贴近了彼此间的距离，贴近之后，即只能将目光错开，俯仰之间假作看天看水，摆出一点矜持的派头。

Xixi Artists' Clubhouse
西溪湿地三期工程艺术集合村 J 地块会所

Location: Hangzhou, Zhejiang Province
Area: 4,000m²
Architect: Zhang Lei, Qi Wei, Zhong Guanqiu, Zhang Guangwei, Guo Donghai
Photographer: AZL Architects
Completion Date: 2011

地点：浙江，杭州
建筑面积：4,000 平方米
建筑师：张雷、戚威、钟冠球、张光伟、郭东海
摄影：张雷联合建筑事务所
建成时间：2011 年

Zhejiang, China
中国，浙江

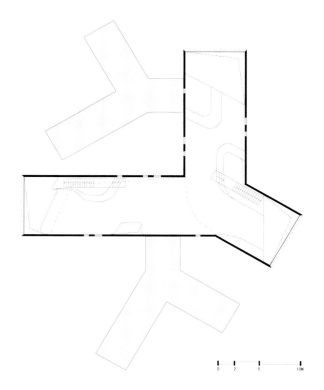

Located in the western part of the Xixi wetlands and near the west perimeter of Hangzhou, the Artists' Clubhouse is laid out like a village with 12 detached plots and each plot is endowed with a 4000-square-metre clubhouse out of 12 designers' hands.

The J plot consists of five units. Each cluster lies on three Y-shaped volumes in two sizes, capped with six- and three-metre-square frameless windows, creating panoramas of the surrounding landscape. Twisting fibreglass installations work with the cubic structure to redefine internal spaces and join walls, floors and ceilings together.

The larger six-metre-tall structure is concrete while the smaller sections sort translucent white PC panels to diffuse direct sunlight. The confrontation between the oblique with the linear, the translucent panels with the concrete and the external shape with the interior installations create a heightened sense of space.

西溪湿地三期工程位于湿地的最西侧，紧临杭州绕城西线。规划以离散式聚落的布局方式将12个地块设定为艺术集合村，每个地块布置约4000平方米艺术家会所，并邀请了12位建筑师分别进行建筑设计。

J地块会所由五组单元构成，每组800平方米大小的单元由一个大Y形和两个小Y形体量组合而成，小Y的尺度恰为大Y长宽高各缩小一半。依据基地四周地形地貌的景观特质，大小Y采用1+2的组合模式沿周围灵活布置，面对湿地采用6米×6米和3米×3米的大尺度无框景窗，获取最大的自然接触面，形成了既遵从生态秩序又有自然变异功能的离散式树状聚落结构，通过有机生长的方式与湿地景观互动，从而形成富有张力的结构和视觉关联。

大Y形为白色水泥和乳白色阳光板表面；小Y形则采用整体玻璃幕墙外饰乳白色半透明阳光板，似灯笼飘浮在树林湿地之间。乳白色阳光板因其漫反射和半透明物理属性，极大地削弱了建筑几何体亮感。

N Plot of Hangzhou Xixi Wetland Art Village

杭州西溪湿地艺术村 N 地块

Location: Hangzhou, Zhejiang Province
Area: 2,500m²
Architect: Wang Weiren
Photographer: Wang Weiren Architectural
Completion Date: 2011

地点：浙江，杭州
建筑面积：2,500 平方米
建筑师：王维仁
摄影：王维仁建筑设计研究室
建成时间：2011 年

Zhejiang, China
中国，浙江

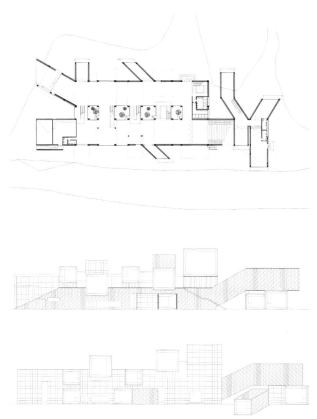

N Plot of Hangzhou Xixi Wetland phase is divided into three narrow long building bases: the north, the middle and the south, all with hill and dune waterscape at the front and back. Visitors can walk along the main road and enjoy different landscape experience formed by mountains, rivers, sky and land. The linear structure of watery places in southern China is totally different from that of the urban blocks in other regions, as the waterway settlement not only serves as the connecting system, but also directs the configuration of the buildings.

Confucius says, the benevolent favours mountains, the wise favours water. The design offers the visual experience of the water pleasure, and more importantly the artistic concept of the enjoyable and livable mountains. Through the arrangement of location, height, spatial patterns and textures, as well as landscape objects, it aims to inspire the visitors to interpret the scenery with different perspectives and viewpoints in different settings and moods.

The architect intends to make the most of the viewers' subjectivity and reframe the landscape during their gaze with conscious movement. Different from the 'enframed scenery' or 'scene changes with every step' concept in traditional landscape design in southern China, this project tends to rediscover the relation between time, space and landscape by the means of restoring films. By reframing the landscape and scenery captured on the move, the original landscape together with the impression and experience of the landscape are re-organised to form 'moving scenery'. The concept of 'enframed scenery' is hereby redefined and enriched.

西溪湿地三期N地块分北中南三条细而狭长的建筑基地，分别面临由前后两方向的山丘与沙丘水景。观者沿着主要道路漫步，山、水、天、地，构成几条线型的不同风景经验。江南水乡的线性织理与中国城市的街廓合院织理形式截然不同：聚落的水道不仅是连接系统，建筑配置更是外向性的以水为主导。

设计不只在提供智者乐水的视觉经验，更在营造仁者乐山、可游可居的空间意境，透过一系列不同的"观景器建筑"的位置、高低、空间形式与质地的安排组合，以及被观的风景对象状态的差异，启发观者对山水景观的情境诠释：视角和视点的不同，以及环境与心境的不同。

设计的意图希望透过观者的凝视间有意识的改变，景观的平面呈现被重新再框，观者的主体性再次被实践。有别于中国江南传统园林设计中叙事空间的"框景"，即"步换景移"的时空关系，这个设计以还原电影的方式，重新探讨时间、空间与风景关系；移动时所见的景观和风景，透过"再框"，重组基地原有景观，重组对基地风景的概念及经验，形成一连串的"动景"。"框景"的概念被重新定义及丰富了。

Xiangshan Campus, China Academy of Art, Hangzhou
中国美术学院象山校区

Location: Hangzhou, Zhejiang Province
Area: Phase I project 70,000m², The phase II project 78,000m²
Architect: Wang Shu & Lu Wenyu, The Amateur Architecture Studio, Contemporary Architecture Creation Study Centre, China Academy of Art
Photographer: Lv Hengzhong
Completion Date: 2007

地点：浙江，杭州
建筑面积：一期工程 70,000 平方米，二期工程 78,000 平方米
建筑师：王书、陆文宇、业余建筑工作室、中国美术学院当代建筑创作研究中心
摄影：吕衡中
建成时间：2007 年

Zhejiang, China
中国，浙江

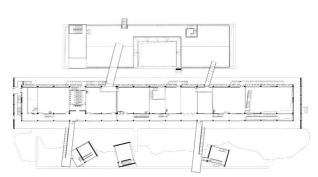

The new campus of China Academy of Art is located around Xiangshan Mountain, Hangzhou. The master plan of its phase I project is a morphological simulation of the natural relationships between mountains. Ten building units imply the trend of the mountains, which is obviously in association with the former villages on the site. Phase I project, which was designed in 2001 and completed in 2004, is occupied by the Public Art Institute, the Media and Animation Institute, library and gymnasium.

The building on the site trends to hide itself, as a metaphor of the art education hiding behind the landscape after contributing itself. The Xiangshan Phase I project is partitioned by courtyards with openings facing the mountain in different angles. The angles, openings and locations are precisely defined. Based on the partition, the form and the detail of the units are made accordingly to interpret the relationship between the site and the scene.

The phase II project at the south of the Xiangshan hill was designed in 2004 and completed in 2007, which consists of ten large buildings and two small ones. It contains the School of Architectural Art, the School of Design, art gallery, gymnasium, students' residential building and dining hall. The new buildings are all arranged at the margin of the ground, which is in the same direction as the hill stretches and similar with the local traditional buildings. Between the buildings and the hill, a large space is vacated, in which the original farm, river and pound are preserved. The form of each building changes naturally along with the undulation of Xiangshan hill.

中国美术学院的新校区位于杭州香山附近。一期工程的总体规划是一个对于山脉间的自然关系的形态模拟。10栋建筑依照山脉的走势而建，明显的与先前该地的村庄联系在一起。一期工程设计于2001年，并于2004年完工，包括公共艺术学院、传媒与动画学院、图书馆以及体育馆。

这一处的建筑倾向于隐藏自身，就像是隐喻艺术教育在做出了贡献后是隐藏在风光背后的。香山校区一期工程通过校园面向山脉的不同开口角度以及校址的所在地进行分区。校区的角度、开口以及位置都经过了准确的定位。基于这一分区，通过建筑的外形与细节的处理来演绎这一校址与景色之间的关系。

校园二期工程位于香山的南面，设计于2004年，并于2007年完工，这一工程包含10栋大型的建筑以及2栋小型建筑。它包含了建筑艺术学院、设计学院、美术馆、体育馆、学生宿舍楼以及餐厅。这一新的建筑群被规划在场地的边缘，与山脉的延伸方向一致并且和当地传统的建筑非常的相近。在山脉与建筑群之间，有一大块空置着的空间，保留了原有的农场、河流以及畜棚。每栋建筑的外形都设计成了自然的，与香山的起伏保持一致。

Alibaba Headquarters
阿里巴巴总部

Location: Hangzhou, Zhejiang Province
Area: 150,000m²
Architect: HASSELL
Photographer: Peter Bennetts
Completion Date: 2009

地点：浙江，杭州
建筑面积：150,000 平方米
建筑师：澳大利亚 Hassell 设计集团
摄影：彼特·班尼特
建成时间：2009 年

Zhejiang, China
中国，浙江

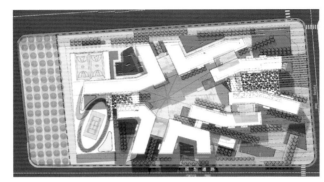

The dynamic campus accommodates approximately 9,000 Alibaba employees and has been designed to reflect the interconnection, diversity and vitality of the company. The master plan principles for the Headquarters are based on the concepts of connectivity, clarity and community – concepts that are also vital to Alibaba's e-commerce business. These principles guided all design decisions from the single workstation to the greater workplace community.

The campus is arranged around a central open space or "common" surrounded by a cluster of buildings or "neighbourhoods" that vary in height from four to seven storeys. The built form and the designed "spaces between places" are integrated so that each defines the other. The humanised scale of the built form and the long, narrow floorplates help to create a strong sense of place at a legible scale, and establish physical connection throughout the campus. The grand central space is complemented by a series of more intimate gardens that nurture the individual within the larger corporate community. Visual permeability – or the ability to see into and across the major courtyards into other parts of the complex – is also key to achieving the sense of community and connectivity.

The Hangzhou context has been embraced with garden networks and sunshading screens that represent Chinese ice-pattern window screens which are prominent throughout the city's renowned historical gardens. The sustainable design incorporates features to minimise the campus' environmental impacts while maximising its contribution to the health, wellbeing and productivity of its population.

阿里巴巴总部园区内大约有9000名员工，其设计旨在展示公司一直推崇的"互动性"、"多样性"及"活力性"原则。整体规划以"连通、清晰及一致"为主要理念，从单独工作台到公共工作区的设计完全以此为引导，当然这一理念对于公司的电子商务业务同样重要。

整个园区围绕着一个中心开阔空间展开，四周环绕着从四层到七层高低不等的建筑群。建筑样式以及空间设计强调彼此之间的联系，窄长形状的楼板营造强烈的区域感。一系列的小花园使得中央公共空间更具特色，从庭院处可以看见建筑群中的不同部分，在视觉上强调连通性。

杭州特有的地域环境决定了花园的打造及遮阳结构的运用，同时展示了传统的中式窗花样式。可持续发展理念在这一项目中得以体现，旨在减少园区对周围环境的影响，营造一个健康、高效的工作氛围。

Park Block Renovation, Luqiao Old Town

路桥旧城小公园改造项目

Location: Taizhou, Zhejiang Province
Area: 10,500m²
Architect: Tong Ming
Photographer: Tong Ming
Completion Date: 2007

地点：浙江，台州
建筑面积：10,500 平方米
建筑师：童明
摄影：童明
建成时间：2007 年

Zhejiang, China
中国，浙江

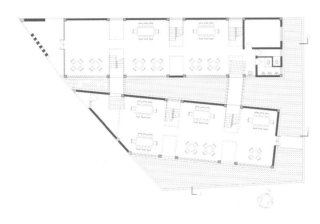

The main object of the renovation project (including five buildings) was to make combination and transition among the different structures and programmes in the city from the aspects of function and volume. On one hand, the design follows the existing typology and street texture to create new architectural volumes; On the other hand, it starts from spatial sequence and patterns on the building surface to create a strong sense of hierarchy visually. Latticed wooden windows, hollow brick walls, wooden partition panels and large expanse of glass differ in material, texture and scale and compliment each other. All those structures were made with simple technology and on site and they combined together to form different patterns, which makes them a complete whole but without interfering each other. Such a design concept reflects the main growing principle of an old town block – individual element can exist independently in a logical way, when combined together, they can coexist in a perfect way as well to bring perfect visual integrity. At the same time, the whole block seems as moving all the time to highlight change and complementation. Then, the spatial complexity and richness come into birth consequently.

The construction work became more operable and elements of material, craft, node and colour were no longer important. This design concept set an example for projects similar to this one.

街区内包含五幢建筑，其意图是要在功能和体量上衔接过渡周围各种不同的城市肌理和尺度。街区的设计一方面依据原有的建筑地形和街道肌理生成新的建筑体量，另一方面则通过空间秩序及建筑表面的纹理来形成视觉上丰富的层次感。由不同材料、不同肌理和不同尺度所构成的花木窗、空心砖墙、木板隔扇以及大片玻璃的相互映射，它们一方面各自按照简单的现成工艺及原则进行制作，另一方面又会构成两个或三个以上的图形层叠，这使得各种要素之间能够相互渗透而不在构造上影响任何一方，于是也就在回应着一个旧城街区的生长原则：各个片段逻辑化的独立生长，相互之间在视觉整体性方面的相互融合。它会形成一种连续性运动，同时也因连续运动而显示相互应和与变化不定，从而展现出一种空间上的繁杂性与丰富性。

由于这样一种空间思考，施工的操作难度变得大为降低，材料、做工、节点、色彩都不再显得重要。这对于一个长远距离且现场质量难以保障的项目是一种切实可行而又易于操作的方法。

Momentary City
瞬间城市

Location: Hefei, Anhui Province
Area: 900m²
Architect: Chen Jiajun, Sunqun / Vector Architects
Photographer: Shu He Architectural Photography, Vector Architects
Completion Date: 2009

地点：安徽，合肥
建筑面积：900 平方米
建筑师：陈嘉俊、孙群 / 直向建筑
摄影：舒赫摄影、直向建筑
建成时间：2009 年

Anhui, China
中国，安徽

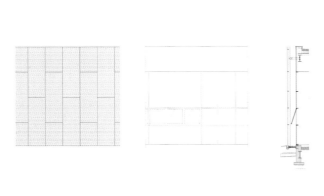

A sequence of courtyards are inserted between the indoor space and outdoor city environment, becoming a visual connection between the building and the city. The ends of these courtyards are enclosed by industrial fibreglass screens which are hung in two layers. On the courtyards' side, the colours of the screens change to reflect the various themes of the courtyards. On the city side, the screens remain clear and translucent. This variation in colour creates a subtle effect when they are viewed from city side. When they are seen from different points of view and distances, the screens' appearances are also shifting. At night, when the courtyards and the trees are lit, the screens glow quietly to cast a faint colour on the sidewalks.

The layout of the courtyards is the foundation of an internal logic that informs the transition of the spaces. Sunlight reaches the interior spaces directly or indirectly through the clerestories on the northern wall. The light made the spaces and time inseparable. With the light changing through the day and the seasons, the spaces take on different expressions. Maybe such ever-changing and irreplaceable moments are the precise definition of eternity.

在室内空间和城市之间嵌入了六个不同主题的院落空间，如竹的院子、花的院子、水的院子等。它们成为提供建筑和城市的视觉联络的媒介。院子的端面是用工业玻璃钢格栅板分两层挂装，在院落的一侧，颜色随院落的主题而变化，而在城市的一侧，玻璃钢无色而半透明。这种细节处理使得这些端面在城市一侧随着视角和远近的不同呈现出不同的视觉差异。在夜间，当灯光把院子里墙和树木照亮时，半透明的双层玻璃钢界面像是一系列有不同颜色的发光体，静静渲染城市的步道边缘。

院落的格局和朝向成为内部空间起承转合的内在逻辑。阳光通过一系列位于北侧墙体上方的高侧窗，直接或间接的渗透到空间中。自然光的充分介入，建立了空间和时间之间不可分割的关联。随着一天中太阳的轨迹变化和四季的更替，空间的光影表情永远在变化，也许正是这些一个个不可再生的瞬间印象，才是对永久记忆的最贴切的表达。

Hefei Art Gallery
合肥美术馆

Location: Hefei, Anhui Province
Area: 16,307m²
Architect: Meng Jianmin
Photographer: Shenzhen General Institute of Architectural Design and Research Co., Ltd
Completion Date: 2009

地点：安徽，合肥
建筑面积：16,307 平方米
建筑师：孟建民
摄影：深圳市建筑设计研究总院有限公司
建成时间：2009 年

Anhui, China
中国，安徽

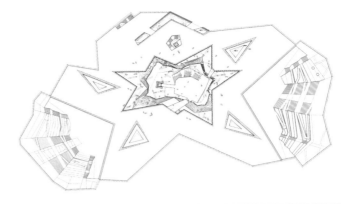

The main function of Hefei Art Gallery is to provide an information platform of artistic communication for the city. Design inspiration of the scheme comes from a traditional children's game of pick-up-sticks. Shafts of various lengths are overlapped and connected, creating an abstract form full of tension. Increasing the number and varying location of connection, the possibilities of form generated also increase.

These intertwined shafts are at once architectural envelope and building structure. This structural model breaks with the conventional one of static load transfer. By substituting the uniform support mode with a dispersed one, beams may bifurcate, columns may turn into beams, and various members are molded into a continuous structural system. It blurs the traditional load transfer model as well as the boundaries among various load transfer members. The knitting in accordance with an order of structural hierarchy generates a complex form. It emerges after an integration of multiple superimpositions. It is a non-linear process.

The design is a process rather than representation. The design is full of dynamic transformations between geist and rationality, chaos and order, and, randomness and control. From the intuitive form in the beginning until the resulting open possibilities, the whole design process is full of dramatics.

合肥美术馆的设计其功能主要为城市提供一个艺术交流的信息平台。方案的设计灵感源自民间儿童游戏棒。长短不一的杆件通过搭接、组合，产生富有张力的抽象形式。随着数量的增多以及搭接位置的不同，形式生成的可能性也越来越多。

杆件在地面按结构逻辑共分四个等级。一级杆件为主受力杆件。二级杆件为次受力杆件。两者共同编织成形式主体的结构体系。三级杆件与一、二级杆件共同组成复杂连续的玻璃折叠表皮。部分主体杆件沿伸至地下，与地下结构杆件共同组成稳定体系。杆件的编织即是建筑的表皮，同时也是建筑的结构。其结构模式打破了常规的静力传递模式，利用分散化的方法代替均等的支撑模式，梁可以分叉、柱子可以成梁，各个元素可以成为连续的结构体系。传统的应力传递模式被模糊，各传力构件之间的界限也被模糊。编织按照结构等级的次序产生一个复杂的形态。这是在不断的叠加合成后出现的。这是一个非线性的过程。

设计是过程而非再现的。设计中充满了感性与理性、混沌与秩序、随意与控制的动态转化过程。从设计起点的直觉形式到设计结果的开放可能性，整个过程充满了戏剧性。

Crossing Battle Memorial Hall
渡江战役纪念馆

Anhui, China

中国，安徽

Location: Hefei, Anhui Province
Area: 15,000m²
Architect: Meng Jianmin, Xing Lihua, Li Jinpeng, Yi Yu, Huang Chaojie / Shenzhen Architectural Design Research Institute Ltd.
Photographer: Shenzhen Architectural Design Research Institute Ltd.
Completion Date: 2011

地点：安徽，合肥
建筑面积：15,000 平方米
建筑师：孟建民、邢立华、李劲鹏、易豫、黄朝捷 / 深圳市建筑设计研究总院有限公司
摄影：深圳市建筑设计研究总院有限公司
建成时间：2011 年

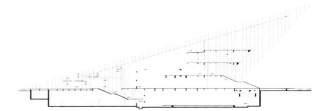

314 / EAST CHINA

Crossing the Yangtze River Campaign Memorial Hall is built to commemorate the important battle in which the communist army made the significant move of crossing the river and later won the civil war. The memorial hall adopts the theme of river crossing and victory, which is delivered with a simple, pictographic way of expression. The gigantic triangle pointing to the front displays overwhelming strength and motion, creating the atmosphere and sense of place related to the theme. A six metre wide 'time' tunnel lies between the two giant triangle structures, connecting the present with history, and providing the experience of 'crossing' and 'landing' which resembled the victory gained in the battle. The huge twin memorial buildings speak to people in a silent way, looking back with a noble spirit of humanity and forgiveness and inspiring the future generations to pursue peace and development.

Chronological sequence and sense of ritual feature the memorial hall interior, with a semicircle sculpture wall of nearly 50 metres as a highlight of the exhibition. The sculptors' passion is perfectly reflected in the works which add to the finishing touches of the ambience. Following the exhibits and story lines of history, the visitors will come to the underwater memorial hall, where they can look up and enjoy an overlapping and layered view of the inside and the outside, merged by the water. Finally people will arrive at the Hall of Military Honours. The huge mural of the river crossing moment and the numerous unnamed gravestones underwater will bring the sentiments to a new climax. One can exit the memorial hall through the Medal Corridor. The whole memorial hall design integrates the landscape and indoor exhibition with the design concept in mind and the theme fully expressed.

渡江战役纪念馆是为纪念中国解放战争中跨江统一全中国的重要战役而设的场馆，该纪念馆以"渡江"、"胜利"为主题，故本创意以简约、象形的表现主义手法表达主题思想，巨大的前倾三角形实体展现出一种势不可挡的力度与动感，从而营造"渡江"与"胜利"的氛围与场所感。在两块巨大三角实体中间空留出一条6米宽的"时空"隧道，将当今与历史贯通于一起，人们通过"渡"与"登"的行为动作体验与感受战争、胜利的隐喻。作为国内战争，两块巨大的纪念体犹如巨碑默默地向后人示意与陈述，以一种崇高的人文精神，包容态度客观地追忆以往，回顾过去，启迪后人追求和平与进步。
纪念馆室内空间注重序列性与仪式感，近50米的半圆雕塑巨墙为陈列展示之高潮。雕塑家对作品的激情注入为空间氛围的营造仿如画龙点睛。沿着展陈流线越过历史的节点与细节，人们会步入水下纪念厅，仰首望去，通过水面将室内室外融为一体，创造出特殊的空间交叠与层感。最后人们进入军功厅，巨型渡江壁画与水下无名碑，再次将人们的情绪推向一个新的高潮，而后经过军功章环廊步出纪念馆。整个纪念馆建筑将室外环境与室内展陈统一创意，理念贯穿如一，使其主题意义得以充分表达。

Anhui Museum
安徽省博物馆新馆

Location: Hefei, Anhui Province
Area: 41,380m²
Architect: He Jingtang, Liu Yubo, Zhang Zhenhui, Liang Weijian/ Institute of Architectural Design at South China University of Technology
Photographer: Zhang Guangyuan
Completion Date: 2011

地点：安徽，合肥
建筑面积：41,380 平方米
建筑师：何镜堂、刘宇波、张振辉、梁玮健 / 华南理工大学建筑设计研究院
摄影：张广源
建成时间：2011 年

Anhui, China
中国，安徽

The Museum is located at the Provincial Culture and Museum Block in the Cultural New District in Hefei. With respect for the local culture and consideration on the site and its urban setting, the architects finally came out with the principal design concept: gathering courtyard as a core. Gathering courtyard is a typical element of local architecture in Anhui Province. In this project, at the core of the building is set a public courtyard, which is divided into complicated spaces by zigzag walls. In this way, a micro-village with traditional Anhui flavour is created. The exhibition hall is a series of continuous yet zigzag spaces, forming a cubic general space open to all sides, defining the basic configuration of the building. The patterns on the exterior walls are derived from the invaluable Tripod of Chu, a piece of exhibit in the museum. They give the architecture a sense of dignity that belongs to the ancient bronze ware. The inner surfaces of the openings on the walls are clad with timber panels, a typical material of Anhui architecture that gives out a natural and elegant air. The zigzag spaces between exhibition spaces become a flowing public route. The circular gallery along the inner side of the zigzag spaces is a blend of different architectural elements such as terrace, bridge and cantilever. It is indirectly linked with the gathering courtyard in many ways. For example, the bridge is built above the courtyard; the terrace is a good place to overlook the courtyard; and you can also view the courtyard through the hollowed-out windows. In this way, the central core is visually extended on all sides, closely related with public spaces.

安徽省博物馆新馆（以下简称新馆）坐落于合肥市政务文化新区内的省文化博物馆园区。基于对地域文化的发掘与理解，结合对建筑基地与城市环境的思考，设计立意构思确定为——四水归堂、五方相连。"四水归明堂，归水亦宏扬"，庭院在安徽传统建筑文化中，具有中心突出的位置。取意"四水归堂"，建筑核心部位形成向天开敞的公共大厅，其空间通过一系列转折勾连、错动交叠的通高片墙进行多样化的限定。展厅是一系列连续转折的实体体量，构成一个四面通透的方体，确定建筑的基本外轮廓。实体外表面覆盖的纹理抽象于镇馆之宝楚大鼎鼎身的兽面纹，突出远古青铜器般厚重浑雄的分量感；实体通透部位切开的内表面覆盖木质墙板，通感于徽派建筑内部的大小木作界面，雅致亲切。展厅之间通透的部分形成连续转折的虚体体量，成为活跃流动的公共空间系统。虚体空间内侧的环廊结合平台、连桥、挑台等建筑元素，与"四水归堂"大厅取得半开半合、柳暗花明的多样化联系，有时是连桥飞跃，有时是平台眺望，有时又是隔着镂花窗隐约可见，形成一个从核心空间向四面伸展，沟通上空与四方外界的"五方相连"的公共空间系统。

Jinan Sports Centre
济南奥林匹克体育中心

Location: Jinan, Shandong Province
Area: 350,000m²
Architect: CCDI
Photographer: Chen Su, Ji Chengke, Li Yan, Zheng Quan
Completion Date: 2008
Award: McGraw-Hill Construction 3rd Bi-Annual "Good Design Is Good Business" China Awards 2010, Best Public Project

地点：山东，济南
建筑面积：359,000 平方米
建筑师：中建国际设计顾问有限公司
摄影：陈溯、籍成科、李岩、郑权
建成时间：2008 年
所获奖项：麦格劳-希尔公司《建筑实录》、《商业周刊》第三届"好设计创造好效益"中国奖项、2010最佳公共建筑

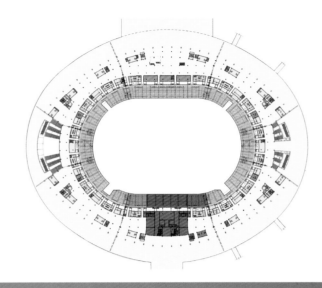

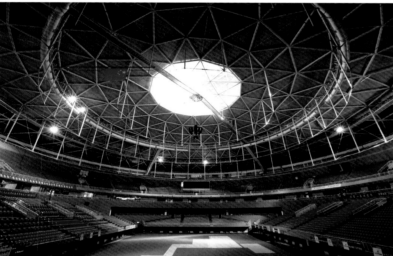

Jinan Sports Centre is not only an important venue for the eleventh National Games of China, but also an initiating project for the construction of the new town in the eastern part of Jinan City. The Sports Centre aims at holding all the sport events of the National Games, and meeting the requirements of General Association of International Sports Federations to provide modern, world-class venues for body building and entertainments.

The plan maximises the plot ratio and distribution of the venues (a triangle in plan). A sense of balance among spaces, and future business during and after sport events are carefully thought. The stadium and the training venue are situated in the west, while in the east are another stadium, a pool and a tennis centre. The western part feels more magnificent, while the eastern layout is much more concentrated, with the central circular stadium and the symmetrical pool and tennis centre. The latter two "harbour" the stadium, forming a biaxial symmetry with the stadium in the west. The western, eastern, and southern parts (Administration Centre in the south) occupy three corners of the triangle, making a harmonious and steady layout. The central part is the plaza of the National Games, and would become a landmark of Jinan City after the Games. The underground part is devoted to shops, services and parking facilities. The plot, with the south on a higher level than the north, and the east and west higher than the central part, is carefully studied on the site to set an appropriate level for the construction in order to maximally meet the functional requirements of the project. Furthermore, the local culture of Jinan is embedded in the complex.

济南奥林匹克体育中心不仅是第十一届中国全运会的举办场馆，同样也是作为济南东部新城区建设的启动项目。济南奥林匹克体育中心的建设目标是承办全运会的所有比赛项目，并且达到国际单项体育联合会总会的要求，提供现代的世界级的健身以及娱乐设施。

这一规划实现了建造比例以及场馆分布最优化（呈三角形规划）。在设计上仔细思考了在空间和未来的商业用途以及在比赛过程中和比赛后的平衡感。体育场和训练场馆坐落在西边，同时在东边建造了另外一座体育场、一个游泳池以及一个网球中心。西边的场地感觉上更加的宏伟，同时东边的布局规划更为集中，集合了中央圆形体育场和对称的游泳池以及网球中心。后者的两个"港口"体育场与西边的体育场形成了一个双轴对称的态势。西边、东边以及南边的部分地区（位于南部的行政中心）分别占据了三角形的三个角，营造出一种和谐又稳固的布局。中心部分是全运会广场，它将在全运会结束后成为济南市的地标性建筑。地下的部分将被用作商店、服务业以及停车设施。这一处的地势，南部要略高于北部，东部西部要高于中心地区，对这一地形仔细的研究后确定了一个合适的建筑高度，以最大化的满足这一工程的功能需求。此外，在综合设施中还融入了济南的本土文化。

Qingdao Grand Theatre
青岛大剧院

Location: Qingdao, Shandong Province
Area: 60,000m²
Architect: gmp Architekten – von Gerkan, Marg and Partners
Photography: gmp Architekten – von Gerkan, Marg and Partners
Completion Date: 2010

地点：山东，青岛
建筑面积：60,000 平方米
建筑师：gmp 建筑设计事务所
摄影：gmp 建筑设计事务所
建成时间：2010 年

Shandong, China
中国，山东

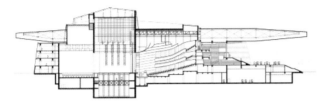

The Qingdao Grand Theatre includes an opera house that can accommodate up to 1,600 people, a concert hall for 1,200 and a multi-function hall with 400 seats. In addition, there are a museum and a hotel with 300 rooms. Because of the unique situation of the massif directly by the sea, the Laoshan ridge is often wreathed in cloud, which creates a unique setting. The style of the building echoes the enchanting landscape, with the massif and the lightness of the clouds reflected in the appearance of the opera house. It rises from the landscape like a mountain and a cloud-like roof seems to wreathe the four buildings. The raised terraces in the surrounding park are reminiscent of a mountain plateau, and take their bearings both from the sea and the mountains.

The opera house itself lies at the northern end of the ensemble of buildings, offering direct views of the coast. Audiences can enter the auditorium on two levels, either via the lower level, giving access to the large cloakrooms and the lower circle, or via the main foyer above with its large glass façades and view of the terrace, park and Yellow Sea. The walls and floor of the foyer are clad with local stone, which emphasises the affinity of the building with the mountain.

青岛大剧院包括可容纳1600人的剧场、1200人的音乐厅以及400人的多功能厅。此外，这里还设有博物馆及带有300间客房的酒店。项目选址在崂山脚下，这里常年呈现多云天气，为建筑营造了独特的背景。设计风格旨在体现迷人的自然景观，整幢建筑犹如从山间"升起"，如云朵一般的屋顶将四个不同的结构"笼罩"。周围公园被抬升的露台让人不禁联想到高原。

剧场建筑位于北侧，可欣赏到海滨的景致。观众可通过底层入口直接进入到衣帽间和圆形广场，或者经由带有全玻璃外观的一层主厅进入。主厅地面及墙壁采用当地石材饰面，让人不禁联想到建筑本身与山的关联。

Sailboat Hotel
帆船酒店

Location: Rizhao, Shandong Province
Area: 50,000m²
Architect: Hong Zhongxuan
Photographer: Chen Zhong
Completion Date: 2009

地点：山东，日照
建筑面积：50,000 平方米
建筑师：洪忠轩
摄影：陈中
建成时间：2009 年

Shandong, China
中国，山东

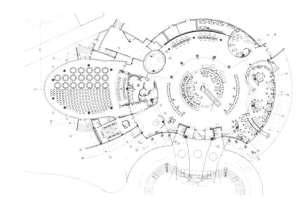

Adjacent to Rizhao Sailboat Base on the Wanpingkou Coast, a hotel with sailboat as its theme, Junhao Hotel, is like sailing in the sea from a long distance.

The lobby is a round architectural space; the roof adopts two layers of canvases spreading tensely; the soft lamplight shed onto the canvases reflected into the whole lobby area, creating a warm and comfortable atmosphere. The most interesting is that the top of the background walls of the lobby has the ceiling made of turfs. A "boat" put to the shore, with the overlapping scenes, together with the flowers, grasses and trees in the central hall, creates a scene of primitive forest. In fact, under the turfs hide air-conditioners, representing the comprehensive design competence of the hotel. The air-conditioning pumps and equipment for the whole are hidden very skillfully, not affecting the beauty of the hotel. The guestrooms continue the style of sea; the blue and white carpet is fresh and bright and full of the smell of sea, creating a "blue dream" atmosphere for guests.

The whole hotel mainly adopts LED lamps and has made a revolutionary break in energy saving. Few traditional wood finishes are used, reducing air pollution. In terms of environment protection, the hotel is leading in the hospitality industry. The hotel design taking metals as main dominant materials provides us a cool, healthy and personalised space.

滨邻万平口海滨山东日照帆船基地，有一座以帆船为主题的酒店——君豪帆船酒店，远远看去犹如航行于海中。

大堂是一个圆型建筑空间，天顶采用双层帆布张拉而成，帆布上的灯光逆照而上，柔和的灯光反射到整个大堂区域，营造出一种温馨舒适的氛围。最有意思的便是大堂主背景墙上方用草皮做成重迭的天花，即将靠岸的船只，看到层迭的风景，与中厅的花草树木共同编织了一副生态原林的景象。其实在草皮的下方隐藏着空气调节设备，这是整个酒店最突显综合设计能力的地方了，整个酒店的空调管道设备隐藏得非常巧妙，未造成美观的影响。客房延续了海的格调，蓝白相间的地毯清新爽朗，它弥漫着海的气息，为休憩的客人制造一个"蓝梦"的氛围。

整个酒店采用以LED灯为主导，在节能上有了革命性的突破，传统的木饰面所使用的比例较低，减少了气味污染，环保的方面也在酒店行业中引导了先锋，以金属为主材的酒店设计给人们一个充满凉爽、健康、人性化的空间。

Mount Taishan Peach Blossom Valley Visitor Centre
泰山桃花峪游人中心

Location: Tai'an, Shandong Province
Area: 7,685m²
Architect: Cui Kai, Wu Bin
Photographer: Cui Kai
Completion Date: 2010

地点：山东，泰安
建筑面积：7,685 平方米
建筑师：崔愷、吴斌
摄影：崔愷
建成时间：2010 年

Shandong, China
中国，山东

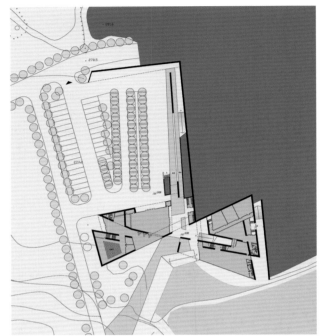

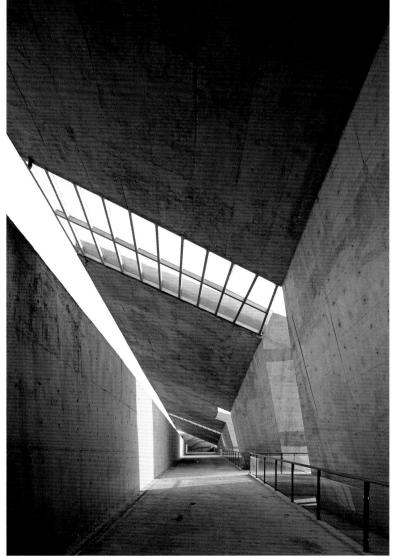

The project is located in the Peach Blossom Valley in Tai'an city to the west of Tai Mountain. It lies along the motorway to Tai Mountain, with South Matao Reservoir to the south formed by the water from the Caishi Stream, villages and army sheds to the east, and roads to the reservoir hotels to the west. The existing public car park and private car park are kept, and by using the terrain elevation, long ramp roads are built to orgaise the traffic. With cloverleaf intersections, drivers can see and communicate with each other without interference, and the traffic to the Visitor Centre is basically well managed. There are bus shelters in the uphill area, ramp roads and canopy between two parking lots. The office area and VIP reception are located on the other side of the road, with the building over-crossing the street to form a gate, as is convenient for management. As the visitors walk in the open air, they can enjoy the beautiful scenery of lakes and mountains at any time.

本工程位于泰安市泰山西侧桃花峪景区内，在通往泰山道路的旁边。南侧为南马套水库，由上游彩石溪溪水汇流而成，东侧为村落和部队平房，西侧紧邻通往水库对岸旅馆的道路。保留现有的社会停车场和内部停车场，充分利用现有地形的高差，用长长的坡道将上山和下山游客的流线组织起来，立体交叉，互不干扰但能进行视线交流，满足游客中心人流的基本功能要求。在上山候车区设置候廊，两个停车场之间设置坡道和挑棚。办公、贵宾接待等放置在路的另外一侧，建筑跨过马路形成关口，有利于管理。游客在室外空间行走，随时可以看到优美的湖光山色。

International Tourist Passenger Terminal, Xiamen Port

厦门港国际旅游客运码头

Location: Xiamen, Fujian Province
Area: 81,274m²
Architect: Wei Yeqi
Completion Date: 2007
Award: HKIA Merit Award of the Year Outside Hong Kong 2007

地点：福建，厦门
建筑面积：81,274 平方米
建筑师：韦业启
建成时间：2007 年
获奖：2007 年香港建筑师学会境外优异奖

Fujian, China
中国，福建

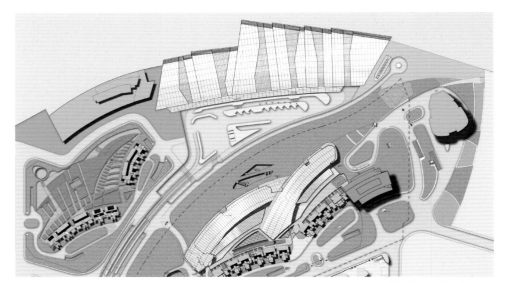

The project takes inspiration from fluctuation of waves. The curvy roof panel of the building brings distinctive effect visually. At the same time, it can make full use of natural light and provides fine ventilation to create a comfortable environment for passengers as well as to achieve the energy-effective result.

Located at the west of the urban city, the building has become gateway to the city. Thus, it would be a necessary ingredient of the urban landscape. Despite of the eye-catching roof, the volume of the building is intentionally reduced to set off the cruise ship.

The flyover above the east ferry road links with the ground towards the terminal and then rises gradually, leading to the main entrance of the building. The steps in front of the south façade of the building are moved intentionally ahead and the appearance of the building becomes clearly as one goes upstairs.

客运码头大楼的设计意念源自海中波浪起伏的形态，覆盖大楼客运廊的屋面盖板一层叠一层，营造出别树一帜的效果。同时设计又能充足利用自然采光及通风，为客运廊带来舒适的环境，并达到节能效益。

客运码头大楼坐落于厦门市以西，成为进入市内的门户，所以其建筑设计将会构成市内景观特色的其中一项重要元素。虽然客运码头大楼的屋顶设计较为突出，但整体的建筑体量已被刻意调低，务求衬托邮轮的非凡气派，使邮轮成为目光的焦点。

在交通流线方面，东渡路的高架桥可通往大楼，高架桥进入码头范围后斜向地面层，形成一小段地面道路，以连接码头用地其他地区，后段逐渐升高，直抵客运码头大楼正门，表示正式抵步。大楼南面的楼梯位置刻意移至前面，令人甫踏上楼梯便能感受抵步的惊喜，大楼形貌也随着上行的步伐逐步展现于眼前。

Fujian Provincial Electric and Power Company

福建省电力公司大楼

Location: Fuzhou, Fujian Province
Area: 63,301m²
Architect: MulvannyG2 Architecture
Photographer: Shen Zhonghai
Completion Date: 2007

地点：福建，福州
建筑面积：63,301 平方米
建筑师：美国 MG2 建筑设计公司
摄影：沈中海
建成时间：2007 年

Fujian, China
中国，福建

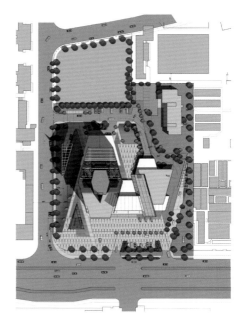

328 / EAST CHINA

The building is set away from the main street, allowing a green, open space at the entrance. This openness extends through the lobby and connects to a quiet courtyard on the other side of the building. With its roof gardens and sky lounges, the project is a vertical oasis in an urban setting.

Unique to the building's design is a stylised bolt of lightning discharging from the building in the form of a 226-foot-high communications tower. The material of choice is stainless-steel metallic fabric. North and south curtain walls shield a crystal-like tower. By using ceramic frit, the lightning motif is integrated into the building's glass panels, appearing as hundreds of transparent triangles. The tips of the curtain shields are made of translucent white glass, which glows at night.

The building's south-north orientation, multi-level sky garden spaces, cost-effective revolving natural ventilation system, and energy-efficient low-E silkscreen glass are some key sustainable design features. Multi-level lounge areas are provided for employees every two to three floors. Transparent and translucent materials such as combinations of clear glass and metal screenings are used in the lobby. A large water wall further magnifies the feeling of transparency and an open flow with nature.

建筑远离街道，在入口处形成了一个绿色的开阔空间，这一开阔感一直穿越大厅，延伸到建筑另一侧的庭院内。屋顶花园以及空中观景廊的设计将整幢建筑打造成城市中的绿洲。

建筑的特色即为69米高的类似信号塔的闪电形状尖顶，不锈金属纤维作为主要的材质。南北两侧的幕墙结构营造了水晶般的造型，陶瓷熔块的运用使得尖顶嵌入到建筑玻璃板中，看似犹如数百块透明三角形结构拼接而成。幕墙顶端采用半透明白色玻璃镶边，在夜晚闪闪发光。

这一设计注重体现可持续发展理念，如南北朝向的设计、多层空中花园、自然通风系统、节能丝印玻璃等全部体现出环保特色。每两层到三层设计有观景廊供员工使用。大厅内多采用透明或半透明材质装饰，如玻璃及金属丝网等。一面宽大的水墙更加强调出空间的通透感以及同自然的密切联系。

Bridge School
桥上书屋

Location: Zhangzhou, Fujian Province
Area: 240m²
Architect: Li Xiaodong, Chen Jiansheng, Li Ye, Wang Chuan, Liang Qiong, Liu Mengjia
Completion Date: 2009

地点：福建，漳州
建筑面积：240 平方米
建筑师：李晓东、陈建生、李烨、王川、梁琼、刘梦佳
建成时间：2009 年

Fujian, China
中国，福建

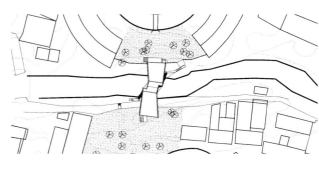

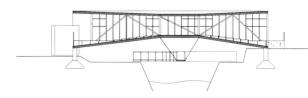

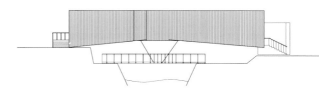

The Bridge, the School, the Playground, the Stage. Located at a remote village, Fujian Province, China, the project not only provides a physical function – a school + a bridge, but also presents a spiritual centre. There's a local legend saying that the two castles in the village used to be enemies and thus built a creek in between. The Bridge School connects the two castles across the river. The main concept of the design is to enliven an old community (the village) and to sustain a traditional culture (the castles and lifestyle) through a contemporary language which does not compete with the traditional, but presents and communicates with the traditional with respect. It is done by combining a few different functions into one space – a bridge which connects two old castles cross the creek, a school which also symbolically connects past, current with future, a playground (for the kids) and the stage (for the villagers).

A lightweight structure traverses a small creek in a single, supple bound. Essentially, it is an intelligent contemporary take on the archetype of the inhabited bridge. Supported on concrete piers (which also has the function of a small shop), the simple steel structure acts like a giant box girder that's been slightly dislocated, so the building subtly twists, rises and falls as it spans the creek. The colour of earth yellow is perfectly integrated with that of the castles, whose circular shape sharply contrasts with the rectangular structure, creating continuity and harmony.

桥上书屋是一座桥梁，一间学校，一个操场，一处舞台。平和县下石村的中心有两个圆形土楼，中间横跨一条溪水，传说旧时两个土楼的家族互为仇敌，遂划渠为界，互不往来。李晓东的桥上书屋就在土楼之间，溪水之上。设计理念旨在为古老的下石村注入活力，通过现代的建筑语言使传统文化（土楼以及土楼式生活）永葆生机——这种现代的建筑语言不与传统争锋，而是带有敬意地与传统交流、融合。建筑师将四种功能融于单一空间内：一座桥梁横跨溪流，连接两座土楼；一间学校，不单履行学校的职能，更重要的是象征着过去、现在与未来的联系；一个操场供孩子们嬉戏；最后，一处舞台为住民提供活动场地。

桥上书屋以细密的桉树木条包裹住方筒式的建筑，横亘于溪水上，下方用钢索悬吊着一座轻盈的折线形钢桥。钢桥采用混凝土柱子支撑，桥下的空间同时也用作小商店。这种结构以现代的方式巧妙地借鉴了下石村的建筑形式。轻盈的建筑结构悬于溪上，仿佛轻微脱位的关节，随着人的运动而发生扭动。土黄的颜色与土楼相融在一起，强烈的方圆对比由此显得柔和而贴切。

Lanyang Museum
兰阳博物馆

Taiwan, China
中国，台湾

Location: Yilan, Taiwan
Area: 7,681m²
Architect: Kris Yao, Artech Architects
Photographer: Zheng Jinming
Completion Date: 2010
Award: Green Building Award, Taiwan EEWH 1st Prize, 7th Annual Far Eastern Architectural Design Award Museum 1st Prize, 2010 Taiwan Architecture Award Superior Construction Quality Award, 10th Public Construction Golden Quality Awards for Outstanding Public Architectural construction, 2011 FIABCI-Taiwan Real Estate Excellence Award – Construction Quality Category Excellence Award, 2011 Golden Project Award - 2011 United Nations Environment Programme (UNEP) International Award for Livable Communities, (THE INTERNATIONAL ARCHITECTURE AWARD 2012) 2012 The Chicago Athenaeum Museum of Architecture and Design, The European Centre Architecture Art Design and Urban Studies: The International Architecture Award.

地点：台湾，宜兰
建筑面积：7,681平方米
建筑师：姚仁喜，大元联合建筑师事务所
摄影：郑锦铭
建成时间：2010年
获奖：第七届远东建筑奖杰出奖2010，第32届台湾建筑奖首奖2010，第十届公共工程金质奖 – 公共工程质量优良奖建筑类特优2010，2011国家卓越建设奖最佳施工品质类卓越奖，2011 "联合国环境规划署"（UNEP）认可之国际宜居城市大会人造环境类别金奖，2012芝加哥雅典娜建筑设计博物馆+建筑艺术设计与城市研究学院"2012国际建筑奖"

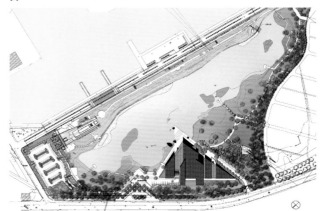

To design Lanyang Museum is like to design a gateway for Lanyang Plain. To design the visiting experience of Lanyang Museum is to design a process for people to gain the ability to know, to introspect, and to face Lanyang when participating at the Yilan tourism, the educational tours or the local community leisure activities. To design the landscape of Lanyang Museum is like to design a museum experience. The design experience of the museum's site comes from utilising the latent natural resources of the site and the historical relics of Wushih Harbour to conduct an inspirable spatial imagination.

To exhibit the cultural resources of the site is to use the Wushih reef located at the water area of the Wushih Harbour as an introduction to show the flourishing, the down fall, and the rebirth experience of the site. The building mass is centralised and extends the orographic trend of the cuesta coastal mountain. It inserts into the water front ground at a 20-degree angle, presenting the local cuesta rock terrain with a low key expression; hence to preserve more natural wetland area and to maintain a better relationship to the natural terrain. The crevice separating the mass cleverly becomes the source of the interior natural lighting and the programme zoning device; it is also the spiritual axis that connects to Guishan Island at a distance, and builds a spiritual tide visually, geographically, and in terms of spatial imagination. Under the slanted roof, there is the 27-metre large span steel truss and load-bearing wall system, which constructs the column less continuous interior exhibition spaces, and also constructs an interpenetrating Yilan landscape experience of the "Mountain-Plain-Sea".

When visitors look back at the spatial scenes of the sails at the Wushih Harbour, the air-raid shelters of the counterattack period, the fish farms, the graveyards, and the faraway mountain next to each other, experiencing the different dramatic sceneries of the site, the Wushih in the water shows a strange yet familiar appearance. The new Wushih is like a mirror, reflecting today's dream and participating in the rebirth of a site.

对兰阳博物馆的设计就像是为兰阳平原设计了一扇大门。为兰阳博物馆设计的参观体验是一个设计过程，通过它来让人们获得认知，去进行反思并且在参与宜兰旅游的时候了解兰阳，这种旅游可以是教育性质的旅行或是当地的社区休闲活动。设计兰阳博物馆的景观是在设计一场博物馆体验一般。通过使用当地潜在的自然资源和乌石港的历史文物来设计博物馆体验，为人们带来能够激发出灵感的空间想象力。

展示基地的人文资源，是将乌石港遗迹水域中的乌石礁作为引子，展示基地自繁盛、荒颓复又重生的经验。建筑的量体集中，延续海边单面山的山势，以20度角插入水边，低调地展现海岸乌石礁地形，以保有较多的自然湿地面积，以及较好的与自然地貌的关系；分开量体的隙缝，巧妙地作为内部的采光和功能区隔，并且连接远方龟山岛的精神轴线，在视线上、地形上，和空间的想象上建立精神的联系。斜屋顶内部27米大跨距的钢桁架及承重墙系统，构成室内无柱的连续性展示空间，构成山—平原—海相互贯穿的宜兰地景经验。

当参访者回顾石港春帆、鱼场及坟墓、远山与乌石并置的空间场景，体验基地戏剧化的不同情境时，水中的乌石展现既陌生又熟悉的外貌，新的乌石似乎一面镜子，同样地并置一个今日的梦，参与一个基地的重生。

Hunya Chocolate Museum
宏亚巧克力博物馆

Location: Taoyuan, Taiwan
Area: 3,967m²
Architect: J.J. Pan & Partners, Architects & Planners
Photographer: Vesper W.S. Hsieh
Completion Date: 2012

地点：台湾，桃园
建筑面积：3,967 平方米
建筑师：潘冀联合建筑师事务所
摄影：谢伟士
建成时间：2012 年

Taiwan, China
中国，台湾

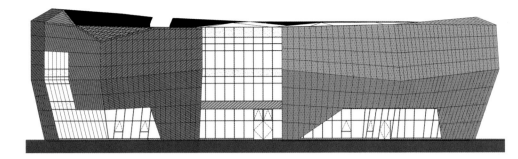

The site is surrounded by agricultural land and commands an open view with local residences encircling the peripheral area. The museum sitting on the north-eastern corner of the site and the factory on the south-western side make room for a largest possible outdoor garden. Experience in the museum provides tourists with knowledge of chocolate, and a tour around the factory familiarises them with the chocolate-making process. The string of activities in the museum, plaza, and factory presents a most fulfilling trip to the world of chocolate.

Hunya is a renowned food brand in Taiwan with a long history. To transform the chocolate brand image and inspire innovations, the building adopts solid volume cut in different angles with the chocolate-coloured exterior to convey the imagery of chocolate. Through several major cracks in the appearance a totally distinct, flowing inner space is seen. A void space is created by the three-storey atrium through the centre of the volume, with a sky-bridge and staircases crossing within. The terminal point of tour is a greenhouse of cacao trees. Various exhibitions take place in the spaces on both sides of the greenhouse, like chocolate, a symbol of ever-lasting taste with surprising fillings, embodying the daring, innovative brand spirit.

基地周围为一般农业用地，视野开阔，再外围则是一圈民房。博物馆配置在基地东北角，与西南方的工厂间留出最大的户外广场范围。博物馆提供有关巧克力的相关知识，巧克力工厂的导览则让游客了解巧克力的体验制作过程。通过参观博物馆、广场、工厂的活动彼此串联，以期给游客最完整的巧克力体验之旅。

多年来宏亚是台湾传统且极具口碑的知名食品品牌。为因应巧克力品牌形象的转型以及创新，本案外观设计上采取厚实、被切削的量体感以及巧克力色外墙等手法来传达巧克力意象的概念；而透过外观上几个主要大型切口则能看到内部截然不同且流动的空间，三层楼大挑空形成的虚空间从中划过建筑物，空桥、楼梯穿越其中，视线的端点则是植有可可树的透明大温室。形形色色的展览活动发生在两旁空间里，象征隽永的巧克力，包覆令人惊喜的馅料，体现大胆创新的品牌精神。

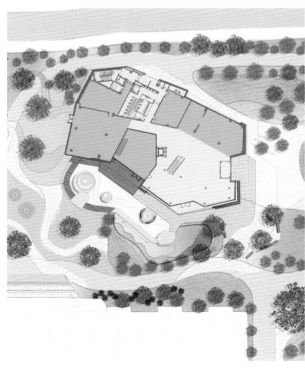

Bellavita
宝丽广场

Location: Taipei, Taiwan
Area: 52,640m²
Architect: P&T Architects
Completion Date: 2009
Client: Chun Yee Co., Ltd.

地点：台湾，台北
建筑面积：3,967 平方米
建筑师：P&T 建筑师事务所
建成时间：2009 年
业主：俊逸有限公司

Taiwan, China
中国，台湾

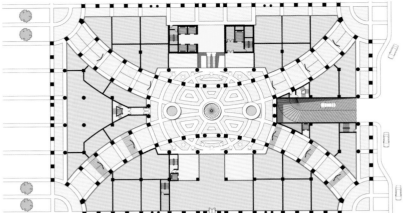

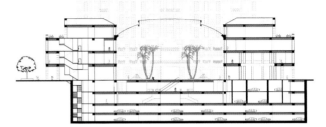

336 / EAST CHINA

The focus of this rectangular mixed-use development is an elliptical atrium which serves as a public plaza and centre, with four curvilinear glass-roofed gallerias sweeping outwards to entrances in each corner. The remainder of the three-storey retail podium is taken up by high-end retail flagship stores, as well as F&B outlets arranged in a symmetrical layout.

A crescent-shaped tower on the north side rises another six storeys and contains several restaurants as well as luxury apartments above. Characterised by half round pediments, double height columns and large windows, it complements the podium which features arched colonnades at ground level, round windows and a rhythm of freestanding and engaged double height columns above, topped by a recessed pitched rooftop.

Brazilian granite, bronze anodised metal works and grey tinted glass enclose all architectural element and produce a visually balanced and elegant solution. Grey granite and white Italian marble and granite are used to all paved areas to create a complimentary and uniform colour palette to the project.

宝丽广场是一个综合性项目，呈现长方形格局，椭圆形的中庭构成整个建筑的核心。四个曲线玻璃屋顶造型的商业空间分别朝向四角的入口，三层的零售空间内安排着高端旗舰店以及对称排列的餐饮店。

6层高的结构从北侧半月形的塔楼上"升起"，内部包括餐厅及公寓。半圆形的山墙，双层高度的梁柱以及开敞的大窗使其与下面的零售空间（圆形的窗户、顺序排列的柱子以及尖屋顶构成其主要特色）相得益彰。

巴西花岗岩、镀铜金属部件以及灰色玻璃窗构成建筑外观，营造视觉平衡感。此外，灰色花岗岩以及白色大理石地面与建筑整体结构相互呼应。

Glassware on Water
亲家爱敦阁接待中心

Taiwan, China
中国，台湾

Location: Taipei, Taiwan
Area: 590m²
Architect: Shu-Chang Kung, Josh Wu, Lisa Chen / AURA Architects & Associates
Photographer: Zou Changming
Completion Date: 2007
Award: 2007 TID Award (The Special Jury Award)

地点：台湾，台北
建筑面积：590 平方米
建筑师：龚书章、吴建森、陈丽雪 / 原相联合建筑师事务所
摄影：邹昌铭
建成时间：2007 年
获奖：台湾室内设计大奖评审特别奖

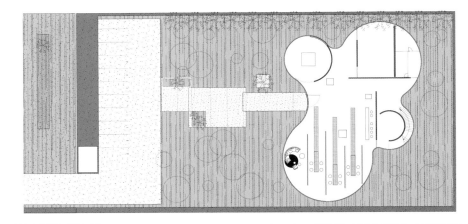

338 / EAST CHINA

The translucent glass bricks created totally different day and night views in this city. During day time, the light inside the space is natural and during the night, using lights inside the building. It turns into a light vase. To determine this project's serenity and noisiness, the architects attempt to use solid recyclable materials such as Charcoal to define this boundary. The large amount of glass bricks created curved wall that turns into interior curved walls. The reflection explains the void and solid of the building but also the relation of the space and view.

The interior trend is defined by the curve of the glass brick wall. When it comes to the evening, the lighting from the outside and from the inside is defined by the trend of the curved wall. This tells two different stories that happen in the same space. Paper tube is used to from the ceiling, which responds to the curved wall of the main building.

清透的玻璃砖材料让此建筑在白天与夜晚呈现出不同的城市地景表情——白天所引领带入室内的是外部的自然光线，夜晚则是利用室内灯光均匀照明，创造出夜晚自本体发亮的光盒，在这都市之中跳脱开来独自成形。从外到内企图利用实质的自然可回收的木炭材料与外在环境界定出之间的喧嚣与宁静。外部大量的玻璃砖创造出的曲线也转换成室内的曲形墙面，前方大片静态中的水池倒影也说明实虚量体存在于空间与景观之间的关系。

内部的动线也因外在玻璃砖的弧形关系清楚地分割出来，当接近傍晚时外部的光线与内部的光源被弧形墙面与动线综合界定出来，说明两种不同的关系却同时存在一个空间中。室内天花部分加入了纸管，企图排列出波浪造型以呼应室内的弧形玻璃墙面。

Star Place
台湾大力精品购物中心

Location: Kaohsiung, Taiwan
Architect: UNStudio
Photographer: Christian Richters
Completion Date: 2008

地点：台湾，高雄
建筑师：UN 建筑师事务所
摄影师：克里斯汀·里希特
建成时间：2009 年

Taiwan, China
中国，台湾

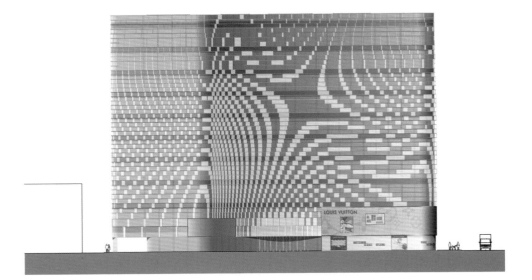

A vibrant new landmark has appeared in the fast and modern city of Kaohsiung: the luxury shopping centre Star Place. Both outside and inside, the building radiates dynamism and the kind of bright perfection that the architect refers to as the "Made in Heaven Effect". Everything about the building moves the eye.

Positioned at an urban plaza with a roundabout, the building occupies a triangular lot, giving it a wide and open frontage. Curving inwardly, the building embraces this position and opens itself fully to the city. For UNStudio the question of the building began with the façade as an urban manifestation. However, the chosen solution of a "deep" front elevation, with a prominent pattern made by the application of protruding elements, was immediately reconnected to the internal arrangement of the spaces around the atrium, the circulation through the atrium and the views from the inside to the outside. As a result, the project now consists of a tight package of inside-outside relations.

The open and transparent glass façade is patterned with projecting horizontal, aluminium-faced lamellas and vertical glass fins that together form a swirling pattern. This pattern breaks up the scale of the building, which, from the outside has no legible floor heights as a result of the one-metre spacing between the horizontal lamellas.

The façade pattern wraps around the complete building, thus providing the closed rear façade with the same identity, but in a more simplified design. At night, coloured lighting replaces the optical effects produced by the depth embedded in the façade motif, with a fluid layer of changing hues and tones. The dots on the laminated glass fins pick up the colours distributed by LED-lights which are integrated at the bases of the fins. The minimised light fittings contribute to the luxurious appearance of the façade.

一座充满生机的新地标——台湾大力精品购物中心，出现在了高雄这一快速发展又现代的城市中。这栋大楼由内而外的散发着活力与明亮的完美感，设计师想要表达的是"天堂效果制造"。这栋大楼的一切都是那么的吸引人的目光。

购物中心位于环形的城市广场中，大楼占据了一块三角形的区域，使得临街地界宽敞而开阔。由于造型上的向内弯曲，这一大楼呈现出环拥地势并且使自身向城市完全开放。对于UN建筑师事务所来说，大楼的问题出现在作为城市的表现形式的建筑外墙上。然而选择的解决方案，一个由突起元素组成的鲜明图案的"深度"正立面，会立即使人将它与内部环绕着中庭的空间布局通过中庭的循环通道以及从内向外的视角联系起来。这样带来的一个结果就是，使得目前这一工程存在着一个室内与室外的紧密联系。

这一开放透明的玻璃外墙图案由横向突出的，铝表面材质的薄板以及垂直的玻璃片组成了一个涡流的图案。这一图案打破了大楼的比例，水平的薄板间的1米距离带来的是，从外部看不出清晰的楼层高度。

外墙图案包裹了整栋大楼的每一个角落，由此也运用相同的设计手法建造了封闭的背部外墙，但却使用了比正立面更加简洁的设计。到了夜晚，彩色的灯替代了由外墙图形的深度嵌入带来的视觉效应，形成变幻着色彩与色调的流动的光影外层。位于玻璃夹层中的圆点聚集了LED发出的灯光，将它们汇集在玻璃层的底部。这些小型灯饰的装配更加有助于体现外墙的豪华外观。

Kelti International Group Corporate Headquarters

克缇企业总部大楼

Location: Taipei, Taiwan
Area: 2,340m²
Architect: Kris Yao, Artech Architects
Photography: Zheng Jinming
Completion Date: 2009
Client: Jing Yung Gi Investment Co., Ltd

地点：台湾，台北
建筑面积：2,340 平方米
建筑师：姚仁喜、大元联合建筑师事务所
摄影师：郑锦铭
建成时间：2009 年
业主：金永基股份有限公司

Taiwan, China
中国，台湾

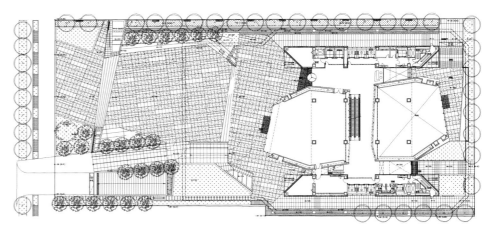

342 / EAST CHINA

The project is located in Xinyi District as a mixed-use complex with corporate headquarters, retail and commercial spaces, and high-end boutique hotel. Adopting the existing site condition which is a narrow rectangular form with east-west orientation, the building sets back from the street on the west side, creating a plaza where green lawn, patio, and platform stairs are arranged to create a cosy space. The dynamic exterior of the building is formed by the overall composition of two masses at the lower level, a suspended volume at the upper level, and the flanking two service cores which support the massive frame. The irregular forms of the masses of the lower level rise from the grade, intersected in the frame's lower portion, and together with the frame create a void space as the main entrance lobby of the building. The alternating dark and light colours on the glazing of crystal-like upper level form juxtapose the stone façade of the service cores to create a delightful rhythm.

The lower parts of building are programmed with exhibition centre and restaurant. The 3rd to 7th floors are designed for product showrooms. The 8th floor and above are allocated for the hotel. This arrangement deliberately keeps the hotel on the top portion of building to create a quiet, peaceful image. The natural sunlight is introduced to the restaurant at the basement from the enormous space at the first floor lobby by an atrium. The service cores are located on the north and south sides respectively to prevent interference to the circulation of the exhibition services and hotel.

本案位于台北信义计划区，是一栋结合企业总部与展示中心的商办大楼。配合东西向长形基地，建筑自街面退缩留设前方广场，搭配植栽、平台与阶梯，穿透建筑连接后方公园绿带，营造开阔连续的都市休憩空间。建筑主体框架以石材营造厚实量感，交错贴覆的水磨石材随季节光线及气候的时序差异，赋予立面纹理细腻的丰富质感。底层量体自地底垄起，抬高地面层高度在框架底部形成入口大厅，大片玻璃延伸视觉深度塑造建筑前后空间通透感。框架内不规则玻璃量体采用无框料玻璃帷幕系统，挪移对扣悬挑其中；深浅玻璃轻盈动态对比石材静定沉稳，显现建筑蕴含的内敛气度与优雅华丽的时尚质感。

建筑地上共15层，一楼作汽车展示中心，B1另设餐厅；设计运用高度变化塑造空间焦点，抬高展场压低餐厅，前后之间高低光影凸显空间属性又能营造场所氛围，入口挑高部分以结构玻璃支撑，让视线延伸至户外景观。三楼以上为企业总部，3楼与14楼为挑高楼层，门型框架结构塑造14楼室内无立柱的挑高空间；建筑由内而外、由下而上密切搭配，满足商办大楼多样复合的需求。

Hsinchu Xiang Duan Village Community Centre

新竹香村段闲谷小区中心

Location: Hsinchu, Taiwan
Area: 1,993m²
Architect: Tao Architects and Planners
Photographer: Li Yanyi
Completion Date: 2011

Taiwan, China

中国，台湾

地点：台湾，新竹
建筑面积：1,993 平方米
建筑师：十方联合建筑师事务所
摄影：李彦仪
建成时间：2011 年

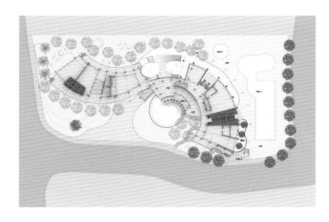

344 / EAST CHINA

This is another hillside villa cluster community project of Chang Yih Construction after the completion of the Chang Yih Construction Headquarter. It is located in the Hsiangshan part of Hsiangshan District, ten miles from the local railway station. Hsiangshan is surrounded by sloping preservation area of hill lands. Most houses are two-in-one open-ended villas whose residents are mostly from downtown Hsinchu and Hsinchu Science Park.

Looking from the southeast corner of the site, the middle of the community inclines slightly towards the southwest, surrounded by rolling hills. One can overlook the hills in the north and enjoy an open view. The design follows the terrain and forms the winding outline of a deck chair. The configuration flows with the contours, zigzagging in the central area of the site, forming the shape of an S. It collects the scenery of the community entrance and at the same time creates a paradise where the residents can build a dream and count the stars. The first arc façade can be found in the direction of the gum tree road, here you can enjoy the scenery on the other side of the entrance axis. In the opposite direction, the waterscape crossing the second central axis of the site (empty space) and the greenery penetrating axis three (empty space) contribute to the inverse 'S', establishing the main concept of building configuration.

With the 'S' configuration, the site is divided into three sections which serve as the major public living space and open area, holding possible community events. The green visual axis links the front and back gardens, perfect relaxing places for the elderly and the young. A simple kitchen is also available here, so the people involved can socialise and communicate through activities like cooking, sports and film watching.

There is an aerobics classroom and community office on the first floor of the building. The sloping board over the pool is designed for outdoor cinema in the evening. Wandering on the roof at night, watching the distant lights from the community, it could really feel like a scene from *Fiddler on the Roof*.

继昌益建设总部大楼完成之后，本案是另一个昌益建设将推出的山坡地别墅小区的俱乐部建筑。本案位居新竹市香山区香山段，距离新竹火车站车程约10公里处。香山附近为山坡地保育区带，多为丘陵地形。小区内多为双拼式透天住宅别墅，居民多来自于新竹市区以及新竹科学园区。

由本基地小区东南隅的小山丘上，纵览小区，小区的中间向西南侧缓缓倾斜，四周环绕着连绵的小山丘，越过北边的山丘，向下俯揽远眺，视野辽阔。循此概念顺着地形，勾勒出曲折类似躺椅的轮廓；沿着等高线的走势，以近似"S"的形状，蜿蜒匍匐在小区的中央地段上，一方面收揽小区入口的端景意象，一方面创造一个可以让小区居民筑梦及数星星的小天地。由枫香树道的视线方向建立起第一片圆弧立面，可以看到小区入口轴线另一方向景观。顺势反曲，借由穿越基地中央轴线的水景（虚空间），以及视觉的轴线三绿色穿透（虚空间），挥洒出S形反曲线，成为建筑的主要配置概念。

借由S形建筑配置，将基地分为三个区带，成为小区的生活场域及开放空间，容纳了许多小区的活动。绿色视觉轴线连接了前后绿园区，是为老人及小孩活动的场所。并且设立了简易厨房，供老人、孩童、亲子，借由料理、下棋、运动、交谊、看电影、生活连结了人与人之间的记忆。

建筑物二楼设有韵律教室及小区办公室。事实上在水池上方的斜板上，是设计可以在夜晚架设投影电影院。夜间徜徉于屋顶上，看见小区星星点点的灯光，颇有屋顶上提琴手的感觉。

Huga Fab III and Headquarters Building

广铔光电中科三厂暨总部大楼

Location: Taichung, Taiwan
Area: 31,860m²
Architect: J. J. Pan and Partners
Photographer: Chun Chieh Liu
Completion Date: 2009

地点：台湾，台中
建筑面积：31,860 平方米
建筑师：潘冀联合建筑师事务所
摄影：刘俊杰
建成时间：2009 年

Taiwan, China
中国，台湾

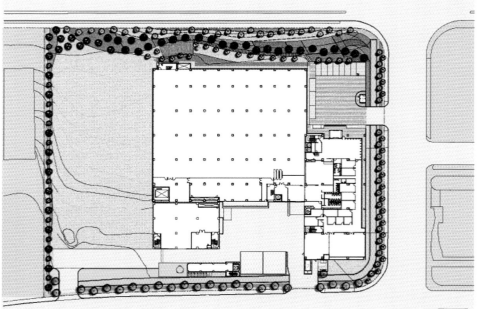

346 / EAST CHINA

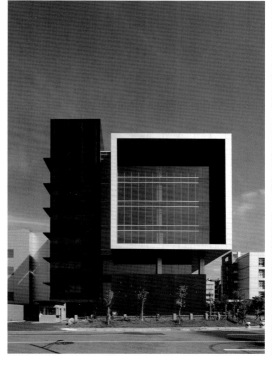  

As a signature image located near the entrance to the Central Taiwan Science Park, The Huga Optotech, Inc. Headquarters Building sets an example for industrial buildings with its simplistic beauty. The two major building volumes are configured as an abstraction of the LED (light emitting diodes) manufacturing process, resembling the electric discharge between electrodes. The landscape elements featuring rounded corner fonts of the "HUGA" corporate image signage transform into in the public area.

The loading zone and parking entrance are located at the lowest point of the parcel nearby the southeast corner of the site in order to take advantage of the six-metre height difference. Fabrication buildings are set back and away from the northeast border, creating a generous entrance.

The ground level, with nine-metre ceiling height, is used for manufacturing LED epitaxial wafer and crystalline grain. The ornamental strips with lighting fixture enrich the fully enclosed façade, symbolising the encapsulated LED products. The six-level office building orients north-to-south; its insulated aluminium wall panels and glass curtain wall achieve excellent energy performance while maximising the view overlooking the scenic Dadu mountains nearby.

The composition of the building's four elevations feature deep setback windows and sunshades. Vertical and horizontal grills are carefully studied for the proper position of frames, columns, joint and alignment to express the rational and meticulous professionalism of the corporation.

本案基地位于中部科学工业园区北端进口西侧，临40米林荫大道，北端临近清泉岗机场。

楼体配置之挑檐语汇源自LED发光二极管制作过程中正负极电子间的相互反应。

企业LOGO中交错弧线和倒圆角字型则转化成公共区域景观设计的元素。

基地南北向高差约6米。将东南侧低处作为停车场入口及服务卸货区；西北侧退缩为作业厂房生产区。厂房地面层楼高9米，二三层楼高5米，生产发光二极管（LED）磊晶与晶粒制造。

办公区共六层，主建筑采取南北向配置，以节约能耗。透过主办公室可看到大度山下之优美景观。办公楼外墙主体采用金属帷幕，厂房采用浅灰及深灰色调搭配的复合板。

建筑的四个朝向根据不同的日照条件，分别采用深窗、遮阳板、纵横百叶等手法构成立面的韵致。通过格构、柱位、板缝、百叶等元素的严整对位，塑造出高科技厂房理性而严谨的意象。

W Hotel Taipei
台北 W 酒店

Location: Taipei, Taiwan
Architect: G.A. Design International Ltd., London
Photographer: G.A. Design International Ltd., London
Completion Date: 2010

地点：台湾，台北
建筑师：伦敦 G.A. 设计国际有限公司
摄影：伦敦 G.A. 设计国际有限公司
建成时间：2010 年

Taiwan, China

中国，台湾

W Taipei is an electrifying sanctuary of serenity and energy in the heart of a bustling neighbourhood, reflecting the surrounding natural beauty of Cising Mountain and Yangmingshan National Park, juxtaposed alongside Taipei's vibrant, modern cityscape. Designed by renowned architects G.A. Design International Ltd., London and encased in radiant glass that elevates to 31 storeys tall, W Taipei is designed to reflect the union of urban city and nature. A perfect example of this union is The Chain – made of the strongest stainless steel with a mirror finish – symbolically anchoring the W Taipei building into Taipei city, as well as a gigantic green wall filled with plants organically grown in Taiwan. W Taipei presents collection of design installations throughout the hotel.

W Taipei's 405 guestrooms and suites have been designed as private sanctuaries inspired by nature with spectacular urban views. Warm-coloured stones, burnished wood and lush electrified floral carpeting contrast with modern, subtle lighting inspired by Chinese lantern boxes. The guestrooms feature an interior balcony/play pad, the signature ultra-comfortable W Bed with its 350-thread count linens and a sleek, modern white work station with a leather ergonomic chair. An oversized vacation-style islander tub will take command in the bathroom, set against red or chartreuse-coloured subway-inspired wall tile and a wooden partition.

W Taipei has multiple dining structures and countless square metres of meeting and event space including: the Mega Room at W Taipei – the largest pillar-less ballroom in the city – featuring a symmetrical lattice ceiling and unique tubular lighting with more than 1,039 square metres of space and, W Taipei's signature restaurant YEN – on the 31st floor with floor-to-ceiling windows – that is finished in a combination of deep midnight purple lacquer and high gloss Macassar ebony and showcases bold, original art pieces.

台北 W 酒店位于时尚国际之都充满活力信义区最美丽的交叉路口，充满着标志性的迷人天然元素和时尚氛围，汇聚了这座城市卓越时尚、动感夜生活和繁华商业的无限魅力。由著名的伦敦 G.A. 设计国际有限公司操刀设计，酒店如同被套在一个晶莹剔透的玻璃方盒中，有 31 层高。台北 W 酒店的设计体现了城市与自然的结合。其中一个完美的展现是入口处用最坚固的钢铁铸成的"枷锁"，表面光洁，象征将台北 W 酒店锚定到台北这座城市，同时还拥有一座巨大的绿色墙体，种植了在台湾地区有机生长的植物。W 淋漓尽致的诠释了酒店的设计理念。

台北 W 酒店的 405 间客房和套房，以自然元素为设计灵感并且结合了令人陶醉的城市美景。暖色调装饰石料、抛光实木、绿意盎然的植物雕花地毯与中式灯笼罩子中充满现代感的精致灯光，形成了令人沉醉的奇妙对比。所有客房全部配备了台北 W 酒店特色的特级舒适睡床，以及 350 纱织密度亚麻丝柔床单和现代的白色书桌搭配符合人体工学的高级皮椅。超大的度假风格的岛民浴缸占据了整个浴室，使用了设计灵感来自于地铁的壁砖和壁板，配以高反差度的红色或黄绿色。台北 W 酒店拥有多种餐饮结构以及包含了超多的会议和活动空间，台北 W 酒店大厅 (Mega Room) 将是全市最大的无立柱宴会厅，超过 1039 平方米的空间以其对称的离子晶格天花板和独特的管式照明为特色，位于台北 W 酒店 31 层的 YEN 签名餐厅拥有全景落地窗，仿佛是以深紫色漆面和光亮的黑檀木以及大胆的展示组合而成的原始的艺术品。

Northstar Delta Project Exhibition Centre

北辰长沙三角洲项目展示中心

Location: Changsha, Hunan Province
Area: 3,000m²
Architect: Zhang Lei, Zhou Suning, Wang Liang
Photographer: Yao Li
Completion Date: 2010

地点：湖南，长沙
建筑面积：3,000 平方米
建筑师：张雷、周苏宁、王亮
摄影：姚力
建成时间：2010 年

Hunan, China
中国，湖南

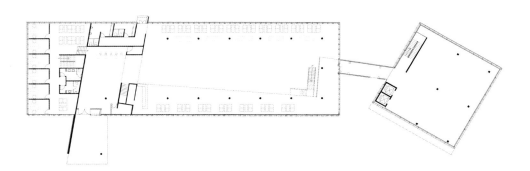

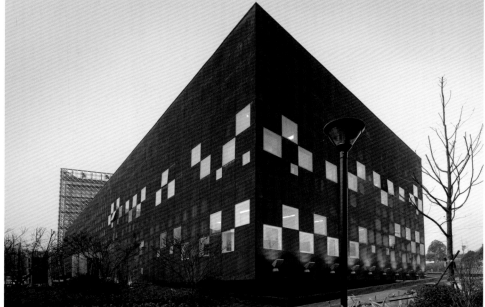

Northstar Delta Project, located in the flat area to the east of Xiangjiang River and to the south of Liuyang River, is an urban complex programme with a site area of one square kilometres and a gross construction area of over 5000,000 square metres. The northern part as well as the future development of Changsha City will be endowed with a new face after the realisation of this Project.

The exhibition centre is located opposite to the west of Xiangjiang River and unfolds to the city centre. The sunshade system on the western façade is a requisite in a city where the summer is extremely hot. The west elevation of the horizontal volume as well as the four elevations of the slanting structure besides is required to install with LED media wall, which poses a visible contrast with the concept of "overlooking the Xiangjiang River". To solve this problem, acrylic sunshade screen is adopted. What's more, the application of translucent sunshade media wall and Corten steel plate enhances the juxtaposition and contrast between space form and material language and expresses the authenticity and complexity of the material itself. In such a way, experience in the architecture becomes more interesting and mysterious.

北辰长沙三角洲项目是一个占地1平方公里、建设量超过500万平方米的大型城市开发项目，位于当地著名的湘江以东和浏阳河以南一片平坦的区域。它的实施将极大地改变长沙城市北部的面貌并影响到未来长沙城市的发展格局。

展示中心面对西侧湘江，同时也是面向城市主要的展开面，西向遮阳设置对于像长沙这样夏季炎热的城市是必不可少的，横向体量的西面和斜塔的四个面按照要求同时必须设置面对城市的LED媒体墙，建筑遮阳与媒体墙这些使用要求和西立面基本的观江理念之间的矛盾显而易见。白色亚克力（Acrylic）遮阳格扇在立面的采用巧妙而有效地解决了这一难题。展示中心通过外部倾斜的几何关系与内部正交格构，半透明遮阳格扇媒体墙和锈蚀耐候钢板的运用，强化了空间和材料语言之间的形态并置和对比，表达了其内在的真实性和复杂性。建筑体验自此变得更有乐趣和充满遐思。

Hallelujah Concert Hall
哈利路亚音乐厅

Location: Zhangjiajie, Hunan Province
Area: 4,970m²
Architect: Yu Kongjian
Photographer: Yu Kongjian
Completion Date: 2010

地点：湖南，张家界
建筑面积：4,970 平方米
建筑师：俞孔坚
摄影：俞孔坚
建成时间：2010 年

Hunan, China
中国，湖南

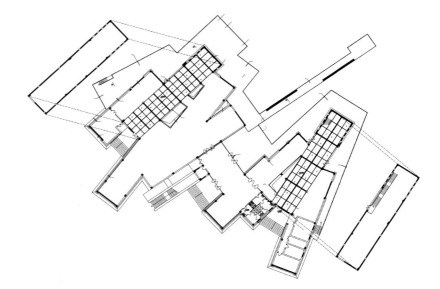

The "Hallelujah" Concert Hall, situated at the entry of the Ecological Square in Yellow Dragon Cave, "a magical karst cave" in Zhangjiajie and with Suoxi River flowing along its south, embraces the paddy fields. Large part of the building was "buried" underground with the volume above ground relatively small in order not to "interrupt" the beautiful natural environment.

Featuring harmonious ecological environment, unique and novel design, and attractive and exquisite appearance, "Hallehujah" boasts the most beautiful surroundings.

In addition, original eco-cultural performance in combination with advanced 3D technologies is to be staged everyday at the concert hall, introducing a graceful, attractive and mysterious Zhangjiajie to tourists from home and abroad.

项目位于举世闻名的风景区湖南省张家界武陵源黄龙洞的洞前广场，前有索溪河，背靠峭壁，因剧场的大部分功能被安排在地坪以下，使得地上部分的体量减少，以避免对风景区的视觉干扰。整体形态是地层结构的剖面，呼应武陵山砂岩峰林核心区周边山地的单斜地壳构造。整体效果和环境相得益彰，并融入环境。夜晚来临，层间玻璃透出暖光，几片弯曲程度不同的岩石飘在空中的感觉更加强烈。

这种形式对于内部功能的意义在于，它把观众厅、舞台、后台、侧台以及舞台上方的挂幕布的空间很好的结合起来，各部分空间都得到充分、恰当的利用。比如外形翘起的高端正好是舞台上方最需要高度的地方，巧妙地解决了传统剧场体型上的难题。斜坡至地面的屋顶是绿化种植，即有节能效果，同时与东侧的稻田景观融为一体。环绕建筑周边的水面也是为了建筑的微气候调节设计。

向上反曲的屋面能够把舞台上传出的声音均匀地反射到观众席，尤其是有助于后排观众获得宝贵的前次反射声。

Maoping Village School
毛坪村浙商希望小学

Location: Leiyang, Hunan Province
Area: 1,168m²
Architect: in+of architecture, Studio Wang Lu, Tsinghua University
Photographer: Christians Richterm, Wang Lu
Completion Date: 2008
Award: 5th Architectural Society of China Architectural Creation 2008, honorable mention; 2010 International Architecture Awards, Chicago

Hunan, China
中国，湖南

地点：湖南，耒阳
建筑面积：1,168 平方米
建筑师：壹方建筑、清华大学建筑学院王路工作室
摄影：克里斯汀·里克特、王路
建成时间：2008 年
获奖：第五届中国建筑学会建筑创作优秀奖，2008；2010 年国际建筑奖，芝加哥

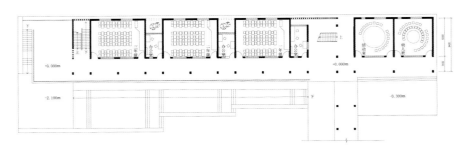

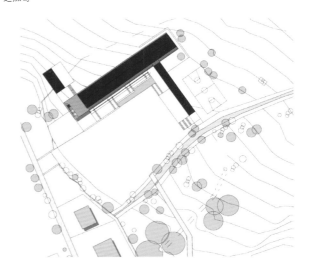

The site of the primary school is on the slope in the northeast of the Maoping Village. The two-storey school building stands on a terraced ground that is embedded in the slope. The configuration, cross-section, materials and colours of the building are basically isomorphic to local houses, and the scale of its gables is largely commensurate with the surrounding houses. The division of the structure by small sky-wells that correspond to teachers' offices and staircases renders the whole building resembling a cluster of local houses; through the breaking-up of the whole, the school amicably blends with the local environment. In order to keep the construction cost under control and to adapt the project to the local construction techniques, bricks are still employed as the main building materials: red bricks are used for the building so as to have a better dialogue with the surrounding houses; whereas the limited amount of those surviving large grey bricks are applied to roads, paths, and open grounds.

The northern brick façade has a few brick lattice works piercing through each of the wall, a measure of architectural treatment that was derived from the tradition of local houses, where this technique had been applied in order to reduce deadweight of the wall and to ensure ventilation. The largest wall with lattice work of this kind on the north side of the lobby becomes the only "decoration" for the lobby space, and entering the lobby, one is presented with a digitalised scene of the outside landscape, making the space distinctive.

The southern façade, with wooden framework screen as its integral part, similarly borrowed the language of local architecture, so that the building was instilled with certain symbolic significance. Like an unfolded role of bamboo slips for writing, the façade gains an air of scholarship for the primary school building. The corridor on the first floor is thereby distinctive: when one looks out into distance, it seems as if the landscape is present behind a stretch of woods, and the building therefore is not only an architectonic structure but also a toy with intersected light and shadow, which children can enter, and with which remain the special memories of living in Maoping.

小学基地在毛坪村东北的一块坡地上，建筑的体形、剖面、材料、色彩与当地的民居基本同构。整栋建筑通过对应于教师办公和楼梯间的小天井划分，像是一组民居的集合；通过这种化整为零的办法，校舍友善地融入了环境。为了控制造价，适应当地施工工艺，砖仍然是建筑的主要材料：小红砖用来砌筑建筑，与周边民居更好的对话；存留不多的大青砖用来铺砌道路、广场。

砖砌的北立面，有几处镂空的砖砌花格墙，其做法来源于当地民居。当地民居中，为了减轻自重，保证通风，采用这种镂空砖墙的砌法。这里位于门厅北侧最大的一堵花格墙，成为门厅唯一的"装饰"，入于厅中，外面的风景被像素化地呈现，空间也有了特色。

木格栅的南立面，像展开的简牍长卷一样，使小学的建筑获得了一个有些书卷气的立面。二楼的走廊也因此与众不同；向外望去，风景仿佛展现于一片树林之后，建筑不单是一栋房子，还是一个小朋友可以进入的玩具，光影交织，留下童年在毛坪生活的特殊记忆。

Jishou University Research and Education Building and Huang Yongyu Museum

吉首大学综合科研教学楼及黄永玉博物馆

Location: Jishou, Hunan Province
Area: 25,727m^2
Architect: Zhang Yonghe / Atelier FCJZ
Photographer: Atelier FCJZ
Completion Date: 2004

地点：湖南，吉首
建筑面积：25,727 平方米
建筑师：张永和 / 非常建筑
摄影：非常建筑
建成时间：2004 年

Hunan, China
中国，湖南

356 / CENTRAL CHINA

The project is located at the campus of Jishou University in Jishou City, Hunan Province, and is mainly concerned with two important issues relating to the site: one is the relationship between architecture and its surrounding environment, and the other is how to establish relationships with local architectural tradition and culture.

The entire campus of the University was built on a hilly area, and nearly all the buildings were built against the hill. The site of the project sits on the south of the artificial lake at the campus centre, where it was formerly a slope which was later terraced. The Research Education Building and the Museum form a wedge-shaped composite section that inserts itself into the land. The building mass, multiple roofs and integrated windows blur the vertical and horizontal forms of walls and roofs, which in turn contributed to rebuilding and reestablishing the physical presence of the site.

Respect for local architectural culture has been developed into two types in Jishou: one is the preservation of "specimen buildings" in the old town, and the other is the duplication of local single residence regardless of structure, material, function, scale, etc. of the new building. Under such circumstances, the architects conformed to contemporary architectural conventions, and decided to keep the immense volume. They tried to bring the pattern of local residential groups into the new building, visually establishing a relationship between the new architecture and local architectural culture. Therefore, conceptually the architecture is both a "hill" and a "village".

建筑位于湖南省吉首市吉首大学校园内,由两部分组成——综合科研教学楼和黄永玉博物馆。设计主要关注两个问题:一个是建筑如何重组建筑与周边物理环境的关系;二是建筑如何与当地原有的建筑文化传统建立积极的联系。

校园建在山地上,几乎所有的建筑都是依山而建。建筑的基地位于校园中心的人工湖南侧,原是坡地,后被削平。教学楼与博物馆形成的整体以楔状的剖面形态插入基地,用建筑的手段恢复了基地物理环境的秩序。裙房部分的屋顶与高层部分的北侧外墙在剖面形态上构成两个不同斜率的连续的表面,从而模糊了屋顶与外墙两种不同功能的建筑构件在形态上的差异。屋顶与外墙上相似的开窗方式进一步加强了这种混淆,使建筑整体加入"造山运动"。

对建筑文化传统的尊重在吉首地区分化为两种模式:一种是对老城区内建筑物"标本"式的保护;另一种是新建的建筑物不顾结构、材料、功能、尺度等方面的巨大差异,对传统民居单体形式的"戏仿"。建筑师以保持当代建造逻辑并接受新建筑的大尺度为前提,尝试将传统民居村镇聚落的肌理带入建筑的形式系统,从而在视觉上建立起新建筑与当地建筑文化传统的呼应。因此,在概念上,这栋建筑既是"山"又是"村"。

Wuhan CRland French-Chinese Art Centre
武汉华润中法艺术中心

Location: Wuhan, Hubei Province
Area: 1,500m²
Architect: Zhang Ke, Zhang Hong, Hao Zengrui, Han Xiaowei, Yang Xinrong, Liu Xinjie, Li Linna, Jing Jie, Lin Lei, Han Liping
Photographer: Cheng Su
Completion Date: 2005

地点：湖北，武汉
建筑面积：1,500 平方米
建筑师：张轲、张弘、郝增瑞、韩晓伟、杨欣荣、刘新杰、李琳娜、经杰、林磊、韩立平
摄影：程苏
建成时间：2005 年

Hubei, China
中国，湖北

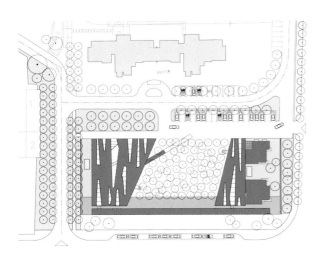

The site of the Art Centre is located near the Tanhualin historic area in Wuchang, about one mile away from the historic Huanghelou tower. The expectation of the building was an important public space for the city and a monument for both the past and the ongoing transformation of the city. The fact that many famous Chinese intellectuals lived in the Tanhualin area inspired the architects. They were interested in testing the possibilities of building something out of the ancient Chinese intellectual practice of ink and water. The Art Centre was conceived as an urban container, within which art objects, events, acts, concepts and activities flourish. In this case the container is made out of intuitive images of ink-and-water. While the site conditions also take part in the formation of spaces: since the site is cut into a half by an unexpected urban infrastructure (a flood pipeline), the 30-metre-wide outdoor space became the central courtyard for spatial organisation, around which seat the east and west exhibition hall and the floating bridge linking the two parts. In the 80-metre-long concrete bridge, the ink-water stroke texture coincides with the necessary structural elements for the 5.5-metre-high concrete hollow beam. This becomes an interesting moment when an image merges seamlessly with a structure.

艺术中心基址隔街紧邻武昌最重要的历史文化街区——昙花林，距离著名的黄鹤楼不到两公里。设计之初，人们期望艺术中心既能够为城市提供一个重要的公共活动场所，又能够成为正在进行的城市变迁和城市历史的纪念碑。

中国近代历史上许多文化名人（包括郭沫若等）都曾经在基地紧邻的昙花林居住过，这给以建筑师最初的一些冲动，建筑师好奇地想试验一下利用中国传统文人的水墨游戏进行建造的可能性。空间布局同时也受到基地条件的影响：由于一条市政排洪沟从中央穿越了基地，从而产生了30米宽的中心院落，围绕院落的空间组织形成了东部和西部分立的展厅以及漂在水面之上联系两个展厅的廊桥。80多米长的廊桥将水墨笔触的肌理与5.5米高、2.5米宽混凝土空心梁的内外孔洞联系起来，建筑在此归结为毫无修饰的结构之美。

Emperor of Ming Dynasty Cultural Museum 明代帝王文化博物馆

Location: Zhongxiang, Hubei Province
Area: 6,200m²
Architect: Shan Jun, Lu Xiangdong, Tie Lei, Wang Xin, Sun Xian, Luo Jing, Liu Si
Completion Date: 2011

地点：湖北，钟祥
建筑面积：6,200 平方米
建筑师：单军、卢向东、铁雷、王鑫、孙显、罗晶、刘思
建成时间：2011 年

Hubei, China
中国，湖北

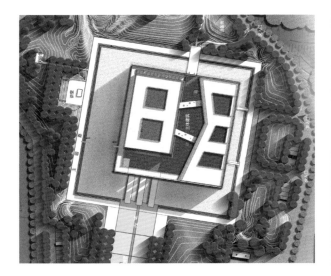

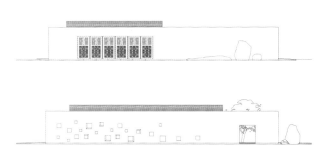

Located in Zhongxiang, Hubei Province, Emperor of Ming Dynasty Cultural Museum stands north to the Xian Mausoleum of Ming Dynasty, which is listed as the World Cultural Heritage, and west to the Mochou Lake. The whole museum is situated within a fantastic environment, surrounded by splendid mountains and shadowed by a lush forest.

In order to highlight the theme – Jiajing emperor and Ming dynasty, the architects proposed a general layout simulating the Chinese character "Ming", which is composed of two characters meaning "sun" and "moon" respectively. The site plan takes the geometry of the Chinese character of "sun" as the main exhibition hall while "moon" is secondary to it, and the space between the two halls together with the loggia provides public and outdoor exhibition spaces that people can freely walk through, making a perfect combination between the character's structure and the museum programmes. The building complex, with a base area of 6,400 square metres, is located in a square pool of 14,400 square metres. By melting into the surroundings, the architecture revealed the poetic conception of "garden as well as yard" resembling traditional Chinese gardens and courtyards.

Chinese character is also used in the three stones that symbolize the word "mountain" in front of the museum, and the skylight symbolizing the word "dragon" in the exhibition hall. The former one is the finishing touch of the landscape theme, while the latter creates an abstract image of caisson, which used to be widely used in traditional Chinese architecture.

With an attempt to mix Chinese traditional garden into contemporary regional architecture, the architects hope to reinterpret the vocabularies such as loggia, window, pavilion and yard in a modern context, reaching a state of "designing with nature" as suggested in the book named "The Art of Garden", which has been taken as a classic treatise on Chinese garden design. A magnificent white wall acts as an icon of the museum, with square openings creating ever-changing shadows in the interiors. A giant copper gate is placed at the main entrance, bringing out both the feeling of the imperial palace and an artistic air.

Due to the limited budget, most areas of the museum's façade are painted white. Indeed, the method of blank-leaving, together with the hues of black, white and grey, reproduces the image of Chinese landscape painting.

明代帝王文化博物馆位于湖北省钟祥市，北邻世界文化遗产——明显陵，西临莫愁湖，周围有绿树成荫起伏的山丘，自然景观优美。该博物馆既是钟祥市博物馆，又是以明代嘉靖皇帝及相关展陈为主体的主题馆。

为突显出生于钟祥的嘉靖皇帝的帝王主题和"明代"的历史主题，设计采用汉字"明"的总体布局，"日""月"分别对应主馆和次馆，两馆之间以及围绕主馆的长廊是可供人穿越漫步的公共开放和室外展陈空间，整体汉字结构的疏密错落与博物馆的功能流线相契合。建筑以方形的园墙围合展馆，80米见方的建筑群体置于120米的方形水池中，与周围景观融为一体，宛如自然山水画中的印鉴，形成介乎中国传统院落空间和园林意象之间的、"亦院亦园"的诗情画意。

汉字的运用还体现在建筑正立面三个景石形成的"山"字组合，以及主序厅12米见方的"龙"字天窗等设计中。前者是对山水主题的点睛之笔，后者则旨在抽象地塑造一种中国传统建筑的藻井空间意象。三块姿态各异的叠石作为白墙的前景，通过墙上的光影和池中的倒影在极简中产生变化；叠石取意"癸山敦（殷）"的山字刻文，与水景共同寓意山水意象。"锺聚祥瑞"为嘉靖皇帝所赐，嘉靖十年承天府因之得名"钟祥"。博物馆主次连廊分别设计有"锺聚"和"祥瑞"两组刻字变形的景窗，通过远近组合形成"锺聚祥瑞"的题名。

设计者试图探讨一种将中国传统园林意境融于当代地域建筑中的设计理念，希冀通过对廊、窗、榭、院等传统园林语汇的现代诠释，达到明代计成所著的中国造园经典《园冶》中所谓"窗棂无拘，随宜合用；栏杆信画，因境而成。……纳千顷之汪洋，收四时之浪漫"的境界。建筑外部白墙随景窗而透出内部墙体的阴影变化，在主要出入口处级巨大铜门，在园林气氛中体现帝王气象和亦庄亦谐的艺术效果。

由于低造价的制约，建筑外立面选择了大面积的白色涂料，而大面积的留白，以及黑白灰的色调则是对中国传统山水画意境的整体再现。

Wuhan New Railway Station
武汉火车站

Location: Wuhan, Hubei Province
Area: 120,000m²
Architect: J.M. Duthilleul, E. Tricaud / AREP Architects
Photographer: T.Chapuis
Completion Date: 2009

地点：湖北，武汉
建筑面积：120,000 平方米
建筑师：J·M·杜地耶尔、E·特里郭 / AREP 建筑师事务所
摄影：T·查布斯
建成时间：2009 年

Hubei, China
中国，湖北

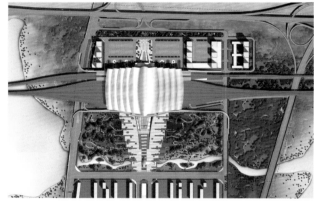
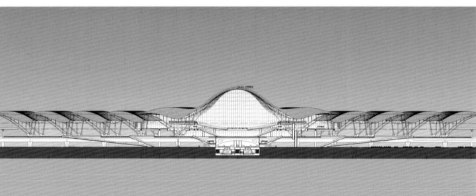

The new station in Wuhan stands as a symbol both of this historic past and its current spectacular development. The sweep of the roofs, which are systematically organised in two large wings, on either side of a central concourse running east to west, illustrates one of the most famous Wuhan legends and alludes to the flight of a legendary bird, a yellow crane, whose return to the country announces an era of prosperity and happiness. Steeped in symbolism, the roof is in nine sections, echoing the city's location at the crossroads of nine provinces. Thus, the image of the station appropriates an ancient component of the Chinese tradition – "a sweeping roof that challenges the sky, built on a massive foundation rooted in the earth"– While at the same time projecting it into a highly-contemporary idiom.

As the station affirms its strong architectural identity, form expresses function. The body of the bird, extending out from the alignment of the urban development, contains the access, reception, waiting and service functions. Access routes to the platforms are central to the composition. The interplay of access routes and levels creates a three-dimensional space beneath the roof canopy, animated and punctuated by the movements of trains and passengers. The wings extending out from either side of the body completely roof in the platforms. They are composed of roof sections created out of a succession of partially overlapping wavelets, which provide views of the sky. They protect the platforms very effectively while at the same time allowing the natural light to filter through. The general, west-facing composition marks the direction of the city.

武汉新火车站已成为这一区域的历史及如今大规模发展的象征。蜿蜒的屋顶被系统地分成两扇巨大的"翅膀"分别位于中央广场的东、西两侧，诠释出武汉最著名的神话，寓意着"黄鹤"的到来，宣布这一时代的繁荣与喜庆。此外，屋顶由9部分结构叠加而成，意指武汉位于九个省的交叉点上。武汉新火车站以其全新的形象展示了中国传统建筑语言"飞向天空的屋顶"，同时赋予其现代象征意义。

"形式体现功能"，其独特的建筑外形决定了其内部的功能。"黄鹤"的身体略微突显出来，内部设计着入口、接待台、等候室及服务台等。通往月台的通道构成整体设计的核心元素——通道与楼层交叉错构成了一个立体感十足的空间，穿行其间的乘客更是为其带来了活力。"两扇"翅膀延展出去，下面建造着月台。透过层叠的波浪外壳可以看见天空的景象，在保护月台的同时将自然光线吸引进来。此外，火车站朝西向的设计标志出了整个城市的方向。

Zhongxiang Culture and Sports Centre

钟祥市文化体育中心

Location: Zhongxiang, Hubei Province
Area: 12,900m²
Architect: Shan Jun, Lu Xiangdong, Wang Xin, Tie Lei, Sun Penghui
Completion Date: 2010

地点：湖北，钟祥
建筑面积：12,900 平方米
建筑师：单军、卢向东、王鑫、铁雷、孙芃卉
建成时间：2010 年

Hubei, China
中国，湖北

Facing the Xian Mausoleum of Ming Dynasty, one of the World Cultural Heritage sites, and Emperor of Ming Dynasty Cultural Museum across the lake, Zhongxiang Culture and Sports Centre is located on the west bank of Mochou Lake, which is to the northwest of the old city.

According to the programmes the culture and sports centre is divided into two parts: the high main part and the low auxiliary part. The main part is the gymnasium, located at the eastern end near the lake and connected with the lake by successive steps; the auxiliary part is a terrace of 230 metres long, 90 metres wide and 5 metres high. A group of seven sunken courtyards is "embedded" in the terrace, for the use of tennis, basketball, volleyball, swimming pool and other training sites, all connected to the terrace with stands and steps. Outside the terrace, there are commercial types as cafés and bars, enhancing the social and cultural functions during the post-game period.

The main part is of cone shape, placing on the high-terrace upside down. Its gradient triangle patterns of the metal curtain surface are abstracted from the traditional local instruments. The centre is surrounded by the urban scenery with alternating natural beauty, forming into a "see and seen architecture". Mochou Square on the south is the most significant public square in Zhongxiang City. There are four sets of curving rope light around the centre with dynamic effect, which signifies the fluttering ribbons of Ms. Mochou, a female dancer of Chu State in the Warring States period of Ancient China. The square and the gymnasium co-create a special artistic effect of dynamic and static combination.

Zhongxiang Culture and Sports Centre is not only one of the main venues for Hubei Provincial Games in 2010, but also the city's most important civilian places for celebrations and public events, such as the large-scale dance party organised by CCTV. In a word, it is a new landmark of Zhongxiang City.

钟祥市文化体育中心位于旧城西北部的莫愁湖西岸，与东岸的世界文化遗产——明显陵和明代帝王文化博物馆隔湖相望。

根据使用功能，建筑的总体布局为一高一低、一主一次两部分。凸起的主体建筑为体育馆，设于紧邻湖水的东端，通过自由连续的台阶与湖水相连；辅助建筑为一个长230米、宽90米、高5米的一层平台，将篮球场、网球场、游泳池等训练场地相融合，形成一组下沉院落群，"镶嵌"于大平台中，通过台阶看台与屋顶平台相连。平台建筑外侧设置酒吧等商业文化业态，以提升其非比赛期的社会文化功能。

主体建筑以倒圆锥体的造型落于高台上，表面是从传统地方器具纹样中抽象出简洁渐变的三角形图案。建筑本身既是莫愁湖畔的最主要景观，沿建筑四面环廊又可一览周边城市和自然景色的交替变化，形成"景观·观景建筑"。南面的莫愁广场是钟祥市最主要的市民广场，设计者以战国时期楚国舞者莫愁女舞蹈时飘动的绸带为母题，通过环绕中心的四组曲线地面灯带展现动感，与体育馆形成一动一静的艺术效果。

建成后的文体中心，不仅是2010年湖北省运会的主场馆之一，也是中央电视台举办大型歌舞晚会及全市最重要的庆典和市民活动的公共场所，是钟祥市的新城市地标。

Zhengdong District Urban Planning Exhibition Hall
郑州郑东新区城市规划展览馆

Location: Zhengzhou, Henan Province
Area: 8,450m²
Architect: Zhang Lei, Qi Wei, Guo Donghai, Cai Zhenhua
Photographer: Yao Li
Completion Date: 2011

地点：河南，郑州
建筑面积：8,450 平方米
建筑师：张雷、戚威、郭东海、蔡振华
摄影：姚力
建成时间：2011 年

Henan, China
中国，河南

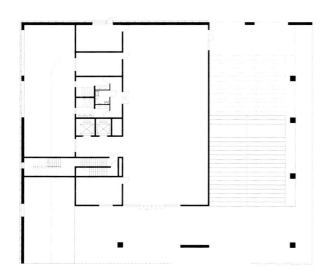

366 / CENTRAL CHINA

Zhengdong District Urban Planning Exhibition Hall, sited close to the completed school building on the north and east and the main arterial roads of the city on the south and west, is a rarely seen landmark building located in a street corner. The specific location defines the open structure of the building to show the welcoming gesture to the city. The ground floor was elevated to form a void space between the city and the building, bringing the lively atmosphere of the street to the site. The building was divided into two parts – the open exterior and the secrete interior. The interior space was wrapped by the solid walls while the exterior space seemed to be insubstantial. Inside, the independent circulation is actively linked to the dynamic one defining the exterior space. The entrance was designed in the traditional manner, making visitors coming and exiting gesture become more interesting. The building itself resembles the gardens and squares in the city.

郑东新区城市规划展览馆基地东面和北面紧邻已建成的学校，南面和西面是主要的城市道路。这是一幢不多见的位于街角的标志性公共建筑。角部开敞是基地在特定位置向城市开放所表现的姿态，展览馆首层局部架空形成了城市与建筑之间的灰色区域，通过它很快将街道的活力注入场地。公共性的策略将建筑切分成反差极大的两部分，私有化的内部展览被墙体的包裹而固化、公共性的漫步因为开放而虚空。展览馆内部空间在具备独立完备动线的同时，其核心功能与穿越建筑体量的开放动线始终存在着积极的交织，传统的标志性独立入口被分解成更多有趣的进入方式。开放和漫游使得内外动线因为界面开合视线互动而成为不确定行为的事件空间，建筑如同立体主义式的城市街边闲散游园和小广场。

Luoyang Museum
洛阳博物馆新馆

Location: Luoyang, Henan Province
Area: 42,000m²
Architect: Li Li / Tongji University Architectural Design and Research Institute
Photographer: Yao Li
Completion Date: 2009

地点：河南，洛阳
建筑面积：42,000 平方米
建筑师：李立 / 同济大学建筑设计研究院（集团）有限公司
摄影：姚力
建成时间：2009 年

Henan, China
中国，河南

Luoyang Museum, located in the south coast of Luo River (one of the tributaries of Yellow River) and adjacent to the Luoyang Ruins of Sui and Tang Dynasty, is the landmark building on the central axis of Luoyang City. Nowadays, the area where it is sited links the old and new city together.

The concept is based on "asymmetric spatial structure"; courtyard and light well are sited at the transition area, which takes inspiration from garden design, in order to create dynamic balance in spatial layout. The interior design boasts the conceptual theme of "void" by linking different rooms together to correspond with the exterior waving typology. The reproduction and imitation of the 13 archeological sites on the roof shows the profound cultural spirit of the old city and combines the site quality and architectural concept together. The concept goes through the entire design to highlight the characteristics of modern museum and local cultural feature to create a complicated architectural form. The form highlights fusion of building and space, coexistence of exterior solidity and interior void, combination of traditional axis and asymmetrical spatial organisation as well as the intervention of light and space. In addition, the close appearance and open typology contrast each other to boast memorial and public functions.

洛阳博物馆新馆位于洛河南岸，毗邻隋唐洛阳城遗址，是洛阳城市中轴线上极为重要的标志性建筑，以其核心的文化设施集中区将成为衔接、缝合洛阳新老城区的城市重要功能片区。

设计构思以非对称的空间结构为支撑，借鉴园林手法在方形流线的转折位置设置庭院和采光天井，使空间布局达成动态的均衡。设计在外部建构了大尺度的起伏地形，内部则通过建构相对应的一系列空间的连接来暗示"虚空"的概念主题，并通过屋面开放的13个遗址考古场景的再现，深刻地揭示了洛阳这座千年古都的厚重内涵，将场地特质与建筑概念融为一体。设计概念贯穿室内外整体，具有强烈的现代博物馆建筑特征和鲜明的地域文化特色，最终形成了概念复合的建筑特征：建筑形体彰显与空间的沉静融合，外部的凝重与内部的虚空共存；古典的轴线与非对称的空间组织融合，光与空间交织成内在的园林意向；封闭的外表与开敞的地形塑造融合，将纪念性和公共性并置呈现。

Culture Museum of Zhengzhou
郑州中原文化博物馆

Location: Zhengzhou, Henan Province
Area: 3,445m²
Architect: Wu Gang
Photographer: Shu He
Completion Date: 2010

地点：河南，郑州
建筑面积：3,445 平方米
建筑师：吴钢
摄影：舒赫
建成时间：2010 年

Henan, China
中国，河南

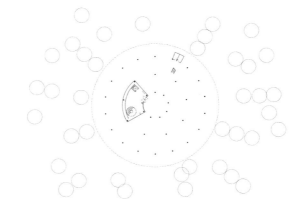

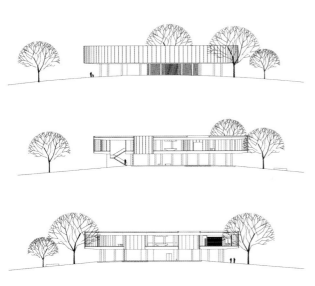

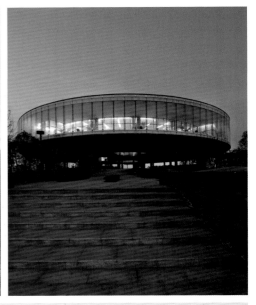

The Culture Museum is located in the prolongation of Tengfei Road in Zhengzhou. It links the north and the south plot in the masterplan, and it is the focus point of this area. The museum echoes the landmark building named "Gate to Zhengzhou" at a distance and emphasises the sequence of the axes.

Surrounded by a large area of urban green space the site neighbours the Qili River and it is rich in landscape resources. The impact of nature is minimised by lifting the building above the landscape.

Being in the main showroom and the offices on the first floor, there are beautiful views into the landscape. Beside the supporting columns, only the reception and the necessary circulation core are on the ground floor. There are separated entrances for staff and visitors. The building appears to be a jade floating above the landscape.

Among the four inner courtyards, some are on the ground floor as the circulation core and day-lighting courtyard; some are on the first floor as landscape courtyards. Thanks to this arrangement, the natural atmosphere is brought into the interior and the indoor ventilation and lighting condition are also optimised. Double-skin façade guarantees a good insulation effect of the building while the space (2.1m in width) between the two layers serves as the closed-corridor exhibition area. Sentences selected from Book of Odes (Shi Jing in Chinese) are carved in the glass façade and have different density according to the functional requirements, concentrated in the structural part and scattered in the area where visitors can see it clear. An interesting interplay of light and shadow is produced by the dramatic shape of the daylight opening on the roof.

This project extends along the Tengfei Road axis of Zhengzhou City. The design concept is derived from Zhongyuan culture, which has been the ancient Chinese culture centre for thousands of years. The building resembles a white jade stone entrapped in a green expanse lying between north and south plots. The Tengfei Road axis becomes emphasised as a local landmark building, and the Zhengzhou Gate, is echoed. The glass façade building appears light as it is nestled into the landscape and embraced with antiquated poetry from floor to ceiling; altogether allowing the interior to be softly lit and the views outward to be clearly enjoyed.

作为建筑师更关注建筑与城市和这片基地的关系。建筑的个性来源于这片地，或者说来源于这个城市以及最后的使用者。这是一片城市中的公共用地。甲方没有什么太多的要求，只是希望它可以和中原的文化，和这个城市的环境有所关系。

中原古代都城的建造秉承"惟王建国，辨方为正，以为民极"的思想，城乡规划上都采用中轴对称的平面布局。中原文化博物馆位于郑州市腾飞路中轴线的延长线上，在布局上正好使南北地块发生联系，犹如一枚晶莹剔透的白玉，镶嵌在南北地块之间大片的城市绿坡上，是处于连接与过渡的关键视觉高潮，与整个地块的标志性建筑"郑州之门"遥相呼应，强调了轴线上的序列感。设计师的目标旨在充分利用周边大块的城市绿地，充分营造新中原文化氛围，使之与总体规划相协调，整个设计旨在体现轻盈的体形，传统的建造，怡人的环境。项目周边为大片的城市绿地，临近七里河，拥有良好的景观资源。设计师的想法是希望尊重这片地，不想让这座建筑阻挡后面大片的景观。所以想到要把建筑举起来，让其成为一个"飘浮的博物馆"。把地抬高，把建筑托举，这样当人们走在这个空间之中，会体会到建筑是与周围风景融合的。通过把地面抬高，把建筑升起，底部架空，降低了对基地植物的破坏，保持了景观的完整性，建筑在占据了这片地的同时又把这片土地还给了整个空间，这是一个很优雅的态度。升起的圆代表了聚会、宏观、希望和圆满。

中原文化非常厚重，观看河南各个博物馆给人的印象也大都如此。设计师认为博物馆里展品才是最重要的，而不是这个建筑。因此这个建筑要"透"，让里面的东西"透"出来。所以设计师把建筑的外面做成了玻璃，玻璃上篆刻《诗经》、具有百叶窗的效果，整体形态犹如一枚晶莹剔透的"悬玉"。

博物馆的展览空间沿周边布置并环绕形成四个不同尺度的开敞式空中庭院。所有的展览空间在不同的方向均可获得开阔的视野及良好的光线，并提供给参观者巨大的视觉享受。

四个内庭院有的位于一层作为交通核和采光庭，有的位于二层作为景观庭，这种立体式庭院将自然气息引入建筑内部，同时也优化室内通风和采光条件。建筑采用双层立面的方式达到良好的保温效果，双层立面之间2.1米的间距利用为环廊式展区。外立面玻璃上的篆刻《诗经》随着功能要求进行疏密变化：在上下建筑结构的部分密集，在中间尤其是人视区域则稀疏。屋顶设计了放射状的采光口，产生丰富的光影效果。

主要展示区以及办公室置于二层，360度都能观赏到外部优美的风景；一层除支撑钢柱外，仅设置了接待区和必要的交通核，并为参观与办公人员分别设置了出入口。

Futian Sports and Entertainment Complex

福田文体中心

Location: Shenzhen, Guangdong Province
Area: 69,059m²
Architect: Nadel Architects
Completion Date: 2008

地点：广东，深圳
建筑面积：69,059 平方米
建筑师：美国纳德华建筑设计公司
建成时间：2008 年

Guangdong, China
中国，广东

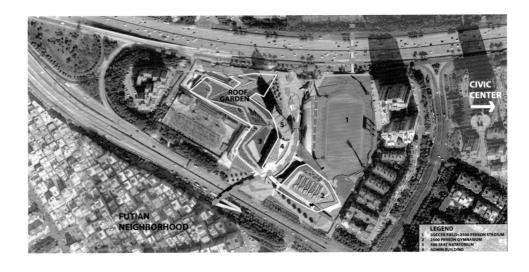

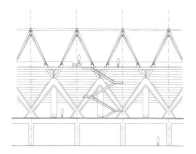

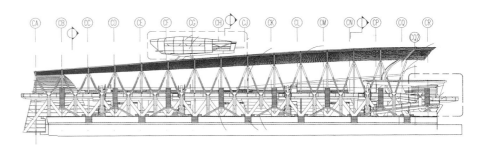

Nadel's core vision for Futian was to provide a connection between two active, dense neighbourhoods, while providing visual drama for the triangular site by utilising a unique set of shapes and forms. The resulting Z-shaped plan restores the urban fabric with juxtaposed rigid and curvy forms that emphasise connections between the neighbourhoods and various venues. Thus, the dynamic interplay of the architectural shapes provides the backdrop for an iconic entertainment and cultural centre for its citizens and tourists.

Since Futian caters to young, athletic people, the design needed to reflect this in its vibrancy. The space-age appearance, with its silver tones and shimmering glass, clearly makes a visual impact. The robust forms also encourage exploration and serve as a reminder that this is a highly accessible community space. Exposed structural design elements, such as steel and concrete supporting glass, bring an unrefined edge that appeals to younger generations who will use the facility. Finally, touches of technology – specifically the giant video screen on the front façade of the hotel, which is used to display athletic events nightly – cap off the youthful sense.

Although sustainability wasn't one of the main goals, the design did benefit from green features. The rooftop park space serves as insulation, regulating temperatures below. The venues take advantage of natural light, and operable windows let air flow through the buildings.

这一项目的设计目标即为在两个充满活力、人口密集的社区之间打造一幢建筑将其连结。通过运用不同的造型及样式，在这一三角形地块上建造了"视觉惊喜"。最终建筑呈现"Z"造型，连续弯曲的造型恢复了城市的地域特色，同时强调了社区之间的紧密联系。总之，相互交错的建筑样式为市民及游客营造了一个独特的运动休闲场所。

鉴于体育中心多吸引年轻及热爱运动的人，因此设计中还应体现出动态感。玻璃材质的银色系外观突出现代风格，注定带来强烈的视觉冲击；裸露的钢筋及水泥等元素带来粗糙的质感；现代化技术设备如酒店建筑外观上的巨型显示屏用于展示体育赛事更是符合年轻人的需求。

尽管可持续发展并不是本项目的目标，但其设计中却体现出绿色特征，如屋顶的公园可行使隔热功能，调节温度；场地充分利用自然光线，而调节的窗户更确保了自然通风。

Bao'an Stadium
宝安体育场

Location: Shenzhen, Guangdong Province
Area: 88,550m²
Architect: gmp Architekten – von Gerkan, Marg and Partners
Photographer: Christian Gahl
Completion Date: 2011

地点：广东，深圳
建筑面积：88,550 平方米
建筑师：gmp 建筑设计事务所
摄影：克里斯琴·加尔
建成时间：2011 年

Guangdong, China
中国，广东

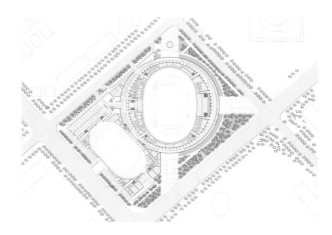

The extensive bamboo forests of southern China were the inspiration for the design. The bamboo look serves two purposes. Firstly, it reflects the character of the region, and thus creates identity. Secondly, it serves as a structural concept for both the load-bearing frame of the stadium stands and the supports for the wide-span roof structure. The outermost part of the stadium unites façade, structure and overarching architectural theme in a single feature. The natural look of the bamboo forest, together with the interplay of light and shadow between the trunks, is interpreted structurally through rows of slender steel supports, as outsize, abstract versions of the bamboo shape.

The stadium is located in the immediate vicinity of a sports arena and swimming bath, which have already established an east-west axis. The stadium and the attached warming-up place fall in with this existing urban axis. The choice of a pure circle for the geometry of the stadium was a decision not to introduce any other geographical orientation into the urban-planning situation, and to emphasise the central character of the sports venue. Appropriately for the uses of the building, the stadium stands on a grassed plinth, which incorporates on the inside the lower tiers of seating and internal functional areas.

Though the supports for the roof structure stand inside the rows of stand supports, they are completely separate from the concrete structure in order to cater for independent movements in the large roof. The steel tubes, which are up to 32m in length, differ qualitatively according to their load-bearing behaviour and function. In diameter, they range from 550mm to 800mm, varying in accordance with their differing static loads. The horizontal stiffening of the structure and drainage of the roof membrane is likewise provided by special supports.

场馆的设计灵感来源于极具华南地区风情的竹林场景，其在重现了华南地域特色的同时还构成了看台以及大跨度屋面的结构支承系统。建筑的外表皮将建筑立面、主体结构以及所运用的象征性建筑语汇整合为一个极具表现力的整体。修长的钢柱在光影中参差交错，如同抽象放大的竹枝，赋予建筑竹林的意象。

体育场紧邻体育中心和游泳馆，位于现有东西轴线之上。体育馆以及其附属的预赛场地的建成更加契合了城市的景观轴线。体育场圆形的几何形式强调了其作为主赛场的核心地位的同时，又避免了在市政规划秩序中引入更多的街道元素。体育场坐落于一个被抬起的平台之上，赛场的下层看台以及一部分内部功能分区被安置于这个平台之内。

建筑屋面的支撑结构位于看台支柱的内侧，构成独立于混凝土看台之外相互独立的结构体系，从而实现了屋面的相对位移，增强了其抗震能力。同为32米长的钢管由于其承重各不相同，管径设计为550~800毫米不等。这种特殊支撑结构系统不仅可以很好的传导水平方向应力，还可一并解决屋面膜结构的排水问题。

Universiade 2011 Sports Centre
深圳 2011 年世界大学生运动会体育中心

Location: Shenzhen, Guangdong Province
Area: 870,000m²
Architect: gmp Architekten – von Gerkan, Marg and Partners
Photographer: Christian Gahl
Completion Date: 2011

地点：广东，深圳
建筑面积：870,000 平方米
建筑师：gmp 建筑设计事务所
摄影：克里斯琴·加尔
建成时间：2011 年

Guangdong, China
中国，广东

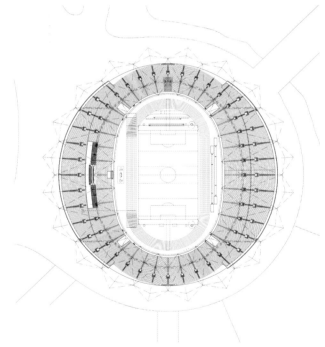

The objective is to create a culturally significant, symbolic project for Shenzhen. Equal importance will be placed on the public facilities in the sports park. When no events are on, they are intended to be available to the public for leisure and recreation.

The new city district offers various sports facilities, residential areas, and leisure and shopping facilities. Both the sports venues – the stadium, multi-functional hall and swimming hall – and the urban areas blend with the landscape at the foot of Mount Tong Gu Ling in the centre of the region. The design is inspired by the surrounding undulating landscape. This enables topographical modulation in the sports centre area, with flows of people on various levels. An artificial lake connects the stadium at the foot of the mountain with the circular multi-functional hall in the north and the rectangular swimming hall west thereof. The central plaza is accessed via a raised promenade from the individual stadia.

The overall complex is laid out as an extensive landscaped park with typical elements of a traditional Chinese garden. Watercourses and plants symbolise movement and development, while crystalline structures in the form of stones and rocks represent continuity and stability. The dialogue between the fluid landscape shapes and the expressive architecture of the stadia constitutes the conceptual framework of the design. The crystalline shape of the three stadia is additionally emphasised by the illumination of the translucent façades at night.

目标即为打造一个具有文化及象征意义的建筑，同时注重中心内公共设施的打造，旨在赛事结束之后为公众提供一个休闲娱乐的场所。 运动中心属于新城区整体规划（运动中心、住宅区、休闲中心及购物区）的一部分。

设计灵感源于四周连绵不断的自然景观，运动馆、多功能厅及游泳中心与其他区域一样与其融为一体。体育中心依地势变化而变化，便于人群流动。体育场位于山脚下，通过人工湖与北侧的多功能厅及西侧的游泳中心连接。中心广场通过被提升的人行道进入。

体育中心犹如一个巨大的景观公园，带有典型的传统中式园林特色。水道以及植物象征着运动与变化，造型好似石块般的透明建筑代表着连续与稳定。流畅的景观与造型感十足的建筑相互"对话"，延续了整体的设计理念。夜晚，建筑外观在灯光的照耀下，更具特色。

Century Lotus Sports Park

世纪莲花体育中心

Location: Foshan, Guangdong Province
Area: 200,000m²
Architect: gmp Architeken– von Gerkan, Marg and Partners
Photographer: Christian Gahl
Completion Date: 2006

地点：广东，佛山
建筑面积：200,000 平方米
建筑师：gmp 建筑设计事务所
摄影：克里斯琴·加尔
建成时间：2006 年

Guangdong, China

中国，广东

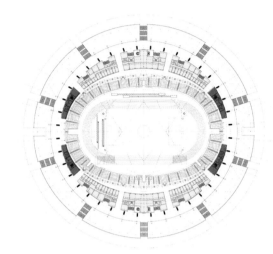

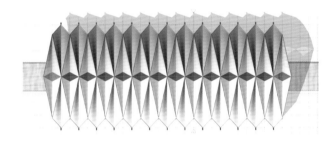

The circular stadium dominates the silhouette of the sports park with its huge, white membrane roof. Being situated on a green hill it looks like a lotus blossom. The stadium's spokes-wheel roof construction with the folded membrane covering measures 350 metres in diameter and covers not just the stands but the outer concourses as well. Above the field the roof can be opened and closed according to demand. With a grand and generous gesture it links the stadium bowl to the surrounding park and turned into a symbol for the games of 2006.

The second venue in the sports park, the swimming pool, reflects the architectural language of the stadium, allowing the two buildings to appear as an ensemble without undermining the unquestioned dominance of the stadium. The translucent membrane roof appears to hover unsupported above the water, and the stands and pools are embedded into the plateau of dike topography. The pools are constructed in a series and set into ground which has been built up to a height of one or two storeys. Unlike the stadium, the folding membrane over the swimming complex is held up by a linear structure of more than 70m long diagonal steel supports, whose tension is taken up by triangular abutments. Supporting cables stretch between these abutments ensuring the stability of the membrane roof.

体育场位于青山之上，看起来犹如一朵盛开的莲花。圆形的造型以及巨大的白色薄膜材质屋顶使其成为体育中心的特色。屋顶呈现车轮形状，直径达350米，一直延伸到广场上方，场地上空的部分可根据需要开启或闭合。宏伟的气势以及大气的形态使其将体育场本身与周围的公园连结起来，并成为2006年体育赛事的象征。

作为体育中心内的另一场馆，在设计上延续了体育场的建筑语言。如此一来，两幢建筑构成一个整体，相互呼应。半透明的白色薄膜屋顶好似"悬浮"在水面之上，而看台以及泳池则恰似嵌入在高原上，泳池由一系列泳道构成。同体育场不同，游泳馆上方连绵的薄膜板通过70多米长的对角线钢结构支撑，而三角形的支柱控制着张力。支柱之间的支撑钢索却抱着屋顶的稳定性。

Atlas of Contemporary Chinese Architecture

Guangdong Xinghai Performance Group Office

广东星海演艺集团办公楼

Location: Guangzhou, Guangdong Province
Area: 13,623m²
Architect: America Teamzero Design & Planning Group
Completion Date: 2006

地点：广东，广州
建筑面积：13,623 平方米
建筑师：美国天作
建成时间：2006 年

Guangdong, China

中国，广东

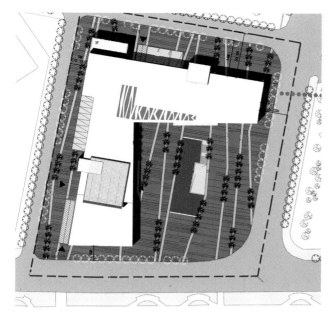

As a landmark on the Er Sha Island, the project is located at the centre of the city. Though as an outbuilding of Xinghai Music Hall, the structure was endowed with characteristics of public cultural architecture and inspired by the "bamboo" in traditional Chinese paintings.

The design of the shape and elevation of the office building started from urban development concept and the surrounding cultural architectures became its background. In addition, architects paid more attention to harmonise with the contour of the Music Hall and buildings on the northern bank of Pearl River visually. The design seeks difference and innovation from the unified structure to make the office to be combined with the Music Hall; the design starts from the texture of the surroundings to emphasise the integrity between form and content, the difference in level and height between individual buildings, and the relationship between exterior skin and interior space. By this, the elevations of the buildings come into shape in a natural way. What's more, the building was given touches of modern artistic style by its abstract and rational structure as well as boasting sculptural sense and strong logic entireness. The common materials were employed. The continuity and solidness of the main structure contrast strikingly with the unique skin resembling the gesture of bamboo created by ancient Chinese scholars.

该项目位于广州的城市中心，同时也是二沙岛上的标志性建筑——星海音乐厅的配楼，具有公共文化建筑的形态特征。方案的艺术构思源于中国传统山水画中"竹子"的形态。

办公楼的整体造型和立面设计从城市设计的角度出发，以周边文化建筑区域环境为背景，重点考虑新建筑与星海音乐厅及珠江北岸建筑轮廓线的关系。在造型和谐中求变化，统一中求创新，使新建筑与音乐厅相互依托、交相辉映。设计从环境肌理出发，注重形式与内容的统一，注重建筑群与建筑个体本身的高低错落，注重建筑表面与内部空间的关系，使建筑立面自然生成。建筑造型以抽象而理性的构图塑造建筑的现代艺术气质，同时注重雕塑感，整体逻辑性强，虚实对比明确，利用普通的建筑材料形成丰富的质感。连续、飘浮的实体的表皮取意于中国传统文人画竹子的抽象形态，形成强烈的标识性。

R&F Centre
富力中心

Location: Guangzhou, Guangdong Province
Area: 157,580m²
Architect: Wei Yeqi
Completion Date: 2007
Client: Guangzhou R&F Properties Co., Ltd.

地点：广东，广州
建筑面积：157,580 平方米
建筑师：韦业启
建成时间：2007 年
业主：广州富力地产股份有限公司

Guangdong, China
中国，广东

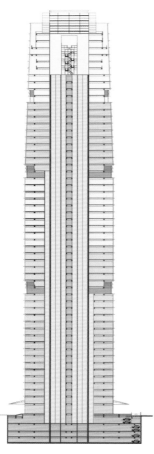

Many architects try to use oriental elements in the design of skyscrapers, but few have succeeded up to now. The project seems to be an exception by taking inspiration from jade craftsmanship with over 2000 years' history in China. The structure resembles jade tower, attaching importance to appropriate scale, spatial balance, simplicity and elegance. In addition, glass as well as welding and sun shading equipment are employed on the exterior wall, rendering the whole building a transparent volume in the day with natural light and at night with lamp light.

There are two-level sky gardens respectively on the 17th and 35th floor while on the lobby of 44th floor, metal louvres are installed on the corners of exterior walls. The roof is planted with greeneries and the glass louvres crown the entire building.

迄今，有不少摩天大厦的设计尝试加入东方元素，但成功的设计案例却不多。富力中心的设计，尝试打破这个定律，从中国2000年历史的玉器工艺中取得灵感，利用建筑展现中国玉花瓶的传统美学比例、型态及质感。富力中心设计成白玉塔，其尺寸比例、平衡、简洁性及优雅，皆是设计重点。同时，大厦亦希望能达到一个通透感，白天因日照而显得明亮，晚间因灯火而变得通明。设计要达到这个效果，外墙巧妙地用了清玻璃、焊接及遮光设备。

18及35楼同时设计了高两层的"空中花园"；四边墙角添加了金属百叶窗，像一条连续的领带，围绕着45楼的空中大堂。顶层设有绿色植物区，玻璃百叶片就像大厦的"皇冠"。

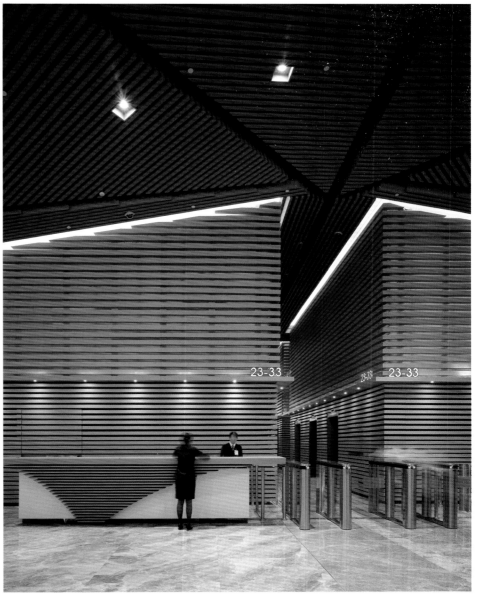

Vanke Centre, Shenzhen
水平摩天楼／深圳万科中心

Location: Shenzhen, Guangdong Province
Area: 121,000m²
Architect: Steven Holl Architects
Photographer: Iwan Baan, Shu He
Completion Date: 2009
Award: AIA New York Honor Award, 2011

地点：广东，深圳
建筑面积：121,000 平方米
建筑师：斯蒂文·霍尔建筑师事务所
摄影：伊万·班、舒赫
建成时间：2009 年
获奖：2011 年美国建筑师协会荣誉奖

Guangdong, China
中国，广东

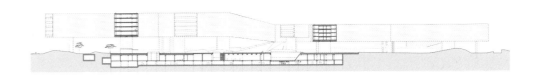

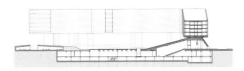 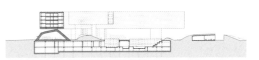

Hovering over a tropical garden, this "horizontal skyscraper" – as long as the Empire State Building is tall – is a hybrid building including apartments, a hotel, and offices for the headquarters for Vanke Co., Ltd. A conference centre spa and parking are located under the large green, tropical landscape which is characterised by mounds containing restaurants and a 500-seat auditorium.

The building appears as if it were once floating on a higher sea that has now subsided; leaving the structure propped up high on eight legs. The decision to float one large structure right under the 35-metre height limit, instead of several smaller structures each catering to a specific programme, generates the largest possible green space open to the public on the ground level.

The underside of the floating structure becomes its main elevation – the sixth elevation – from which "Shenzhen Windows", offer 360-degree views over the lush tropical landscape below. A public path beginning at the "dragon's head" will connect through the hotel and the apartment zones up to the office wings. As a tropical strategy, the building and the landscape integrate several new sustainable aspects. A micro-climate is created by cooling ponds fed by a greywater system. The building has a green roof with solar panels and uses local materials such as bamboo. The glass façade of the building will be protected against the sun and wind by porous louvres. The building is a Tsunami-proof hovering architecture that creates a porous micro-climate of public open landscape.

盘旋在热带花园上空的"水平摩天楼"如帝国大厦高度般的长度，是一个综合建筑群，其功能包括公寓、酒店以及万科集团总部办公室。会议中心、水疗中心和停车场都位于热带花园大面积的绿地下方，景观特点包括土丘下的餐厅和容纳500人的报告厅。

此建筑看似曾经一度漂浮在较高海面上，如今海面已经退去，留下结构高高屹立在八个支撑腿上。在35米限高下抬起一个整体的结构以取代数个小结构体分别满足特定功能，使绿化空间最大限度地敞向地面层公众。

悬浮结构的下方成了它的主立面——第六个立面——深圳之窗在此提供了360度的全景视野，鸟瞰下方苍翠繁茂的热带景观。一个起始于"龙头"的公共通道连接酒店，并由公寓楼划分出办公楼翼。为了适应热带的气候，这一建筑以及景观融合了若干最前沿的可持续性设计策略。由混水处理系统作为支撑的冷水池形成了一个微气候。这栋大楼利用了绿色屋顶、太阳能板以及当地的材料如竹子。大楼的玻璃外墙配置有遮挡太阳以及防风的百叶窗。整个建筑的设计具有抵御海啸功能的盘旋架构，创造了一个多孔的微气候的公共开放景观。

Gemdale Meilong Town, Shenzhen

深圳金地·梅陇镇

Location: Shenzhen, Guangdong Province
Area: 425,200m²
Architect: WSP Architects
Photographer: Yao Li, Shu He
Completion Date: 2007
Award: The International Competition, Implementation Plan; 2006 Architectural Design Innovation Building Award, Planning Innovation Building Award; 2006 Century Architectural Planning and Design Awards, Monomer Design Award, Architectural Style to Create Honourable Mention

地点：广东，深圳
建筑面积：425,200 平方米
建筑师：维思平建筑设计事务所
摄影：姚力、舒赫
建成时间：2007 年
获奖：国际竞赛一等奖，实施方案；2006 建筑设计创新楼盘奖、规划创新楼盘奖；2006 百年建筑规划设计大奖、单体设计大奖、建筑风格创作设计优秀奖

Guangdong, China

中国，广东

This project is located at the Meilong Town 10km away from the central part of Shenzhen. In the overall planning of Shenzhen City, here will be the important accessory life area serving the central part and an important link of Shenzhen central section development axis.

This project will be composed of two plots different in size and complicated in shape. Both the Bulong Road in the north side and the Meilong Road on the west side are the traffic trunk lines of Shenzhen City. The wide Minan Road is between north plot and south plot, and the whole plot is 6-8m higher than surrounding road surface and it's currently a vegetable land.

The design key is to construct a whole dwelling, recreation, and business area. A platform constructed according to the relief will be able to provide space for above three different functions. The edge of the platform will be an open business space, and concave inside courtyard and walk streets in between will provide abundant and proper business opportunity. At the same time, it will be convenient to enter from city roads and community inside. There is a whole "central park" on the plane far from noises of the city, and buildings constructed on stilts will be able to guarantee the continuity of landscape.

High dense residential district adopts thin and deep slab-type apartment building and can provide the uniformity to the maximum and guarantee continuity and maximization of landscape between buildings.

The detailed design of houses has fully considered local lifestyle such as courtyard garden, overhanging balcony, and French bay window and can provide more free spaces for tall-storey residents. The outside flat window is composed of double-layered glass, LOW-E film, and Al alloy sunshade shutter and large-area light colour wall is adopted in consideration of local climate. Partial interspersing colour will make the whole project colourful and also adapt to the future users and make the building mode endure during development.

项目位于深圳市中心区，正北方10公里处的梅陇镇。在深圳市总体规划中，这里是服务于中心区的重要配套与生活区域，是深圳中部发展轴的重要一环。

项目由南北两个大小不等、形状复杂的地块组成，北侧的布龙路和西侧的梅龙路都是城市交通主干道，南北地块之间是一条较宽的民安路，整个地块高出四周路面6~8米，现状为菜地。

设计的核心是建造一个完整的居住、休闲、商业区域。一个依地势而建的平台，提供了以上三种不同的空间：平台边缘是面向城市开敞的商业空间，凹入式的内院和贯穿其间的步行街提供了丰富适宜的商业尺度。同时便于从城市道路和社区内部进入。平面上是完整的"中央公园"，远离城市的嘈杂，建筑底层架空也保证了景观的连续性。高密度的居住采用了薄进深的板楼方式，提供了最大限度的均好性，同时也保证了楼间景观的连续和最大化。

住宅部分的细节设计充分考虑当地的生活习惯，入户花园、挑空露台、落地飘窗，为居住在高层的人提供更多自由的空间。外平窗双层玻璃，LOW-E膜，铝合金遮阳百叶，大面浅色墙体都出于当地气候的考虑。局部点缀的色彩缩小了整个项目的尺度，也和将来使用者的改造融为一体，使建筑的形式在生长中得以持久。

Shangri-La Hotel, Guangzhou
广州香格里拉酒店

Location: Guangzhou, Guangdong Province
Area: 43,000m²
Architect: HBA
Photographer: Shangri-La Hotels and Resorts
Completion Date: 2008

地点：广东，深圳
建筑面积：43,000 平方米
建筑师：HBA 建筑师事务所
摄影：伊万·班、舒赫
建成时间：2008 年

Guangdong, China
中国，广东

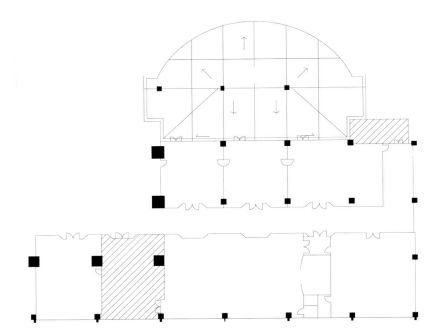

The only hotel located next to the Guangzhou International Convention Centre, Shangri-La Hotel, Guangzhou combines spectacular views of the Pearl River with elegant landscaped gardens, providing an urban oasis for guests' comfort and pleasure.
This luxury hotel offers guests an oasis of warm comfort as they indulge in the lavish accommodation and dining options presented alongside the renowned Shangri-La hospitality.
The Shangri-La Hotel, Guangzhou offers 704 exceptionally spacious and luxurious guestrooms and suites. Suites at the Shangri-La Hotel, Guangzhou, offer some of the largest and most luxurious accommodations in the city, and feature the very highest level of Shangri-La service and elegance.
Shangri-La's Signature CHI, the Spa gives you the luxury of space and timelessness. Eleven spacious and private spa treatment suites are where the ultimate experience of your treatment begins, bringing your senses to a whole new level of rejuvenation. Horizon Club rooms offer an enhanced level of service, comfort, and convenience for busy travellers, including late check-out and a personal concierge.

广州香格里拉酒店是唯一一家坐落在广州国际会议中心附近的酒店,将珠江的壮丽景致与典雅的景观花园结合在一起,为客人打造了一个舒适愉悦的城市绿洲。奢华的住宿环境以及丰富的美食让人流连忘返。酒店共包括704间客房及套房,其中套房格外奢华,提供香格里拉酒店特有的高水准服务。
水疗构成酒店的标志特色,让人感受时空的静止。11间宽敞的理疗套房给人带来独特的感受。俱乐部为繁忙的游客提供便利舒适的服务。

Driving Range – Mayland International Golf Resort
美林国际高尔夫度假村

Location: Guangzhou, Guangdong Province
Architect: Patel Architecture, Inc
Photographer: Patel Architecture, Inc
Completion Date: 2009

地点：广东，深圳
建筑师：美国 Patel 建筑设计公司
摄影：美国 Patel 建筑设计公司
建成时间：2009 年

Guangdong, China
中国，广东

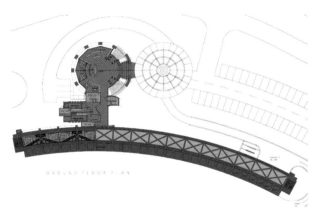

Understanding the sensitive setting for the project, the architect Narendra Patel decided to "reinvent the structure as a landscape pavilion or folly". The idea was to create a light and translucent sculptural building, using tensile fabric structure, instead of solid walls. In addition, the upper floor structure became fairly lightweight, reducing the need for heavy foundations, columns and beams. The architects used concrete for the foundation and second-level deck, but designed the rest of the structure as an exercise in tensile fabric and steel.

The fabric structure designed by G.H. Bruce Inc. is a major feature of the project. In addition to the experiential pleasure that all the Golfers get, it provides shade from the sun. The approach to the tensile structure was to develop sun shade that would attach to the Golf range building and in consistent with the architecture of the entire resort, without having to repeat the same style. Rather, it has its own unique character, more like a sculpture that grew out of the Golf course. The tensile structure design is integrated with the Driving range building. Each module is supported by a cluster of four steel tubes, in such a way that it leaves everything around unobstructed. Modules are interconnected with steel tube arch and the fabric. The play of light and texture throughout the project was inspired by the surrounding mountains, lake, Golf course and its dynamic environment. At twilight, the entire building transforms into a glamorous backdrop for the Golf course, as the whole building comes alive with programmable LED lighting.

设计师在充分了解这一项目所处敏感环境之后，决定打造犹如风景亭或旧时乡间豪华花园中的装饰性建筑一般的结构。最终，他们采用可张拉结构造型代替坚实的墙壁结构，设计了一个轻巧的、半透明雕塑建筑。地基及二层采用混凝土材质建造，而上层结构则尝试采用膜结构和钢材打造，减少梁、柱等承重体的运用。膜结构构成这一设计的主要特色，不仅为球手带来极大的愉悦感同时可用于遮挡阳光。张拉结构的运用旨在提供这样功能，使其与球场建筑以及度假村内其他建筑相连通。除此之外，它更具备自身特色，犹如从球场内"生长"出来的雕塑。每一个模块结构都由一系列钢管支撑，确保四周视野开阔。模块之间通过钢管及织物结构连结。光线以及纹理设计受到周围的山、湖、球场以及其他环境的影响。黄昏时分，整幢建筑形成一个灯光闪烁的背景，格外吸引眼球。

Sheraton Dameisha Resort Hotel
大梅沙喜来登度假酒店

Location: Shenzhen, Guangdong Province
Area: 60,000m²
Architect: Terry Farrell and Partners
Photographer: Terry Farrell and Partners
Completion Date: 2007

地点：广东，深圳
建筑面积：60,000 平方米
建筑师：TFP 建筑设计公司
摄影：TFP 建筑设计公司
建成时间：2007 年

Guangdong, China
中国，广东

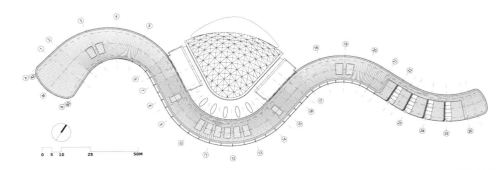

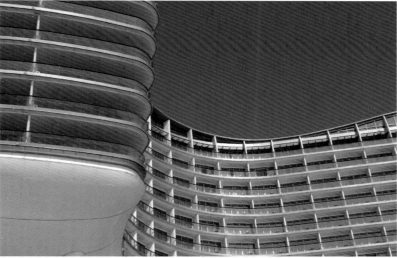

The client's brief is for a destination luxury resort hotel development consisting of 368 guestrooms – each with spectacular ocean views, exhibition and conference facilities, restaurants, wellness spa facilities, fitness centre, swimming pools and two luxury private villas, whilst maximising opportunities offered by the beach topography and magnificent scenery.

The inspiration for the hotel design is derived from the "aura" which is the perceived invisible force surrounding the site. The concept is embodied in the natural and ephemeral qualities of the surrounding geographic and environmental elements which generates a distinctive atmosphere and unique "spirit" and sense of place, referred to in Roman mythology as the protective spirit of a place known as "genius loci".

The natural characteristics of the Dameisha "aura" of the ocean, waves, winds and surrounding mountains act as a framework and are embodied in a building form that has a distinct spirit, a dynamic form, organic shape and a sense of movement, which responds to the client brief, site and in harmony with the natural elements.

The building form, planning and massing follows a sinuous and fluid flowing shape; there are no right angles; everything is set out with curves which create numerous twists and turns with different vistas and views that is reminiscent of a billowing sail, waves, a sine curve and the silhouette of the mountains.

客户要求打造一个奢华度假酒店,其中包括383间海景客房、会展中心、餐厅、健身水疗馆、游泳池以及两幢独立别墅,并充分利用海滩的独特地形以及迷人的美丽景致。

灵感源于其所处场地周围可以感知的元素。设计理念植根于周围地形及环境中的自然因素以及不断变幻的特质,从而营造一种独特的氛围以及地域感,正如罗马神话中提到的"保护地方特色"。

海滩、浪花、微风以及山峦构成了灵感来源的自然元素,并将其作为"主要"框架呈现在建筑造型上,营造出独特的灵魂、活力十足的样式、统一的造型以及动态感,满足客户要求并与周围环境相互呼应。

此外,建筑在造型规划中注重蜿蜒流畅性,摒弃了直角结构。曲线被大量运用,景色随着角度而变化,让人不禁想到前行的小船、跳动的浪花以及山川的剪影。

Dafen Art Museum
大芬美术馆

Location: Shenzhen, Guangdong Province
Area: 17,000m²
Architect: Urbanus Architecture Design
Photographer: Urbanus Architecture Design
Completion Date: 2007

地点：广东，深圳
建筑面积：17,000 平方米
建筑师：深圳都市实践建筑设计事务所
摄影：深圳都市实践建筑设计事务所
建成时间：2007 年

Guangdong, China
中国，广东

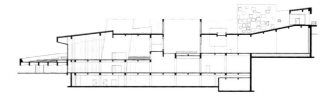

Dafen Village is located in Buji Township, Longgang District, Shenzhen. Best known for its replica oil painting workshops and manufacturers, its exports to Asia, Europe and America bring in billions of RMB each year to the area.

Design concept focuses on reinterpreting the urban and cultural implications of Dafen Village, which has long been considered as a strange mix of pop art, bad taste, and commercialism. A typical art museum would be considered out of place in the context of Dafen's peculiar urban culture. The question is whether or not it can be a breeding ground for contemporary art and take on the more challenging role of blending with the surrounding urban fabric in terms of spatial connections, art activities, and everyday life. Therefore, the strategy is to create a hybridized mix of different programmes, like art museums, oil painting galleries and shops, commercial spaces, rental workshops, and studios under one roof. It also creates a maximum interaction among people by creating several pathways through the building's public spaces. The museum is sandwiched by commercial and other public programmes which intentionally allow for visual and spatial interactions among different functions. Exhibition, trade, painting, and residence can happen simultaneously, and can be interwoven into a whole new urban mechanism.

位于深圳市龙岗区布吉镇的大芬村，是深圳著名的油画产业村，村中遍布油画复制品的创作坊。这里的油画出口到亚、非、欧、美各大洲，每年创造数亿元人民币的销售额。然而，大芬村的油画长久以来是被视为一种低俗艺术，是庸俗品位与商业运作的奇妙混合体。但政府看到这种创意产业的价值，于是在一个似乎最不可能出现美术馆的地方，大芬美术馆出现了。这一决策引发了另一个问题，作为一项政府行为是否能在另一层面上促成当代艺术的介入，并且通过这一公众设施对周边的城市肌理进行调整，使日常生活、艺术活动与商业设施混合成新型的文化产业基地。设计策略是把美术馆、画廊、商业、可租用的工作室等不同功能混合成一个整体，让几条步道穿越整座建筑物，使人们从周边的不同区域聚集于此，从而提供最大限度的交流机会。美术馆在垂直方向上被夹在商业和各种公共功能之间，并且允许在不同的使用功能之间有视觉和空间上的渗透。其结果是展览、交易、绘画和居住等多种活动可以同时在这座建筑的不同部位发生，各种不同的使用方式可以通过不断的渗透和交叠诱发出新的使用方式，并以此编织成崭新的城市聚落形式。

以肯定大芬村油画产业为初衷的大芬美术馆的产生，如果机构化的气氛太浓，或许会不经意地埋葬大芬村油画产业。将美术馆与村落结合，既是一种形式上的调和，也是从本质上让自发的生活形态能在被设计的环境中得以延续和发展的策略。

OCT Art and Design Gallery
华·美术馆

Location: Shenzhen, Guangdong Province
Area: 2,620m²
Architect: Urbanus Architecture Design
Photographer: Urbanus Architecture Design
Completion Date: 2008

地点：广东，深圳
建筑面积：2,620 平方米
建筑师：深圳都市实践建筑设计事务所
摄影：深圳都市实践建筑设计事务所
建成时间：2008 年

Guangdong, China

中国，广东

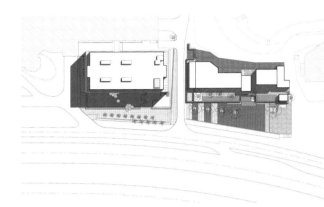

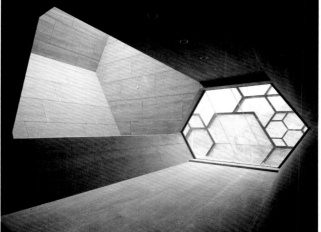

The site has a rather unremarkable history. Originally made for a laundry facility for Shenzhen Bay Hotel in the early 1980s, it is situated along the main road, between a Spanish-style OCT Hotel and the Hexiangning Gallery. Over many years, the warehouse itself remained unaltered while the city around it rapidly transformed. Considering the significance of its location, the owner decides to remodel the warehouse in a meaningful way. For Urbanus, the remodelling of the site poses difficult questions of how to address the existing urban condition, and how new interventions would relate to it. The main architectural gesture is to wrap the entire warehouse with a hexagonal glass curtain wall. The pattern is created from four different sizes of hexagons. As a result, the new wall becomes a lively theatrical screen.

The geometric pattern is more than just surface deep. It is actually a three-dimensional matrix of intersecting elements that project into the gallery spaces, structuring the building's interior design. The result is the creation of delightful and unexpected spatial experiences.

建设中的华侨城洲际大酒店展馆原是建于20世纪80年代早期的深圳湾大酒店的洗衣房。在高速发展的城市中，虽一同并列于深南大道南侧，这座存在于华侨城西班牙风情主题酒店和典雅的何香凝美术馆夹缝中的旧厂房，因其单调的建筑形式早已成为不为人留意的都市残留物。厂房的所属方考虑到其优越的地理位置，决定将其保留并改造为附属于酒店的艺术展馆。它虽是邻近的国家级美术馆为展览空间的延伸，但酒店展馆的特殊定位，决定了改造后展馆的独特性：其设计既要突显个性，与两边建筑风格形成差异性对比，同时也要体现与两端建筑的关系及整体性。改造策略完整地保留了原建筑立面的窗墙体系，加建的立面通过包裹的手法，将单一的原始六边形通过复杂有机的组合形成由实至虚，由小到大，多层次渐变的三维视觉效果。从而，在车辆由西至东快速通过的瞬间，形成强烈的视觉冲击力，通过立面结构的缩小放大，逐层递减，如同面纱般轻轻揭开，最终透出原建筑立面的戏剧性变化过程。

展馆的室内设计再次运用立面所含有的六边形元素作为基本平面形态，在竖向上作90度的拉伸，形成一系列折叠平面互相交叉、互相切入构成的复杂但带着明确功能元素的公共空间。这种表达形式改变了原本单调的立面几何图案，用三维方式生成新的室内空间，这种"突变"形式使设计产生了不可预料的惊喜结果。

Guangdong Museum
广东省博物馆

Location: Guangzhou, Guangdong Province
Area: 67,000m²
Architect: Rocco Design Architects Ltd
Photographer: Almond Chu, Marcel Lam
Completion Date: 2010

地点：广东，广州
建筑面积：67,000 平方米
建筑师：许李严建筑师事务有限公司
摄影：朱德华、马塞尔·拉姆
建成时间：2010 年

Guangdong, China
中国，广东

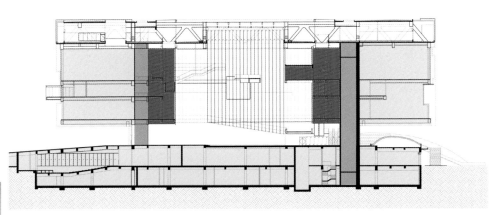

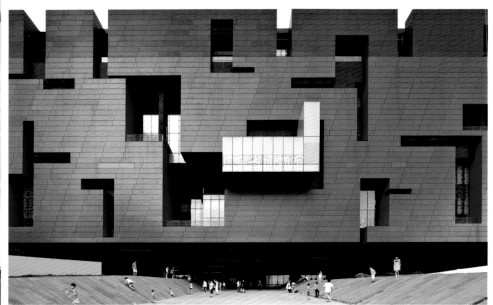

The Museum's spatial arrangement takes its inspiration from the legendary concentric ivory ball carving. Each carving slices through the box and reveals different layers and varying degrees of tranparency within the interiors, forming interesting spatial patterns and luring visitors through its exhibits inside. The interweaving of interior space pockets also reveals the intricate relationship between the visual and physical connections and separation of the atrium corridor, the individual exhibition halls and the back-of-house service areas. This deliberate arrangement not only reinforces the clarity and coherance of the treasure-box concept, but also allows flexibility in planning and operation of all the exhibition spaces. In addition, all the main exhitbition halls are punctured with random alcoves of dynamic spatial geometry. Filled with natural light and served as visual breakouts to the outside, they are also transitions between the exhibition halls which offer visitors initmate and well-balanced resting spaces.

The overall treatment of the main façade is also based on the analogy of ivory ball. Using materials such as aluminium panels, fritted glass and GRC panels, each elevation is uniquely designed with different geometric voids recessed into the building mass. In order to achieve a smooth transition between the museum and the adjoining landscape, an undulating landscape deck is introduced underneath the elevated museum box, metaphorically symbolising a silk cloth unwrapping a much treasured piece of artwork.

广东省博物馆这一项目的空间布局从传奇的象牙球雕刻技术中获得灵感。每个小球从大盒子中"穿过",并展示出其不同的"面"及"透明度",构成趣味十足的表面图案。里面的展品若隐若现,吸引着游客进去参观。内部空间交错连结,在视觉及实体上体现出错综复杂的联系以及中庭走廊、单独展厅及办公区之间的相对独立性。独特的空间格局不仅突出了"珠宝盒"这一设计理念,同时确保了展示空间规划及运用的多样性。此外,主展厅之间通过随意排列而动感十足的"凹室"空间分隔,既能是光线射入,又能与外部空间构成联系,同时为游客提供了温馨的休息场所。

建筑外观设计同样以"象牙球"为理念,采用铝板、玻璃熔块及节能保温板材质打造。不同的立面在设计上呈现差异性,多样的几何造型"凹陷"到主建筑结构上。为使博物馆本身与周围景观环境实现平缓的过渡,设计中在主体结构下方打造了一条蜿蜒曲折的景观带,犹如一块展开的丝绸,将其内部的珠宝呈现。

OCT Design Museum
OCT 设计博物馆

Location: Shenzhen, Guangdong Province
Area: 5,000m²
Architect: Studio Pei-Zhu
Photographer: Studio Pei-Zhu
Completion Date: 2011

地点：广东，深圳
建筑面积：5,000 平方米
建筑师：朱锫建筑设计事务所
摄影：朱锫建筑设计事务所
建成时间：2011 年

Guangdong, China
中国，广东

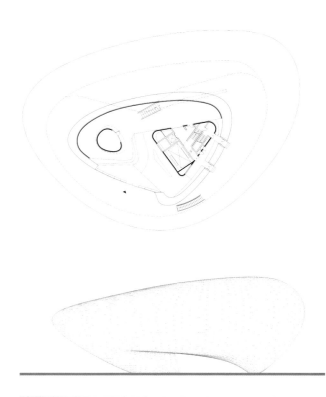

The inspiration for the project comes from both the location being close to the bay and from the needs of the programme, a surreal space for design exhibitions. The OCT Design Museum focuses mainly on fashion shows, product design, and conceptual automotive shows.

The goal was to create a space that is surreal to the subject matter and that brings transcendental experiences. The design of the interior relies on a continuous white curving surface that casts no shadows and has no depth. The result is a surreal borderless space that seems to go on into the infinite, similar to the feeling of a James Turrell installation. The effect is like being in a cloud or dense fog. The building becomes a blank surreal background, with only small triangular windows scattered randomly, as if they were birds in flight. Typically an automobile looks very heavy, but in this limitless space it becomes weightless, letting its curves, shadows, and intense colours become the focal point of the show. The ground floor of the building holds the entry lobby and café, while the first and second is mainly exhibition spaces. Storage space is spread out evenly through the floors, with movable walls allowing the exhibition spaces to be very flexible in scale and function.

The exterior form of the building is a direct reflection of the continuous curving space inside. The smooth organic form has a similar surreal yet transcendental effect when seen outside in its urban setting. Set into its landscape, the building's form seems to float above the ground, as if it was not from this planet. Being 300 metres from the ocean, the architects took inspiration in the smooth stones found along the beach. It is like a purely smooth stone cast into an overly saturated urban setting.

建筑的设计灵感来自于临近海湾的基地位置及其功能需求，一个超现实的空间被用于设计展览。OTC设计博物馆在功能上主要用于概念车、大型产品设计展示、时装表演等。

设计的目的是创造一个超现实主题的空间并且带给人和周围环境超然的体验。室内设计依附于一个连续的、没有投影、没有厚度的白色曲面。从而形成了一个超现实的无边界空间，似乎要走向永恒，让人联想到詹姆斯·特瑞尔的装置艺术。由此产生的效果好似云中雾里。建筑变成了一个极简纯净的背景，仅仅有一些随机散落的三角形小窗户跳跃在其上，好似展翅飞翔的鸟儿。普通认知上的汽车看起来重量感十足，但是在这个无限的空间里它变得没有质量，从而促使曲线、光影和强烈的色彩变成展示的焦点。建筑的首层拥有一个入口大厅和咖啡馆，二层和三层主要用作于展览空间，储藏空间均匀分布在每层之间，利用可移动的墙体来灵活控制展览空间的规模和类型。

建筑的外观直接体现了内部连续的曲线空间。当看到建筑放置在城市布局中时，其光滑有机的造型仍拥有相似的超现实效果。建筑周遭景观就像建筑的形态流动于地面之上，似乎来自外太空一样。在距离海湾300米的基地上，建筑师从散落在海滩上光滑的石头中汲取灵感，建筑就如同一块光滑纯净的石头被投放在过度饱和的城市背景之下。

University Town Library
大学城图书馆

Location: Shenzhen, Guangdong Province
Area: 46,730m²
Architect: RMJM
Photographer: H.G. Esch
Completion Date: 2007

地点：广东，深圳
建筑面积：46,730 平方米
建筑师：英国 RMJM 建筑设计集团
摄影：H·G·埃施
建成时间：2007 年

Guangdong, China
中国，广东

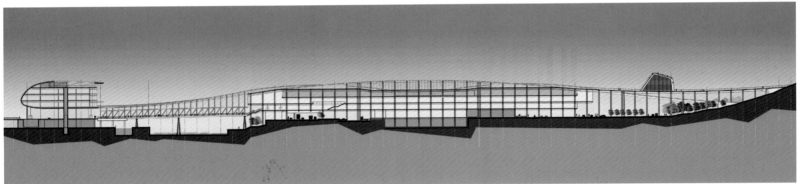

Located in the heart of the campus, the building has the beauty of "crouching tiger and hidden dragon". Taking into account the design concept of "unity and modernity", the long, undulating form of the design echoes the gentle form of the surrounding hills whilst the contemporary materials reflect the erudite language of education and knowledge. The building is at the interface of the universities' campuses and while the entrance plaza, bridge and roof garden provide a series of important public spaces for the entire campus, the structure itself has become a cutting-edge landmark building.

The design promotes a neutral background of light and volumes framing extraordinary vistas of the natural landscape setting. Decoration is kept to a minimum and the space is neatly defined by surface, lines and nodes. The external colours of the building combine white and grey, which makes the building sit harmoniously amongst the green hills and surrounding water. The neutrality of grey allows adjacent colours of the natural environment to stand out gracefully. Modern materials such as glass, aluminum and steel on the exterior, were chosen for their endurance, low maintenance and suitability.

The majority of the building's internal spaces have open plan layouts. Full height curtain walls are utilised on the east and west façades so that natural daylight floods the interior. The interior colour palette takes a combination of grey and white mixed with orange to create a vibrant theme in the library.

图书馆位于园区中心，呈现出"卧虎藏龙"般的美观。其"统一性与现代感"并存的设计理念赋予建筑蜿蜒连绵的造型，映射出周围小山构成的背景环境，当代建筑材质更反映出结构本身的教育性与知识性。图书馆位于大学城内不同校园的连结点上，入口广场、小桥以及屋顶花园为整个园区提供了公共空间，建筑本身则成为了这里的时尚地标结构。

这一设计宣扬素雅的背景以及能够"融合"自然美景的体量结构。装饰被限制在最小程度上，空间由点、面、线等结构打造。外观色彩以灰、白为主，在青山绿水的背景下显得格外宁静。玻璃、铝材及钢板等材质的选用多从其持久性、低维修性以及可持续性出发。

内部空间以开放式格局为主，东、西外立面全部采用玻璃幕墙结构，便于光线照射。室内色彩仍以灰、白为主，些许的橙色增添了空间的动感。

Liberal Art Department, Dongguan Institute of Technology

东莞理工学院文科系馆

Location: Dongguan, Guangdong Province
Area: 9,150m²
Architect: Atelier Deshaus
Completion Date: 2004

地点：广东，深圳
建筑面积：9,150 平方米
建筑师：大舍建筑设计工作室
建成时间：2004 年

Guangdong, China
中国，广东

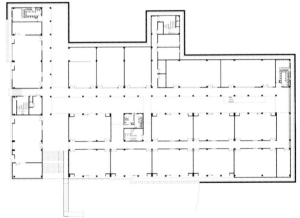

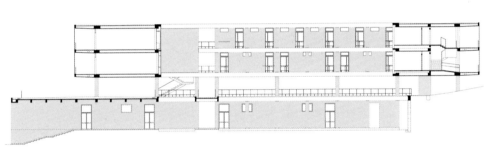

404 / SOUTH CHINA

The ground floor of the Liberal Art Department building is plugged in the hill, and thus the building is melted in the terrain. The ground floor's two sides are opened to the outside and own the nice scenery, while the other part of the building gets the daylight from an "L" shape courtyard. The roof of the ground floor becomes an open platform in rolling hills, while the elevated square building stands above the platform. The platform, courtyard and corridors together create a centralism of outdoor public square. In the main direction of walking, a strong feeling of block building clearly faded away, which reduced the building's pressure to the environment. When eyesight crosses the elevated building and falls on the hills behind, people may recall the original image of this area.

文科系馆首层的建筑被半插入山体，建筑的部分体量由此成为地形的一部分，因为地势的原因，首层建筑仍有两面是开敞的，并面对很好的池塘景观，其他部分则通过一个L形的内院采光。首层屋顶在起伏的地形中形成了一块开放的平台广场，上部建筑则呈正方形围合并架空在平台层上，这样，内院、围廊、平台共同形成了一处具有中心感的公共开放空间。在人流的主要行进方向，建筑的体量被明显化解，有效地减轻了建筑对于环境的压力，当视线穿越架空层而停留在后面或远处隆起的山坡上时，人们大致可以想象或回忆起这处地形原本的模样。

The New Conghua Library
从化市图书馆新馆

Location: Conghua, Guangdong Province
Area: 10,000m²
Architect: Xiao Yiqiang, Liu Huixie, Shi Liang, Qi Baihui
Completion Date: 2010

地点：广东，从化
建筑面积：10,000 平方米
建筑师：肖毅强、刘穗杰、施亮、齐百慧
建成时间：2010 年

Guangdong, China
中国，广东

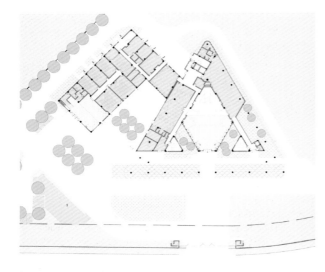

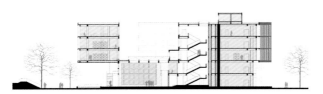

This project is the new city library in Conghua, a distric-level city of Guangzhou. Situated in the northeast of the city, near the highways and the Liuxi River which runs through the city, the library is planned to cover an area of 20,000 m² with the first phase finishing 10,000 m². The usage, construction duration, and sustainability of the functions as well as other elements of the project are taken into thorough consideration, as the construction is going to be carried out in several phases. There is a clear distinction in the architectural forms between the first and second phase of the construction. The first phase focuses on two sections of the structure: the main building of the library and the lecture hall for cultural activities and training purposes.

The construction site is next to the banks of Liuxi River, and the body of the project is designed on the first floor for a better view of the river landscape. The main building of the library has a lobby at its bottom, where the yard overhead and the platform on the roof form the layout of a three-dimensional courtyard. The irregular lands are well utilised as the main building and the lecture hall forming a courtyard. The library building serves as both the traffic centre and the service spaces, providing flexibility and adaptability in case of function changes.

The cost of construction including basic decoration is quoted at only 2,000RMB per square metre. The concept of economy and energy-saving is revealed through the design: the building façade uses double-layer surface, which offers satisfying performance in heat shielding by a combination of vertical and horizontal boards; natural ventilation is achieved by the enclosing yard.

项目为广州区级市从化市的图书馆新馆。位于城市东北部，临近穿越城市流溪河及广东省绿道。规划新馆规模为20000平方米，首期建设10000平方米。因项目分期建设，设计充分考虑了建筑的使用、建设周期及与功能可持续性等因素。将一、二期的建筑形态作了明显的区分。首期建筑包括两部分，分别为图书馆主楼和作为文化活动功能的报告厅及培训功能。

建筑用地紧临高出的流溪河河堤，设计将建筑主体架空于两层以上，使得建筑的阅览空间得到良好的沿河景观视野。图书馆主楼下部设置为门庭，架空庭院与内庭的屋顶平台形成了一个立体化庭院的布局。主体建筑与报告厅形成一个庭院，较好地处理了不规则的用地关系。图书馆主楼平面集中交通核心和服务用房，使得平面具有良好的灵活性，适应建筑的功能变化需要。

建筑含基本装修造价仅为2000元/平方米。建筑师通过设计方式体现建筑的节约和节能。主楼立面采用双层外表面处理，通过水平板和垂直板的组合，形成良好的防热效果；平面围合成内庭，形成良好的建筑自然通风条件。

Guangzhou Opera House
广州歌剧院

Guangdong, China
中国，广东

Location: Guangzhou, Guangdong Province
Area: 73,019m²
Architect: Zaha Hadid Architects
Photographer: Christian Richters, Iwan Baan, Simon Bertrand
Completion Date: 2011

地点：广东，广州
建筑面积：73,019 平方米
建筑师：英国扎哈·哈迪德建筑事务所
摄影：克里斯琴·里克特、伊万·班、西蒙·伯特兰
建成时间：2011 年

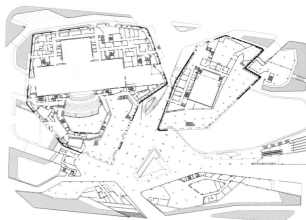

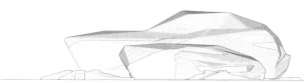

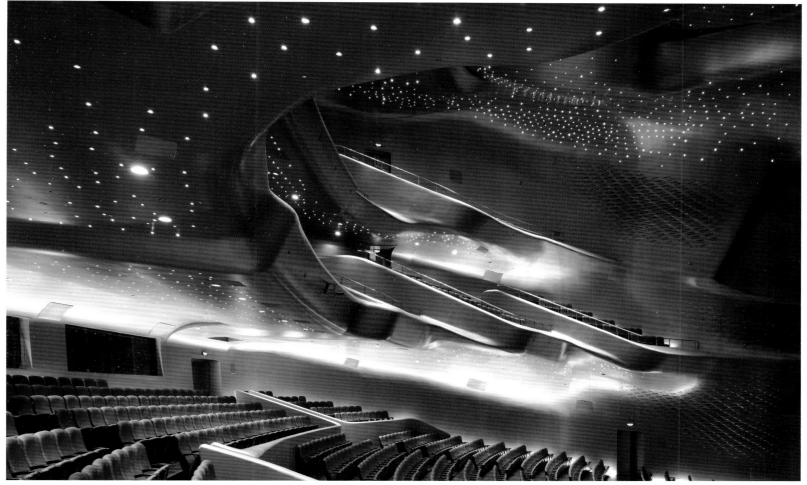

The Opera House is at the heart of Guangzhou's cultural development. Its unique twin-boulder design enhances the city by opening it to the Pearl River, unifying the adjacent cultural buildings with the towers of international finance in Guangzhou's Zhujiang New Town.

The design evolved from the concepts of a natural landscape and the fascinating interplay between architecture and nature, engaging with the principles of erosion, geology and topography. The Guangzhou Opera House design has been particularly influenced by river valleys, and the way in which they are transformed by erosion.

Fold lines in this landscape define territories and zones within the Opera House, cutting dramatic interior and exterior canyons for circulation, lobbies and cafés, and allowing natural light to penetrate deep into the building. Smooth transitions between disparate elements and different levels continue this landscape analogy. Custom-molded glass-fibre reinforced gypsum (GFRC) units have been used for the interior of the auditorium to continue the architectural language of fluidity and seamlessness.

The Guangzhou Opera House has been the catalyst for the development of cultural facilities in the city including new museums, library and archive. The Opera House design is the latest realisation of Zaha Hadid Architects' unique exploration of contextual urban relationships, combining the cultural traditions that have shaped Guangzhou's history, with the ambition and optimism that will create its future.

广州歌剧院位于广州文化开发区中心，其独特的外形如"圆润双砾"，就像置于平缓山丘上的两块砾石，在珠江边显得十分特别，并将与之相邻的文化建筑与珠江新城内的国际金融中心大厦统一起来。

其理念源于自然景观以及建筑与自然的相互联系，运用冲蚀、地质及地形原则。建筑设计尤其受到周围溪谷环境的影响，并着重诠释出溪谷经过冲击侵蚀之后呈现出的状态变化。

景观设计中的"折叠线"确定了剧院内部的形态以及空间分布，形成了流通线、大厅、咖啡厅等区域，同时将自然光线引入进来。独立元素以及不同楼层之间平缓过渡，同样延续了景观设计理念。特制的玻璃纤维石膏板用于装饰礼堂内部，体现了流畅及连续的建筑语言。

广州歌剧院已成为城市文化设施发展的催化剂，更实现了设计师对于城市与环境之间的联系的探索，将广州历史上的文化传统与未来发展理念完美结合。

Canton Tower
广州塔

Location: Guangzhou, Guangdong Province
Area: 114,000m²
Architect: Mark Hemel, Barbara Kuit / Information Based Architecture
Completion Date: 2010

地点：广东，广州
建筑面积：114,000 平方米
建筑师：马克·海默尔、芭芭拉·库伊特 / IBA 建筑事务所
建成时间：2010 年

Guangdong, China
中国，广东

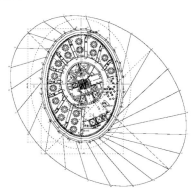
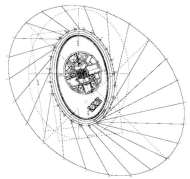

410 / SOUTH CHINA

Besides making a good, nice, functional building, the architects search to capture the essence of contemporary culture. The idea of the tower is simple. The form, volume and structure is generated by two ellipses, one at foundation level and the other at a horizontal plane at 450 metres. These two ellipses are rotated relative to another. The tightening caused by the rotation between the two ellipses forms a "waist" and a densification of material.

The structure consists of an open lattice-structure, built up from 1,100 nodes and the same amount of connecting ring and bracing pieces. The resulting structure can be seen as a giant three-dimensional puzzle of which all 3,300 pieces are totally unique. Recent State of the Art fabrication and computerised analysis techniques allow designers to create much more complex structures than ever before, and therefore wander into a new realm of "informed" structures.

The Canton Tower consists of 88 floors of which 37 floors hold public programme like exhibition spaces, a conference centre, a cinema, two revolving restaurants and one VIP-restaurant, various cafés and observation levels. A deck at the base of the tower hides the giant building's infrastructural workings. These levels support other facilities as well, including exhibition spaces, a food court, extensive commercial space, a 600-vehicle parking and tourist coaches. Panoramic double-decker lifts serve both entrance levels.

设计的目标除了要打造一幢美观、实用的建筑之外，更应该抓住当代文化的精髓，使其在建筑中得以体现。广州塔的设计理念比较简单，其造型由两个椭圆结构决定，其中一个从地面层拔地而起，另一个在平面之上450米高度处与之"会和"。两个结构相对旋转，连结的位置形成建筑的"腰部"。

整幢建筑包含一个由1100个节点以及同等数量的连接环和固定装置打造而成的开放式格子结构，犹如一个巨大的三维迷宫，3300块部件更是别出心裁。现代构造及计算机分析技术让设计师有机会创造更为复杂的结构，带人进入"信息王国"。

广州塔共为88层，其中37层用作展览、会议中心、电影院、旋转餐厅、VIP餐厅、咖啡厅及观景台等公共空间。除此之外，还包括美食广场、商业空间、停车场以及观光服务区等。基础设施全部"隐藏"在建筑基座上的甲板内，双层观光梯供进入塔内人员使用。

Guangzhou South Railway Station
广州南站

Location: Guangzhou, Guangdong Province
Area: 210,000m²
Architect: Gavin Erasmus, Stefan Krummeck
Completion Date: 2010

地点:广东,广州
建筑面积:210,000 平方米
建筑师:加文·伊拉斯玛斯、斯蒂芬·克鲁美克
建成时间:2010 年

Guangdong, China
中国,广东

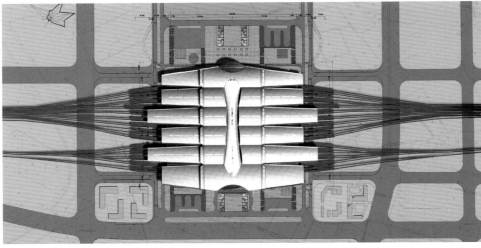

The main building of the station is a four-storey building, which includes one underground floor and three floors above ground. The underground floor covers 930,000 square metres and consists of the parking lot with 1,808 parking positions and the entrances and exits for the subways. Above ground, the ground floor is the transportation level, which mainly handles passengers to/from Wuhan, Zhuhai, and Shenzhen. There are also taxi stations, Port of Entry, Joint Inspection Hall, and the Joint Transportation with Baiyun International Airport. The first floor is the VIP Waiting Rooms, the platform, and the Station Hall. The second floor is a 71,722-square-metre (85,778 square yards) waiting room including a 14,694-square-metre commercial area.

Guangzhou South Station emerges as one of the largest railway stations in the world, roughly three times the size of London's Kings Cross Station. Designed for a daily turnover of 300,000 passengers (by 2030), a clear vertical organisation strategy was adopted to assist in orientation and to accommodate the substantial volume of commuters. Located in-between the Departures Concourse Level and the Arrivals Level are the 28 elevated viaduct platforms for high-speed trains. A contemporary design that incorporates clear wayfinding, new technology and environmental features, Guangzhou South Station's unique and innovative design will also promote rapid development of the regions' economy.

广州南站主楼共为四层，其中地下一层，地上三层。地下一层面积达 93 万平方米，包括 1808 个停车位以及地铁站出入口。地上一层用于交通枢纽，设有出租车乘车站、联检大厅以及白云国际机场乘车站；二层包括 VIP 等候区、月台及候车大厅；三层设有面积达 71,722 平方米的候车大厅，其中包括 14,864 平方米的商业空间。

作为目前世界大规模火车站之一，广州南站在面积上是伦敦国王十字车站的 3 倍。预计到 2030 年，车站每日输送乘客人数将达到 30 万，为此其在设计上强调垂直空间架构，以打造明确的指示标识，同时为大量的通勤游客提供方便。出站层与进站层之间设有 28 个高架桥月台，共高铁乘客使用。集清晰指示系统、全新技术以及环境特色于一身的广州南站将带动这一区域经济的发展。

Dongguan Toy Factory
东莞玩具工厂

Location: Dongguan, Guangdong Province
Area: 15,000m²
Architect: Liu Yuyang, Larry Tsoi, Charles Lam
Completion Date: 2007

地点：广东，东莞
建筑面积：15,000 平方米
建筑师：刘宇扬、蔡晖、林秋雷
建成时间：2007 年

Guangdong, China
中国，广东

414 / SOUTH CHINA

This is an extension of the Hong Kong story: the context is the extremely banal industrial zone in the Pearl River Delta; the subject is the typical factory architecture; participants are local builders; the producer is a pragmatic Hong Kong industrialist; the script writer and director is an architect in search of the extraordinary energy embedded within the banality of the everyday practice – in life and in architecture. The project investigates the possibilities and variations of generic architecture through the research, design and construction of a toy factory building located in Dongguan, PRD.

Located in a city renowned as "world factory", the project aims to design a five-storey building for a world-class toy manufacturer as assembling workshop and storage space. The challenges lie in the requirements that the building should integrate practicability, durability and playfulness as well as in accordance with the overall image (a team composed of diligent administration staff and perfectly-skilled workers). In addition, the construction cost per square metre should not be more than 1,000 RMB.

The project investigates the possible mutations of generic industrial architecture in the Pearl River Delta. Exterior walls are clad with typical tiles, through colours and pattern variations, transforming the perception of cheap materials. A new sky-bridge diagonally inserted between the old and new part of the factory results in an interstitial outdoor balcony with acid-etched glass floor, in-planting in the rational structure momentous flow, suspension, and spatial rhythm that changes with the natural light.

这是一个从香港延伸出去的故事：它的背景是极为普通的珠三角工业区，主体是典型的工厂建筑，参与演出的是当地的施工队，制片是务实的香港实业家，编剧和导演则是建筑师，创作的过程中尝试发掘平凡建筑中的不平凡能量。作品通过研究、设计和建造一座位于东莞的玩具厂房，探讨通属建筑的可能与变奏。

本项目为一家世界级的玩具制造厂提供一栋五层楼的装配与储存空间。基地位于具有"世界工厂"之称的东莞。设计的挑战性在于：建筑需兼备趣味性、实用性、耐久性，而且要符合工厂的整体形象——务实的管理层和熟练的工人所组成的团体，并维持每平方米1000元人民币的低廉造价。

作品探讨珠三角工业城市中，通属建筑的可能与变奏。外立面使用的是极为普通的面砖，但透过在颜色和排列上的差异性，转换对廉价材料的既有印象。同时，新旧厂之间的连接天桥和斜角阳台扰动了原有的矩形结构，为工业厂房重新植入流动、停顿和随着自然光变化的空间韵律。

InterContinental Sanya Resort
三亚洲际度假酒店

Location: Sanya, Hainan Province
Architect: WOHA
Photographer: Ken Seet
Completion Date: 2011

地点：海南，三亚
建筑师：WOHA 建筑设计公司
摄影：肯·希特
建成时间：2011 年

Hainan, China
中国，海南

InterContinental Sanya Resort occupies one of the most beautiful locations in Sanya – Xiao Dong Hai, a mere 5 minutes away from CDF and the city of Sanya and a 25-minute drive from the Sanya Phoenix International Airport. Surrounded by pristine white sands and majestic mountains, this resort offers guests inspiring views from every room. Beautiful water features and water gardens throughout the resort create a serene yet stylish ambience.
InterContinental Sanya Resort offers 343 rooms including 24 beachside villas. All the villas feature an outdoor dining area, a private swimming pool, a pool deck, an outdoor shower and a covered Cabana with day beds and a garden that's just perfect for leaving the world behind. Designed with a wonderfully refreshing and exotic Orchid theme, the 8 double spa villas let one enjoy relaxing outdoor showers amidst beautiful, lush landscaping.

三亚洲际度假酒店选址在风景秀丽的小东海，距三亚离岛免税店仅为5分钟的路程，到达凤凰国际机场只需25分钟车程。酒店被原始的白色沙滩和壮丽的山峰环绕，从每间客房内都能欣赏到外面的美丽景致。此外，漂亮的水景以及水上花园更是营造了恬淡而时尚的氛围。
酒店共343间客房，24栋滨海别墅，每栋别墅都设计有室外就餐区、私人泳池、室外淋浴以及私人花园等。其中的8栋水疗别墅采用清新的田园设计风格为主题，让客人在大自然中享受水疗的乐趣。

Creative Media Centre, City University of Hong Kong
香港城市大学创意传媒中心

Hong Kong, China
中国，香港

Location: Hong Kong
Area: 30,425m²
Architect: SDL
Photographer: SDL Photography
Completion Date: 2010

地点：香港
建筑面积：30,425 平方米
建筑师：SDL 建筑事务所
摄影：SDL 摄影工作室
建成时间：2010 年

The Creative Media Centre for the City University of Hong Kong provides facilities that will enable the University to become the first in Asia to offer the highest level of education and training in the creative media fields. The building will house the Centre for Media Technology and the Department of Computer Engineering and Information Technology.

The distinctive crystalline design will create an extraordinary range of spaces rich in form, light, and material that, together, will create an interactive environment for research and creativity. Internal activity spaces have been designed specifically to encourage collaboration through openness and connectivity. The Centre will also serve as an exciting place for visitors, who will be welcomed to enjoy the facilities as part of an extended public outreach programme of courses and events.

The facility will also include a multi-purpose theatre, sound stages, laboratories, classrooms, exhibition spaces, a café and a restaurant. Secluded landscaped gardens to the north of the building will be available for students and the general public alike.

香港城市大学创意传媒中心将成为亚洲第一个在创意传媒领域提供最高水准的教育及培训课程的机构，包括传媒技术中心以及计算机工程与信息技术中心。

水晶般的外形格外引人注目，内部空间在造型、光线以及材质等方面营造了一个互动性十足的环境，利于学术研究及创意活动的开展。此外，内部空间在设计过程中更注重营造开阔感与连通性，促进学生的合作交流。该中心同时对外开放，让游客体验公共拓展项目课程及活动。

此外，中心内还包括多功能剧院、实验室、教室、展厅、咖啡厅及餐厅等。建筑北侧的封闭景观花园面对学生及公众开放。

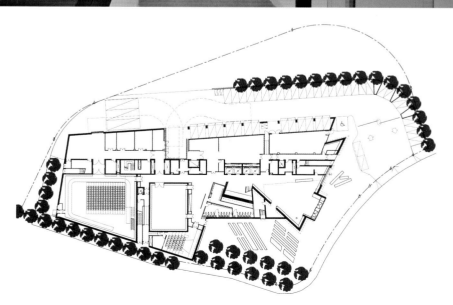

Hong Kong Polytechnic University – Hong Kong Community College

香港理工大学——香港专上学院

Location: Hong Kong
Area: 31,600m²
Architect: AD+RG Architecture Design and Research Group Ltd.
Photographer: Wong Wing Fai, Tang Ka Fai
Completion Date: 2008

地点：香港
建筑面积：31,600 平方米
建筑师：AD+RG 建筑设计及研究所有限公司
摄影：王永辉、邓家辉
建成时间：2008 年

Hong Kong, China
中国，香港

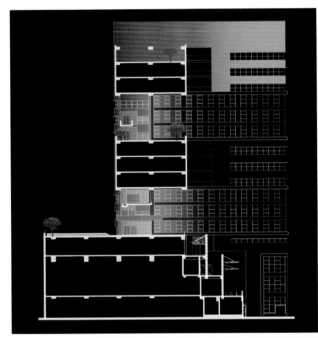

A new interpretation of campus environment in a multi-storey architecture is explored and realised in the project.Due to the limited land resources, the Hong Kong Community College (West Kowloon Campus) aims at experiencing the transformation of the campus communal space design, which is usually, provided in the form of garden on lower floors, into major sky decks similar to high-rise landscape gardens. With associating activities like canteen and student union facilities, campus atmosphere is provided to the sky communal decks. They form the nodal points in the upper campus, which are also proudly visible in the architectural form of the building. The two sky decks are provided to create a sense of community and a place for students' gatherings.

Pre-cast beam and semi-precast slab systems are adopted for the project. This helps to minimise waste during construction and facilitate the construction progress. The twin towers and the connecting structures are carefully articulated. Solidity and transparency are the key architectural manipulation for such articulation. "Red-brick" tile is adopted for the solid twin towers while aluminium cladding with glass wall are employed for the transparent connecting bridge structures. This layering puts a great emphasis on the connection between the two towers. The twin tower design can minimise the plan depth of each tower and so facilitate the penetration of natural ventilation and lighting into the tower campus.

The building design encourages embrace of natural environment and use of environmentally friendly materials. Innovative building technologies are used in the design with close collaboration with academics from the Hong Kong Polytechnic University in order for the building to achieve energy efficiency and a healthy community.

校园善用空间布局，以高层建筑作垂直发展，于课室、饭堂及图书馆等公用设施附近设置平台花园，为专上教育院校带来全新面貌。为了在有限的空间之上，争取最多的可用空间，香港专上学院（西九龙校园）重新设计空间布局，将原本的低层平台花园向高空发展，成为空中花园。并于高层空中花园附近加入饭堂和学生会设施等聚集人气的活动，有效地将人气汇聚到校园的高层平台，符合垂直发展的建筑理念。另外，大楼上的两个空中花园不但能够营造理想校园环境，鼓励学生主动学习，增强归属感，更是师生交流及讨论的好地方。

校园的大楼采用双塔拱形设计，有效减少大楼的深度，一方面可以使室内空间摄取足够天然光线，加强对流通风，另一方面又可以避免景观受到旁边建筑物的遮挡。校园于选材方面经过精心考虑，务求做到实体与透明的视觉对比。两座大楼主要运用红砖外墙以增加实体的感觉；而连接两座大楼的拱桥则以金属外层加上玻璃外墙覆盖，以加强透明感，与大楼的实体布局形成强烈对比，更加强了大楼之间的连贯性。

香港专上学院（西九龙校园）的建筑设计尊重自然环境，采用环保物料，符合可持续发展的建筑理念。校方更与香港理工大学的学者教授紧密合作，运用创新的建筑技术，达到能源效益，共建环保小区。

The Hong Kong Design Institute

香港知专设计学院

Location: Hong Kong
Area: 42,000m²
Architect: The CAAU Studio
Photographer: Sergio Pirrone
Completion Date: 2010

地点：香港
建筑面积：42,000 平方米
建筑师：CAAU 设计工作室
摄影：塞吉奥·皮尔尼
建成时间：2010 年

Hong Kong, China

中国，香港

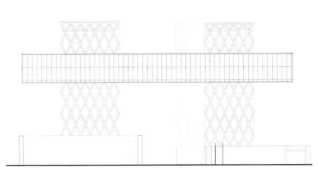

The architectural project metaphor for creativity about to burst forth, the "Blank Sheet" expresses the project's intentions: bringing together and then presenting the multidiscipline nature and targets of the future Institute of Design. In concrete, glass and steel, its radical architecture, light and transparent, invites one to reflect on the combination of multiple and opposing situations: introversion and extroversion, modesty and exhibition, exclusivity and wide accessibility, micro and macro city, classicism and experimentation.

Each functional element, first decomposed, amalgamates and interpenetrates or cuts itself off, by offering the project an immediate clarity from the outside which is very resonant in the city. The flexible and evolutionary plan allows one to envisage future liaisons with the neighbouring campus, LWL.

The base of the building, the giant "urban lounge" favours meetings and exchanges, whilst taking advantage of internal and external green spaces and views of the countryside, thus fulfilling the liaison with the city.

The podium, whose gentle slope stands 7m below the King Ling Road, designed as a landscaped extrusion of the ground, directly linked to the urban environment on two levels – a common space and at the same time an external gallery – is characteristic of Hong Kong infrastructures. Open, sheltered by the platform above, it can host multiple events. The podium is made up of four auditoriums, a café, a space for exchanges with the design industry, a sports hall and an exhibition hall. For the roof, an urban park and sports grounds are available to the students and visitors from nearby. The large auditorium, with capacity for 700 seats, is intended to host conferences, seminars or classical music concerts, but also more recreational activities, fashion shows, pop music concerts, and contemporary dance spectacles.

这一建筑项目隐喻了即将爆发的创造力，"白纸"表达了建筑的目的：汇集并展现香港知专设计学院的多领域的特点与未来的目标。混凝土、玻璃与钢铁组成了这座特立独行的建筑，灯光和通透的设计，多种形式与对立环境的组合引人深思：内敛与张扬，谦逊与昭示，独占性与广泛的包容性，微观与宏观的城市以及古典与探索的风格。

每个功能性的元素，首先进行分解，之后加以混合，嵌入又或是自我割裂，由此为这一工程带来一个直接清晰的外观，与这座城市相呼应。灵活且渐进式的规划使其能够在未来较好的与附近的LWL校园相联系。这栋大楼的基地，那巨大的"城市休息室"可做举行会议与交流互动之用，巧妙的利用内部和外部的绿色空间以及乡村的景色，实现了建筑与城市之间的联系。

建筑的平台，缓缓的倾斜到金陵路7米以下的地方，设计成为延伸到地面上的一道景观，直接的从两个层次上连通到城市环境中———一个公共的空间同时也是室外画廊，这是香港基础设施的特点之一。它是开放式的，上方有平台的遮蔽，能够举办多种活动。平台由四个不同功能的区域组成，它们是一家餐厅，一个设计行业的交流空间，一间体育场以及一个展览厅。而屋顶则可以作为一个城市公园以及操场，向附近的学生以及游客开放。其中的大宴会厅能够容纳700人，原本的设计旨在举行会议、研讨会或是古典音乐会，但同时更多的是举办一些更为娱乐性质的活动、时装秀、流行音乐会、现代舞蹈表演等。

Clinical Science Building at Prince of Wales Hospital for The Chinese University Hong Kong

香港中文大学威尔斯亲王临床医学大楼

Location: Hong Kong
Area: 14,930m²
Architect: WMKY
Photographer: Jermaine Tsui
Completion Date: 2007

地点：香港
建筑面积：14,930 平方米
建筑师：香港 WMKY 设计公司
摄影：杰梅恩·徐
建成时间：2007 年

Hong Kong, China
中国，香港

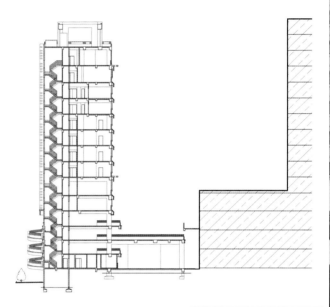

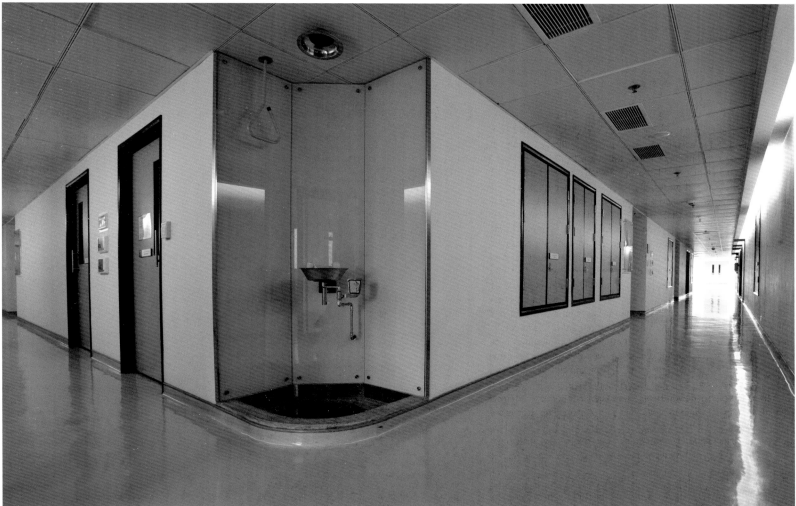

The project is to design an extension to the existing Professional Block of the Medical Department of the University. The new Clinical Science Building is a 13-storey complex of 14,930sqm in floor area accommodating 50 nos. of carparking spaces, seminar rooms, a Pathology Teaching Lab., and various research departments.

The new glass bridge links between the existing clinic science building and the new extension, providing a functional physical and visual transparency of flow between the two buildings. The interaction between inside and outside is enhanced by the use of the glass bridge.

The articulation of the exterior elevation modular spacing levels expresses the common form and structural element without detracting the building from its surrounding environment.

The specific briefing of the required areas of various departments is achieved by the versatile L-shaped rectilinear block form plan, also mimic neighbouring environment and addresses the attached spiral car parking access below, leading up to the internal lobby of the building.

As one by passes via the public road below, the vertical ladder-like feature elevation of the building draws your attention, highlighted and contrasted by the horizontal shading devices.

该工程是为香港中文大学原有的专业医学系大楼设计的扩建项目。新的临床医学大楼高13层，占地面积14,930平方米，容纳了50个停车位、研讨室、一间病理教学实验室及各种研究部门。

一座新建的玻璃桥廊连接了原有的科学大楼和新的扩建大楼，为两栋大楼提供了物理性和视觉通透性上的连接。通过这一桥廊的使用，增强了两栋大楼内部与外部上的互动性。

楼体外部立面的模块间隔度清晰，展现了普遍的形态与建筑上的细节，但又不会被周围的环境削弱了大楼的存在感。

通过使用L形的直线分块布局实现了每个系部所占的区域的明确的要求，同时也仿效了周边环境，在下方建造了螺旋形的停车场通道，连接到大楼内部大堂。

行人从大楼下方的路上经过时，目光会被扶梯状的建筑立面所吸引。垂直的"扶梯"与水平的这样装置相辅相成。

The Arch
凯旋门

Location: Hong Kong
Area: 100,000m²
Architect: Grace Cheng
Photographer: Marcel Lam
Completion Date: 2006
Award: 2008 Quality Building Award (Residential Category) – Winner; 2005 Emporis Skyscraper Award – Runner-up

地点：香港
建筑面积：100,000 平方米
建筑师：格瑞丝·程
摄影：马塞尔·兰姆
建成时间：2006 年
获奖：2008 年质量建筑奖（住宅类），2005 年安玻利斯摩天大楼奖第二名

Hong Kong, China
中国，香港

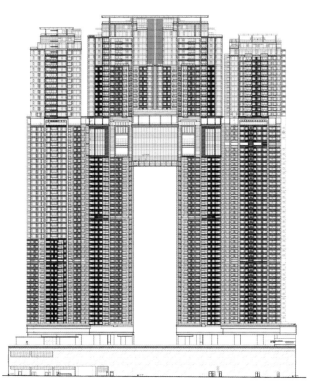

The Arch is a "Unique Icon", as a visionary gateway to the cosmopolitan lifestyle created at Union Square, Kowloon Station.

The location of the development is dramatised by the spectacular view of Victoria Harbour and Hong Kong Island. The aechitects treasure not only this luxury of panoramic view but also the opportunity to blend in the blue sky and the green space in the living environment at different levels of sky living. The four residential towers are integrated into one dynamic architectural form with a strong sense of identity. The Arch Portal in the centre of the building accentuates the sculptural dynamics. The design is further challenged by the client's brief of providing one million sq.ft. residential area with 1,056 units that offers unit size ranging from 500 sq.ft. to 5,000 sq.ft.

"Lifestyle in the Sky" is the planning approach. The Arch Portal signifies a welcoming gesture and provides a sense of homecoming. The city life of this community is celebrated in the Sky Club on sixty-second floor. It adopted the concept of "A City within a City", integrated with extensive Sky Garden of 148m in height, incorporating different outdoor rooms with leisure theming while capturing the postcard views of the Hong Kong Island. The Boutique Sky Apartments below the Sky Club connect the city view in a variety of unit size and orientation. As you rise beyond the Sky Club, you reach the Sky Apartments and Sky Houses that truly appreciate the infinity of sky and the sea.

凯旋门是一个"独特的地标",创造了全球化生活方式的视觉入口,建在九龙站之上。

地段所处的位置由于望向维多利亚港口和香港岛的壮观景象而更具戏剧性。建筑师所珍视的不仅仅是奢华的全方位美景,更是在不同层面上将蓝天和绿地与居住环境融合起来的机会。4个住宅塔楼被整合进了一个具有强烈特性的动态建筑结构。建筑中央的拱形门洞强调了雕塑的动感。客户的要求很具挑战性,100万平方英尺的住宅区包含1056个单位,这样每一户的面积范围就在500平方英尺到5000平方英尺之间。

建筑师的规划一直围绕着"空中的生活"。拱形门展现了一个欢迎的姿态以及为人们提供了宾至如归的感觉。这一社区在62层的天空酒吧举行了城市生活的庆祝活动。它采用了"城市中的城市"这一概念,结合了位于高度148米处的宽敞的空中花园,将不同的户外房间与悠闲的主题相结合去捕捉明信片上一般的香港岛的美景。处在位于天空酒吧之下的精品天空公寓中,提供了不同大小以及方向的户型俯瞰城市的美景。当你到达天空酒吧的高度时,就仿佛是已经触摸到了天空公寓以及天空。在这些房子里可以真正的欣赏到无边无际的天空与美丽的海景。

12 Broadwood Road Residential Development

跑马地 12 号乐天峰住宅开发项目

Location: Hong Kong
Area: 10,563m²
Architect: Philip Liao, Partners Limited / WMKY
Photographer: Jermaine Tsui
Completion Date: 2010

地点：香港
建筑面积：10,563 平方米
建筑师：廖宜康国际有限公司 / WMKY 设计公司
摄影：杰梅恩·徐
建成时间：2010 年

Hong Kong, China
中国，香港

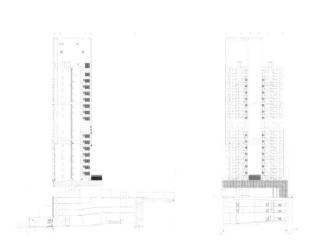

Luxurious 41-storey residential development located at the most prestigious location, Broadwood Road of Happy Valley in Hong Kong.

The design concept evolved around "Simplicity" to enhance the subtly modern lifestyle in the heart of the city. With the mission of being the landmark building of Happy Valley region, the building façade is dressed up with jumbo ceramic tiles and aluminium cladding for residential portion and terre-cotta louvres, steel canvas, stone cladding, curtain wall and feature lighting for non-residential portion.

The layout is designed to maximise the capture of unobstructed panoramic view overlooking the Happy Valley Racecourse for each individual unit, but also able to view the luscious garden and swimming pool.

Broadwood Twelve has also taken leading edge of commissioning sustainable building by adoption of green principles. Therefore, design to reduce energy demands is always inherent in the concept during design development stage. With the use of low-e glasses, heat gain in summer and heat loss in winter would be reduced. Less light reflection also means reduction of light pollution to proximate environment. In addition, double glazing windows further mitigate impacts from external, i.e. thermal transmission and exterior noise reduction. Cross ventilation window to enable effective natural air circulation and obtain maximum natural light. Aluminium features on façade act as shading to control direct solar gain to premise while add extra value on aesthetic sense. Using LED lighting whenever possible to reduce energy consumption and maintenance cost.

奢华的41层住宅开发项目坐落于香港跑马地乐活道最繁华的地区。

这一设计概念围绕着"简单"来展开，在香港的心脏地区提倡一种精致的现代生活方式。目标为将其建造成为跑马地区域中的地标，大楼的外墙装饰有为住宅部分设计的大型的瓷砖和铝板层，以及百叶、钢帆布、石材、幕墙和具有特色的非住宅部分的照明。

在布局的设计上则尽可能实现每一单元都能够最大限度、一览无余的俯瞰跑马地的美景，但同时又能够兼顾美丽花园和游泳池的景色。

跑马地12号乐天峰住宅也采取了前沿的遵循绿色环保的原则的可持续建筑。因此，在这一设计的开发阶段也使用降低能源需求的设计。在材料的选用上，使用低辐射的玻璃，夏天的热量增加与冬天的热量流失都将会有所减少。此外，双层玻璃的窗户能够进一步减轻来自于外部的影响，例如，热传输和外部降噪。交叉通风窗口能够有效的实现自然风的循环以及最大限度的自然采光。外墙铝板的特点则是能够充当阴影的角色，来控制直射的阳光并且为建筑增添了额外的美学价值。使用LED照明尽可能减少了能源消耗和保持了成本。

One Island East, TaiKoo Place
港岛东中心，太古坊

Location: Hong Kong
Architect: Wong & Ouyang (HK) Limited
Photographer: Wong & Ouyang (HK) Limited
Completion Date: 2008

地点：香港
建筑师：王欧阳（香港）有限公司
摄影：王欧阳（香港）有限公司
建成时间：2008 年

Hong Kong, China
中国，香港

One Island East is the latest office development at TaiKoo Place, Quarry Bay, Hong Kong. The basic form of the building is a square plan with central core.

The four corners are rounded. Two corners facing north and south "open up" at the top floors to address the Harbour view. At the base, the two corners facing east and west "open up" to address the open space. The edges of the four façades "sail" beyond, creating a "floating" effect, and giving lightness to the building. Lighting feature is incorporated into the edge of the four façades to enhance the floating effect.

Without a podium structure, the tower sits freely in front of a large landscaped open space to the east. The canopy at the porte cohere is specially designed as a piece of sculpture. The landscape area beyond is designed to have platforms at different levels incorporating water features. This urban landscape gives an appropriate scale as a forecourt to the building and provides a leisure space for both the enjoyment of office workers and for the neighbourhood.

港岛东中心是香港鲗鱼涌太古坊地区最新开发的办公大楼。大楼的基本形态是以中央为核心的方形布局。
大楼的四角呈半球形。楼顶层的两角开放式的面向北方和南方，俯瞰着美丽的海景。在楼体底部两角则开放式的面向东和西方向，以实现开放的空间。立面的四个边缘呈"帆"的形象，创造出了"飘逸"的效果，为大楼赋予了轻盈的感觉。四角与照明功能的结合，进一步增强了这一飘逸的效果。
没有平台的结构，使得塔楼得以无限制的坐落在东面的大块景观开放空间前。将大楼的雨篷特别的设计成一座雕塑的样子。远处的景观带则结合了滨水的特性，被设计成不同层次的平台。这一城市景观可作为与大楼规模相当的前院并且为办公楼里的员工以及附近的居民提供了一个娱乐休闲的空间。

Landmark East
城东志

Location: Hong Kong
Area: 100,000m²
Architect: Arquitectonica
Photographer: William Furniss, Amral Imran, Rogan Coles
Completion Date: 2008

地点：香港
建筑面积：100,000 平方米
建筑师：Arquitectonica 建筑事务所
摄影：威廉姆·弗尼斯、阿莫莱尔·伊姆兰、罗根·高斯
建成时间：2008 年

Hong Kong, China
中国，香港

Landmark East is a Grade-A high-rise, two-tower office development that rises from a dense urban site in the Kwun Tong district of Hong Kong. The slender rectilinear slabs that form the tower volumes are derived from the long narrow plot. The towers respond to the site, rising above the adjacent buildings in a slender, elegantly moving motion reminiscent of bamboo swaying gently in the wind.

The rectangular floor plates of both towers provide efficiency and allow side cores to the north. This in turn orientates the office floors to the south, taking advantage of the privileged views of Victoria Harbour and Hong Kong Island beyond. The orthogonal tower forms have been segmented in section creating a series of interlocking parallelogram slabs. This arrangement emphasises the sensation of movement and conveys lightness to the towers, forming a strong, dramatic, unified identity for the development.

Efficient floor plans are therefore retained whilst providing dynamism to what would otherwise be a regular box derived from a rectangular plan. Where the forms interlock additional corner offices are created taking advantage of the valuable harbour and island views. The car parking and loading areas are restricted to the basement of both towers and a podium structure at the base of tower-two. As a result, the remainder of the site is left open to form an elegant and beautifully landscaped public plaza with sculpture gardens and cafés that provide valuable open space for tenant and local residents use.

城东志是一栋双塔式 A 级高层写字楼复合建筑，伫立在高楼林立的香港观塘区。根据地段的狭长而建的大楼体量呈纤直细长型。塔楼与选址相呼应，以其纤长的身材凌驾于周围建筑物之上，形成一种动感让人联想起在风中轻轻摇曳的竹子。两栋塔楼均为矩形板式结构，这种结构为大楼提供了效率以及使副楼可以向北倾斜。这样就使办公楼层朝向南面，得以尊享维多利亚湾以及远处的香港岛的美景。两个板层塔楼形成相互关联的平面。这种布局强调了一种动感并表现了塔楼的轻盈感，在这一开发项目上形成了一种强烈的、动感的、统一的标识。

因此，为原本的一个从矩形平面中分离出的普通的盒式结构保留了高效的楼层平面设计的同时又注入了活力。拐角处额外创建了办公空间，在上层可看到宽阔的海景。停车场和卸货区被限制在两座大楼的地下室以及 2 号塔楼底部的平台上。因此，剩余的空间是开放的，形成了一个拥有优雅和美丽风景的公共广场，同时配有雕塑花园和咖啡馆，以提供宝贵的开放空间为租户和当地居民使用。

Enterprise Square 5, MegaBox
企业广场第五期红点购物中心

Hong Kong, China
中国，香港

Location: Hong Kong
Architect: Wong Tung & Partners Ltd
Photographer: Marcel Lam Photography
Completion Date: 2007
Award: Certificate of Finalist (Non-Residential Category) for Quality Building Award 2008 by the Organising Committee of Quality Building Award 2008

地点：香港
建筑师：王董建筑师事务有限公司
摄影：马塞尔·莱姆摄影工作室
建成时间：2007年
获奖：2008"优质建筑大奖"非住宅类项目优胜奖

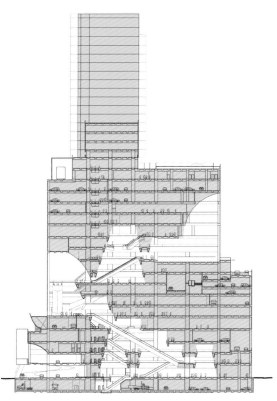

434 / SOUTH CHINA

The Enterprise Square 5 is an amalgamation of shopping, entertainment and office spaces set at a conspicuous waterfront site of Kowloon Bay. Looking from afar, the twenty-storey shopping podium, aptly named as MegaBox, appears to form a single monolithic box. To break down the overall mass, 30 different tones of red are introduced to the façade. An expansive circular glass wall cuts through the front elevation, not only to add visual interest to the building, but also enhance interaction between inside activity and urban fabric. Perched high above on the shopping podium, the 15-storey twin office towers are clad in glass to reflect their different use, while inviting natural sunlight as well as glamorous sea view into the interior space.

The architectural vocabulary of the Enterprise Square 5 stands in stark contrast to the urban fabric in Kowloon Bay. While the office towers are rendered transparent, the shopping podium is conceptualised as a massive opaque box in striking red, dominating the waterfront and creating an iconic new landmark.

企业广场第五期位于美丽的九龙湾岸边，集购物、娱乐及办公于一身。红点购物中心高达20层，远远望去犹如一个独立的大盒子，恰如其名。30种不同的红色色调装饰在建筑表面，打破了大体量带来的厚重感。正面的透明玻璃结构似乎将整个建筑割裂开来，增添了趣味性，同时更将商场内部的活动与外面的城市气息融合。高达15层双子办公楼耸立在商场上方，采用玻璃材质饰面，旨在突出其不同的用途，便于自然光线的射入并将壮丽的海景引入室内。

这一项目在建筑语言上与九龙湾地区的城市结构形成鲜明对比。办公空间突显通透性，购物中心以封闭为特色，点缀在九龙湾岸边，并成为这一区的全新地标。

City of Dreams
新濠天地

Location: Macau
Area: 512,540m²
Architect: Arquitectonica
Photographer: Photo Courtesy City of Dreams
Completion Date: 2009

地点：澳门
建筑面积：512,540 平方米
建筑师：Arquitectonica 建筑事务所
摄影：新濠天地拍摄照片
建成时间：2009 年

Macao, China
中国，澳门

436 / SOUTH CHINA

Four Hotel towers rise from a casino podium that defines a water theme for the development. Each hotel tower establishes its identity while participating in the overall architectural message.

The Hard Rock Hotel is iconic in its own way; playful and young. Its pure circle provides efficiency and reception of room modules. It appears as a glass cylinder with a series of curved octagons cantilevering beyond its skin. These bands are rotated gradually to create a spiraling effect by the movement of shadows created by the horizontal eyebrows. The added movement to the cylinder is almost like a water spout or vertically flowing water cascade. They form a three-dimensional sculpture that moves both vertically and horizontally carrying the eye around the form. The result is simple and complex, fun and interesting, architectural and theatrical.

The Crown Hotel stands in the northwestern corner of the site. This elliptical tower rises from the expansive series of reflecting pools and water features that define the main frontage of the project along the Cotai strip. The pure curved façade of the building is interrupted by a series of long vertical fins that break up as they descend the creating a cascade effect as the façade disappears into the water at its base. A similar glass type is used on all towers, with their striking yet complimentary forms defining their identity.

新濠天地由四幢塔楼组成，该工程建立在一个以水为主题的娱乐平台上。每间酒店的塔楼在参与了整个建筑布局的同时又有着其独特的特性。

硬石酒店有其独有的标志性，好玩和年轻化。它的圆柱形的玻璃塔楼提供了高效的和包容性的房间模式。这些圆柱形玻璃形成了一系列的曲线八角形结构，八角形结构逐层旋转，创造出螺旋形的视觉感受。圆柱形玻璃的旋转带来的视觉感受就几乎是水喷涌而出又或是垂直的像瀑布般倾泻而下。它们形成了一个立体的雕塑，通过这一形式垂直和水平的带动人们的目光的移动。结果是简单而又复杂，好玩又有趣，即是建筑又像是戏剧一般。

皇冠酒店位于新濠天地的西北角。酒店采用椭圆形结构，立于广阔的一系列反射水池和水景之中，酒店的正门沿着金光大道而建。酒店弧形的外墙结构拉长了垂直方向的距离，玻璃立面给人热带暴雨倾盆而下的效果。相近的玻璃结构运用在了新濠天地的所有塔楼上，使人们一眼望去就能够通过出众的外观将它们分辨出来。

Chongqing Library
重庆图书馆

Location: Chongqing
Area: 50,381m²
Architect: Perkins Eastman
Photographer: Gu Zhihui, Tim Griffith
Completion Date: 2007

地点：重庆
建筑面积：50,381平方米
建筑师：珀金斯·伊斯曼
摄影：谷智慧、蒂姆·格里菲斯
建成时间：2007年

Chongqing, China
中国，重庆

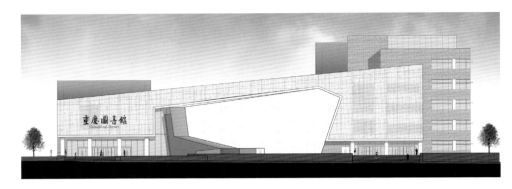

438 / SOUTHWEST CHINA

To convey the importance of this new city landmark, the design team predicated their concept on the notion that reading, learning, knowledge, and the exchange of ideas needs to be free, open, and accessible to all of the public – a basic belief shared by the government and the library leaders. To reinforce this notion, Perkins Eastman sheathed the building almost entirely in glass so that the public could readily see the inside of the building and the actively engaged visitors. The architects inscribed this glass façade with a text pattern that idealises the profound impact of life-long learning on individuals and society. The design team researched and selected quotes from scholars throughout history – rom world leaders such as Chairman Mao and President Roosevelt to renowned philosophers including Socrates and Confucius – all related to the empowerment, knowledge, and freedom one gains from words and books.

The building's form is based on both the Chinese architectural tradition of the courtyard and the interior multi-storey atrium halls of traditional Western libraries. The outdoor courtyard spaces provide a unique oasis within the bustling city, visible from the street level but sunken below grade to achieve separation from the active urban streetscape and only accessible through the secured building's reading rooms. The boundaries between inside and outside are purposely blurred, allowing the building's users to feel that they are in a place of nature and serenity.

为突出图书馆作为新城区地标建筑的重要性，设计团队的理念基于"阅读、学习、求知以及交流应提倡自由、开放、方便公众"这一原则，当然这更是政府及图书馆领导人员所坚持的。为此，整幢建筑采用玻璃材质"包裹"起来，公众可以清晰地看到内部的场景。玻璃外观上的刻字强调了学习知识的必要性，文字内容来自于不同历史时期的名人语录，包括毛泽东及罗斯福，孔子及苏格拉底。

建筑在造型设计上以中式传统庭院以及西方图书馆为基础。室外庭院略低于街道平面，但在街道处可见，与城市街道景观分离开来，同时在喧嚣的都市中营造出一片绿洲。室内外之间的界限刻意被模糊，让人犹如置身于宁静的自然之中。

Huxi Campus Library of Chongqing University
重庆大学虎溪校区图书馆

Location: Chongqing
Area: 37,123m²
Architect: Tanghua Studio / Tanghua, etc.
Completion Date: 2009

地点：重庆
建筑面积：37,123 平方米
建筑师：深圳市汤桦建筑师事务所 / 汤桦等
建成时间：2009 年

Chongqing, China
中国，重庆

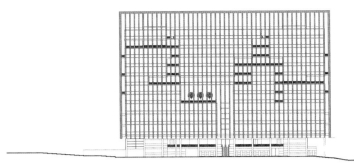

Huxi Campus library is a building located on the side of the central campus axis, with great landscape significance. It faces the Shanshui Park and Huxi City Park to the north and south.

The whole building displays horizontal layers in its structure, having the site, the base and main building from bottom to top. Adopting the method of ground landscape, the building merges with the local landscape on both sides. The base follows the landform and is partly imbedded in the ground, containing the lecture halls, meeting rooms, information centre and computer rooms, etc. Main body of the building rests on the base, partly built on stilts like a stilt house. The structure extends itself facing the Shanshui Park as a compact 'L', simple in façade but contains complex spatial structure and offers various functions.

The landscape engineering comes in two parts. One is the plant management of the landscape, which is different from afforestation, as it is based on firm ground and large square of regular plants and water flows with several gaps suitable for moving around. The other part is the greenery of the architecture. Since the building has several openings on both horizontal and vertical levels, plants from the ground landscape can extend upwards onto the building. A variety of side courts, terraces and wells can also be employed in the design so the plants can reach every corner of the architecture. Then the building will work as an open window, through which the scenery and landscape can be enjoyed with no barrier.

重庆大学虎溪校区图书馆位于教学中心区校园主轴的一侧，是校园中轴线边的重要建筑，南北临近校园中央山水公园和虎溪城市公园。

整个建筑在大的空间结构上呈现出垂直分层的关系，由下至上分别是场地、基座和建筑主体。场地的处理采用地景式的方法，楔入两侧大尺度的景观中去。基座结合地形，部分埋入地下，主要包含报告厅、讲座厅、会议室、网络信息中心和机房等功能。建筑主体承托于基座之上，局部架空，犹如吊脚楼。建筑朝向山水公园展开，是一个由密肋包裹的L形体，外形简洁，其内包含着复杂的空间结构和机能。

绿化系统为两个大的层次，其一是场地绿化景观。场地处理与园林化方式不同，以硬地为底，形成大片的广场，结合规则的植栽和叠水，形成若干尺度适宜的活动空间。其二是建筑中的绿化。由于建筑无论是在水平方向还是垂直方向上都形成了若干开口，这样地面层的绿化景观可以向上延伸渗透，在建筑中还可形成性质不一的边庭、露台和开井，由此绿化可充盈于建筑的每一个角落，整个建筑就像是一扇漏窗，视线和景观可以不受遮挡地穿越。

Science and Technology Museum Chongqing
重庆科技馆

Location: Chongqing
Area: 40,000m²
Architect: Architecture-Studio
Completion Date: 2009

地点：重庆
建筑面积：40,000 平方米
建筑师：法国 AS 建筑工作室
建成时间：2009 年

Chongqing, China
中国，重庆

Located at the northwest of Yuzhong Peninsula, Chongqing Science and Technology Museum will become one of the ten main public facilities in the near future as well as a place open to the world for knowledge training, acdamic research and technological exchange. The architectural shape of the building follows the geometric curve of the typology and is highlighted by exterior walls made of stone and glass in order to compliment with the natural landscape of mountain and river.

The dome cinema is located at the highest part of the site and has become a symbolic landmark of the Museum. The buidling reflects intimate relationship with the surounding natural scenes by the use of effective arhcitecrural language. It merges into the surroundings as well as full of modernity to show the developemnt of science and technology in Chongqing. Moreover, the design takes sustanability as one of the main concepts.

重庆科技馆位于长江口岸渝中半岛的西北，将成为重庆未来的十大主要公共设施之一，面向世界成为知识培训、学术研究和技术交流的互动地。其之所以成为城市不可缺少的一部分，是因为它作为整个地区其中一个重要的有机体，外形的设计结合地形的几何曲线，配合建筑外墙的石材与玻璃的相互渗透，与周围的山水自然景观相得益彰和谐统一。建筑格调融入了科技创新的现代设计理念并考虑到现代的技术更新。建筑构思通过运用各种建筑手法力图使重庆科技馆成为重庆市的象征。

球幕影院位于地块最高的位置，在中国自由纪念馆的纵向延伸线上，它成为了科技馆一个象征性的标志。科技馆通过有力的建筑语言表达了与周围自然景物的关系，它与环境相结合然而又充满现代感，代表了现代重庆科技发展的新气息。为了满足可持续发展的要求，达到节约能源的目的，重庆科技馆的设计和建造也考虑到城市环保的特点。

Chongqing Grand Theatre
重庆大剧院

Location: Chongqing
Area: 100,000m²
Architect: gmp Architekten – von Gerkan, Marg and Partners
Photographer: Heiner Leiska, Modellbau Werner, Hans Georg Esch, Ben McMillan
Completion Date: 2009

地点：重庆
建筑面积：100,000平方米
建筑师：gmp建筑设计事务所
摄影：海纳·雷斯卡、莫德尔堡·沃纳、H·G·埃施、本·麦克米兰
建成时间：2009年

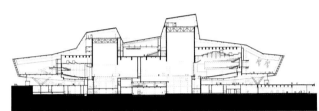

The architectural shell is intended to give structural expression to this extraordinariness and the world of illusion. An expressive sculptural array of parallel double-walled, spaced glass strips that jostle with and nudge up to each other generates in overview and side view a metaphorical image of a ship – an almost dramatic image of a theatre building sailing in a sea of light. The Grand Theatre is close enough to the Yangtze River to seem to "float" on it. Surreal light reflections and light patterns create poetic compositions of reality and fiction comparable to the world of illusion of a theatre performance. The uniqueness of the location and the panoramic view of the imposing skyline of the city in conjunction with the sculptural architecture make this theatre a symbol, a place of international cultural encounters.

A stone platform forms a base for the glass sculpture. Despite the apparent arbitrary expressiveness and the maritime analogies, the ground plan and elevation are strictly in accordance with the functional specifications. The Grand Theatre is not a building with the customary façade of walls and window areas. The light radiating from within and visible from outside, together with the reflections of sun, clouds and water on the multi-faceted and multi-angled glass surfaces, create constantly changing patterns of light – the structure seems to shine and gleam in ever-mutating, enigmatic, almost mystic moods. The upper floors of the theater are organised in accordance with the functional specifications.

这一设计旨在在建筑外形上清晰地展现出其独特的魅力并营造一个幻景般的世界。规则摆放的双层玻璃板条结构"簇拥"在一起，使得整幢建筑无论从整体或侧面看上去，都犹如一艘"大船"。由于剧院坐落在长江附近，从远处望去，更像是停泊在江面之上。灯光的倒影和各种图案融合在一起，营造了一个现实与梦幻共存的诗意环境。独特的地理位置、城市天际线的壮丽景致同雕塑般的建筑样式使剧院本身成为一个象征——国际文化交流的融汇地。

石头平台为玻璃结构打造了一个坚实的基座。建筑本身除了具备象征意义之外，在平面及立面格局规划上注重功能性。与那些外观由墙壁及窗户构成的建筑不同，这一结构通过室内外光线的融合，阳光、云朵以及水波映射到玻璃表皮上的影子营造了不断变化的光影效果。如此一来，剧院本身犹如是"置身"于一个神秘变幻的环境中。此外，建筑上层空间的分配同样遵循功能性原则。

Beity Hot Spring Hotel

贝迪颐园温泉度假酒店

Chongqing, China
中国，重庆

Location: Chongqing
Area: 32,000m²
Architect: Hong Zhongxuan / HHD East Holiday International Design Institute
Photographer: Chen Zhong
Completion Date: 2007

地点：重庆
建筑面积：32,000 平方米
建筑师：洪忠轩 / 假日东方国际酒店设计机构
摄影：陈中
建成时间：2007 年

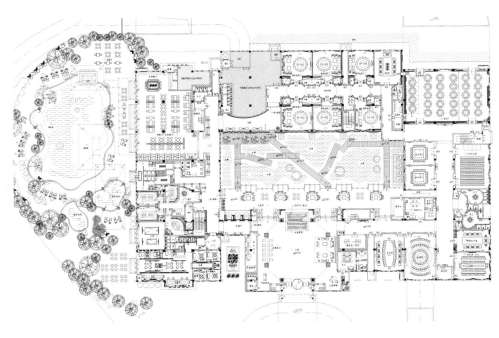

Beity Hot Spring Hotel absorbs the essence of nature to show its complicated beauty of serenity, quietness, smartness and uniqueness with Xuanwu Mountain on the back and Zhongliang Mountain from afar.

"A quiet place in the noisy city" is the real portraiture of Beity Hot Spring Hotel with a 300-acre ecological garden and post-modern architectural style. With the beautiful Yijing Lake, the hotel itself is just a natural "Oxygen Bar" as well as to create a picturesque environment. The design pursues the feeling of "serenity and quietness" to make guests relaxed completely. The linear arrangement expresses order and purity. The water is self-circulated and filtered shows the sustainable feature; the dark tones emphasise "quietness" as well as electricity effective. In addition, the design also aims to create an environment with fragrant smelling to make guests seem to be bathed in clusters of flowers.

背依玄武山丘，远眺中梁山脉，贝迪颐园吸取了大自然的精粹，如写意园林画般，向人们展示着它幽、静、灵、奇的独到之美。

离尘不离城，这是对贝迪颐园最真实的写照。300亩生态园林、后现代建筑风格、坐拥颐景湖，如一个天然氧吧，营造着现代水墨画般的唯美意境。它让人暂离繁忙都市，挥别喧嚣嘈杂，另辟乐活天地，让身心呵护于幽幽翠绿、潺潺玉泉，涤荡于日月精华，喧嚣不再。在这种氛围的感召下，贝迪颐园温泉酒店追求的情绪是"静"。有秩序的线条排列最有利于创造条理和单纯；自然资源之一的水自循环和过滤系统是创造"生态"的最直接方式；暗色调的设计除了表现"静"之外，更重要的是节省电力能源；淡淡的"暖"色调使置身其中的人如浴花丛，追求的是酒店的一种"香"的味觉。

Jianchuan Mirror Museum and Earthquake Memorial

建川镜鉴博物馆暨汶川地震纪念馆

Sichuan, China
中国，四川

Location: Anren Town, Sichuan Province
Area: 6,098m²
Architect: Li Xinggang, Zhang Yinxuan, Fu Bangbao, Liu Aihua
Photographer: Zhang Guangyuan, Li Xinggang
Completion Date: 2007
Award: The Chicago Athenum / Europe International Architecture Awards (2010), shortlist of Weinerberger Golden Brick Awards (2010)

地点：四川，安仁镇
建筑面积：6,098 平方米
建筑师：李兴钢、张音玄、付邦保、刘爱华
摄影：张广源、李兴钢
建成时间：2007 年
获奖：THE CHICAGO ATHENUM / EUROPE 国际建筑奖（2010）、欧洲维纳博格金砖奖提名奖（2010）

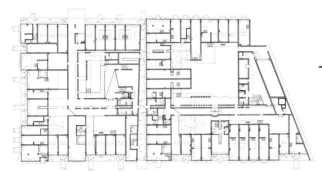

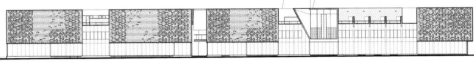

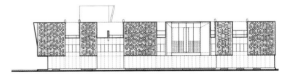

The construction of Jianchuan Mirror Museum paused for a period of time and re-started due to the addition of Earthquake Memorial. After the design adjustment, the museum and the memorial lap over each other and are separately independent in the spaces; the virtual images and realities are mixed and reciprocally referred. Mirror Museum sets a serious of rotatable "Mirror-doors" and creates pure, abstract and changeable virtual-image spaces through which the visitors can simulatively experience the period of unconscious and madness in Culture Revolution. Earthquake Memorial's spaces and exhibits are temporary, crude, concrete and actual through which the visitors can feel the reality of grief and heartquake in Sichuan Earthquake. The museum and memorial both commemorate, exhibit and experience the tragedies of human beings happened in Chinese history, giving later generations warning and caution. The contemporary building in a 1,000-year-old town is peace and calm outside but collects crazy virtual images and heartquaking realities inside, providing people with the prompt experience and commemoration of history and tragedies. Additionally, the design and construction process is devious, changeable and full of reality and dramatic. So it could be called as the current time's "Mirror Museum".

Red & grey bricks and concrete are the two main kinds of materials used. Enrich local bricklaying tradition is embodied by the "Leaky Brick-walls" in various patterns and with different transparency corresponding to different indoor functions at the same bricklaying modulus control, and satisfying the different requirement of light, airiness, view and privacy. A type of "Armor plate Glass Bricks", which are cheap and easily manufactured by local people are also invented following the above modulus to be laying on the empty positions of the "Leaky walls" of the indoor rooms.

原设计建造并停工难产的文革镜鉴馆由于汶川地震的发生加入了地震纪念馆而再造重生，经设计改造后的两个馆在空间上相互叠加并置，虚像和现实相互混合、对照。镜鉴馆通过组合的镜门装置造成纯净、抽象、变幻多端的虚像空间，而让参观者以游戏的方式模拟体验失去理智的疯狂年代；地震馆则以临时、粗砺、具体、真实的空间和展品让参观者感受痛切而震撼的现实。两者以各自的方式纪念、展示和体验着历史上发生的"人祸"和"天灾"两大人间悲剧，给予后人以鉴戒和警示。建筑外部平和而宁静，却混合收藏着内部的虚像狂乱和现实震荡，给予当代人对往往历史和灾难的即时体验与纪念，加上其曲折、多变、富于现实感和戏剧性的设计建造历程，可称得上是这个时代的"镜鉴"之馆。

红、青两色的砖和混凝土是两种被使用的主要材料，当地丰富的砌砖传统在这里体现为在同一砖砌模数单元控制下，不同的室内功能对应不同通透程度的砖砌"花墙"，以满足不同的采光、通风、景观和私密性等要求。设计者还设计发明了符合上述模数的透明"钢板玻璃砖"，造价低廉并易于加工，用于"花墙"上对应室内空间的砌空部分。

Xinjin Zhi Museum
成都新津县芷博物馆

Location: Chengdu, Sichuan Province
Area: 2,353m²
Architect: Kengo Kuma & Associates
Photographer: Kengo Kuma & Associates
Completion Date: 2011

Sichuan, China
中国，四川

地点：四川，成都
建筑面积：2,353 平方米
建筑师：隈研吾建筑都市设计事务所
摄影：隈研吾建筑都市设计事务所
建成时间：2011 年

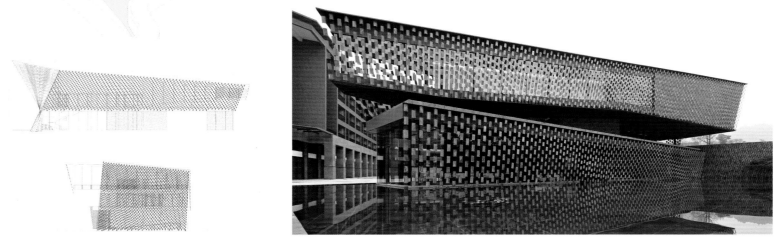

450 / SOUTHWEST CHINA

This pavilion is located at the foot of Laojunshan mountain in Xinjin, to usher in people to the holy place of Taoism, while the building itself shows the essence of Taoism through its space and exhibitions.

The tile used for façade is made of local material and worked on in a traditional method of this region, to pay tribute to Taoism that emphasises nature and balance. Tile is hung and floated in the air by wire to be released from its weight (and gain lightness). Clad in breathing façade of particles, the architecture is merged into its surrounding nature.

The façade for the south is divided into top and bottom and staggered in different angles. This idea is to respond to two different levels of the pond in front and the street at the back, and avoid direct confrontation with the massive building in the south. For the east side, a large single tile screen is vertically twisted to correspond with the dynamism of the road in front. The façade for the north side is static and flat, which faces the pedestrians' square. Thus the tile screen transforms itself from face to face, and wraps up the building like a single cloth.

Taking advantage of the varied levels in the architecture's surroundings, the flow is planned to lead people from the front to the back, motion to stillness, like a stroll type of garden. The exhibition space inside is planned spiral moving from darkness to light. From the upper floor a paramount view of Laoujunshan can be enjoyed. Direct sunlight is blocked by the tile, and the interior of the building is covered with gentle light with beautiful particle-like shade.

该展览馆位于新津县的道教圣地老君山下，引领人们进入道教圣地，它的展览空间有着道教的精髓。

外墙的瓦用当地材料和传统工艺制成。道教强调对大自然的敬畏和阴阳平衡。瓷砖用钢丝挂在空中，在空中飘动，好像没了重量。还有一些瓦嵌入外墙，这种处理手法让建筑与周围的环境相融合。

南立面分为顶部和底部，它们处于两个不同的角度，互相交错。它们对应于前面的池塘和后面的街道，避免在南面出现单一巨大的建筑体量。东立面是扭曲的，以此对应前面忙碌的道路。北立面安静而平坦，对应着步行广场。瓦片的幕布在每个面都不同，像一件衣服包裹着展览馆。

展览馆内部有多层，参观流线是从前面到后面，观众安静的沿着展览馆内部的螺旋的路线，从黑暗走向光明，像是在花园散步。在展馆的顶层可以看到老君山。直射的阳光被瓦片所遮挡，建筑的内部光线柔和，美丽的瓦片网在阳光下像树荫一样。

Relics Exihibition Hall at the Jinsha Site Museum
金沙遗址博物馆文物陈列馆

Sichuan, China
中国，四川

Location: Chengdu, Sichuan Province
Area: 20,190 m²
Architect: Architectural Design and Research Institute of Tsinghua University Co., Ltd / Zhuang Weimin, Ge Jiaqi, etc.
Photographer: Shu He
Completion Date: 2007

地点：四川，成都
建筑面积：20,190 平方米
建筑师：清华大学建筑设计研究院有限公司 / 庄惟敏、葛家琪等
摄影：舒赫
建成时间：2007 年

Jinsha Site is located in the western suburb of Chengdu City, occupying about 29 hectares, including Excavation Hall, Relics Exhibition Hall, and Cultural Heritage Preservation Centre. The Relics Exhibition Hall as a key part of the Site, has a total floor area of 20,190 square metres (including underground parking). Design inspiration comes from the approach of gridding in archaeology. The 10m×10m grids set a good order for the plan. The cubic mass looks as if growing out of the ground – a metaphor of the excavated jade. The height difference between south and north part further makes sharp contrast with the Excavation Hall, establishing an echoing relationship between the two buildings. For the Relics Exhibition Hall, dry-hang travertine is adopted for façade and public space floors and walls. Different from traditional museums where exhibition hall, public space and education space are separated, here exhibition spaces are interactively planned around the central atrium, and the independent hall is combined with the open exhibition terrace. In this way, passive "fixed exhibition" is transformed into flexible "interactive exhibition", and static visiting experience is changed into dynamic multi-media museum experience. Here visitors are the real protagonist, and the concept of humanism is thus fully conveyed. After the Wenchuan Earthquake, the Jinsha Site Museum becomes a shelter for not only the relics, but also the citizens. It has gone beyond the initial intention, interpreting unexpected humanism in a broader sense.

金沙遗址位于成都西郊，占地434亩，包含遗迹馆、文物陈列馆和文物保护中心等配套设施。文物陈列馆作为园区主体，建筑面积20190平方米（含地下车库），以考古中的"探方"作为构思切入点，10米×10米基本模数隐喻文物考察的秩序。建筑形体方正，仿佛从大地中生长出来，隐喻被发掘的玉璋。造型北高南低，与遗迹馆一刚一柔，一实一虚，交相呼应。陈列馆造型简洁大方，外墙和室内公共空间的墙地面全部选用干挂洞石，使得园区建筑浑然一体，统一协调。陈列馆围绕中庭布置展厅，突破了传统博物馆将展厅、公共空间、教育空间割裂的形式，而将独立展厅与开放式台地展区相结合；将被动"固定式"陈列与互动"情景式"陈列空间相渗透；将静态单一参观模式与动态多媒介模式相整合。这种方式破除了人与展品静止对立的格局，人成为陈列馆的主角，充分体现"人本主义"的理念。汶川大地震后，金沙遗址博物馆不但成为了文物的庇护所，还成为广大市民躲避地震灾害的避难所。建筑超越了原设计的意图，提供了更为广阔的人文关怀。

Chengdu Fluid Core-Yard
置信综合办公楼

Location: Chengdu, Sichuan Province
Area: 1,000m²
Architect: PATTERNS
Photographer: Marcelo Spina
Completion Date: 2010

地点：四川，成都
建筑面积：1,000 平方米
建筑师：PATTERNS 建筑设计事务所
摄影：马塞洛·斯宾纳
建成时间：2010 年

Sichuan, China
中国，四川

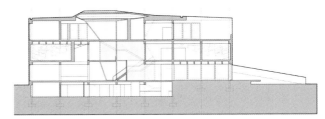

454 / SOUTHWEST CHINA

The project presents a perfect occasion to speculate in a context with very little context; despite the specificity of each new building, everything is also fairly generic. Given those limitations, the proposal aimed at blending the building in with the expected massing schemes, while differentiating it internally in a distinctive, yet insistent way.

A diagonal wedge of circulation that dynamically cuts the plan and connects front and back organises the building. A regular volume is subtracted from it at its corners, generating similarly opposed structural cantilevers that produce a strong sensation of levitation in its mass. Coffered hyperbolic surfaces connecting vertical walls and horizontal slabs further induce the sense of plastic obliqueness throughout the building, linking the front and back visually and physically. The repetition and inversion of the same hyperbolic geometry at opposite ends produces a sense of spatial reciprocity and strange symmetry. Subtly but substantially subverting the generic mass, the inverse repetition of the waffle parabolic surface creates a sense of déjà vu when rapidly moving in and out through the building.

Using hyperbolic paraboloid geometry, the coffered surfaces occupy the underside of the diagonally opposed cantilevered volumes at the front and back façades. Rather than a priori system, the coffers are locally articulated. Indexing the movement of the surface from its cartesian origin, the coffer cells gain depth as the hypar [Hyperbolic Paraboloid] shape moves away from the walls and towards the waffle ceiling. The function of these surfaces is to introduce a radically small scale of articulation in the most public areas of the building.

这一项目所处背景环境极为简约——周围的每一幢新建筑都具备自身的特色，但同时又不失普遍性。因此，设计目标即为与四周环境融合，并个性十足。

对角线通道设计将整体平面切割，并将前后连结起来，构成了建筑架构的基础。长方形的体量似乎从四角"抽离"出来，形成悬臂结构，营造出强烈的"飘浮"感。外观上方曲线格子图案将竖直的墙壁与水平板结构连结起来，加强了建筑造型感，同时更将办公楼前后在视觉上连通。建筑另一侧重复同样的图案，营造对称性。巧妙地设计让人在出入建筑的时候产生似曾相识的感觉。

设计中运用双曲抛物线打造格子图案，装饰在建筑前后成对角线排列的悬臂机构下方。格子图案由来已久，从墙壁到天花板装饰，这一设计旨在建筑公共空间内引进"小规模空间"的理念。

Urban Planning Exhibition Hall at Beichuan Antiseismic Memorial Park

北川羌族自治县抗震纪念园——
城市规划展览馆

Location: Beichuan, Sichuan Province
Area: 2,221.7m²
Architect: Zhuang Weimin, Ren Fei, Cai Jun, Wang Xiaoxia / Architectural Design and Research Institute of Tsinghua University Co., Ltd
Photographer: Zhang Guangyuan
Completion Date: 2010

地点：四川，北川
建筑面积：2,221.7 平方米
建筑师：庄惟敏、任飞、蔡俊、汪晓霞 / 清华大学建筑设计研究院有限公司
摄影：张广源
建成时间：2010 年

Sichuan, China
中国，四川

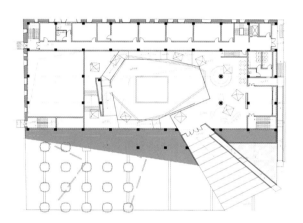

The memorial park consists of Meditation, Heroes and Happiness from east to west. The exhibition hall is a two-storey building with a height of 13.6 metres and a floor area of 2,221.7 m².

The sculptural 'white stone' design responses to the theme and enhance the symbolic meaning the memorial hall. The main structure extends to the square in the Happiness Park and forms a platform of landscape; the main exhibition hall features simplicity and purity just as the 'white stone' implicates, implying traditional Qiang culture in a subtle yet modern way and at the same time sending wishes of 'sacredness', 'protection' and 'good luck' to the new Beichuan; the open landscape platform connects the square of greenery and provides a welcoming public space of harmony, embracing future happiness.

The exhibition area strives to obtain balance in the overall design, therefore an asymmetric layout is adopted and certain receding is made in face of the central area and the waterscape, leaving public space for the visitors on the square. Main entrance for visitors and VIPs to the exhibition hall is accessed at the southeast side. But people can also follow the gentle slope and get to the roof platform. The exhibition is arranged around the sand table along in the main exhibition hall, where people can look around and study the city sand table. There is also a variety of exhibits including multi-media, pictures, models, and objects to provide the diversified and modern viewing experience.

纪念园自东向西分为三部分，依次为静思园、英雄园和幸福园。展览馆为两层，高度为13.6米，建筑面积为2221.7平方米。

建筑以富有雕塑感的"白石"造型设计呼应并共同形成抗震纪念园的主题。建筑主体以幸福园广场的延伸，形成地景平台；主展厅体量简洁纯净，抽象为"白石"的意象，以含蓄现代的手法表现传统羌族文化，暗喻"神圣"、"庇护"、"吉祥"，为新北川祈福；开敞的景观平台与广场绿地相结合，提供亲切、和谐的城市公共生活空间，表达对未来幸福生活的希望。

展览馆建筑单体的布局力求在纪念园整体设计中求得均衡，因此在地段内采用非对称布局，面向园区中心及水面方向进行适当退让，留出广场人群活动的余地。展览馆的主入口设在建筑东南侧，参观人流和VIP流线由此进入。参观人群和市民也可直接沿缓坡行至屋顶平台之上。展陈流线围绕主展厅城市沙盘延缓坡道排列，人流可环绕并俯瞰城市沙盘全景。同时结合多媒体、图片、模型、实物展等类型，创造出丰富、现代化的展览体验。

Nanning International Convention and Exhibition Centre

南宁国际会展中心

Location: Nanning, Guangxi Zhuang Autonomous Region
Area: 130000m²
Architect: Meinhard von Gerkan and Nikolaus Goetze
Photographer: Heiner Leiska, Jan Siefke
Completion Date: 2005

地点：广西壮族自治区，南宁
建筑面积：130,000 平方米
建筑师：迈因哈德·冯·格康、尼古拉斯·格茨
摄影：海纳·雷斯卡、赛风
建成时间：2005 年

Guangxi, China
中国，广西

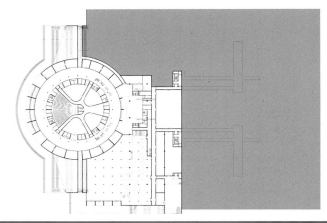

458 / SOUTHWEST CHINA

A multi-functional hall with a folded domical roof, 70-metre-high and 48-metre in diameter, forms the head of the exhibition complex and rises above the city silhouette as a landmark. The roof is conceived as a filigree load-bearing steel structure, which is on both sides covered with a translucent membrane. Due to its central location the circular hall can be used separately from the exhibition and conference operations. As a rotunda it provides an ideal space for exhibitions, large conventions, theatre plays, concerts, and opening ceremonies.

The convention centre comprising of five halls and an adjoining restaurant is located underneath the hall and can be used separately. The exhibition halls adjoining the rotunda are accessible from both sides of a two-storeyed foyer. A stone plinth forms the visual foundation; during events it can be used as an open-air terrace and exhibition area. Reinforced concrete columns, positioned in front of the ascending stone socle, stretch across the complete height of both hall levels and support the distinct roof, which is rhythmically structured by the recesses of the cores accommodating the technical plant and services.

Natural materials and surfaces with their inherent haptic and aesthetic qualities characterise the appearance of the exhibition and convention centre. Externally and internally stone, glass and concrete are the predominately used materials; colours were scarcely used and only applied for the accentuation of special functions.

多功能厅带有一个高70米、直径达48米的折叠状圆屋顶，构成了会展中心的"首部"，使其成为城市的地标建筑。屋顶为承重钢结构，两侧采用半透明薄膜板铺设。由于其独特的地理位置，这一展厅可脱离其他会议及展示空间，单独使用，举办展览、大型会议、戏剧表演、音乐会及开幕式等活动。

会议中心共由五个大厅构成，与其相连的餐厅设置在地下，可独立使用。靠近圆顶结构的展示空间可从两层高的大厅的两侧进入。石头结构在视觉上构成整个建筑的地基，在活动期间可用作露天展示空间。钢筋混凝土梁柱"竖立"在逐级上升的石头台阶前，在视觉上与圆顶屋顶相呼应。

天然材质打造的外观以其特有的触觉及视觉美感构成了设计的特色。石材、玻璃以及水泥等材质被大量运用，色彩的选用旨在强调空间的功能。

Planning and Architectural Design of the Saloon Street, Yangshuo, Guilin, Guangxi

广西桂林阳朔酒吧街规划与建筑设计

Location: Guilin, Guangxi Zhuang Autonomous Region
Area: 17,000m²
Architect: Guangzhou Tianzuo
Completion Date: 2006

地点：广西壮族自治区，桂林
建筑面积：17,000 平方米
建筑师：广州天作
建成时间：2006 年

Guangxi, China
中国，广西

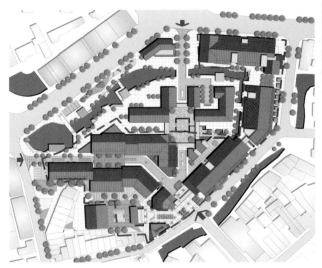

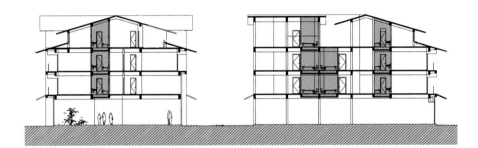

460 / SOUTHWEST CHINA

The bamboo house, as the main dwelling and commercial structure of Yangshuo County, features narrow façade and large depth. The lower part functions as shops while the upper was used as residence. The project inherits the traditional building style and combines modern design method of open space to create distinctive architectural space and to promote the unique and ethnic residential culture of Yangshuo County.

Pile house in Guilin boasts characteristics of lightness and transparency to accommodate the local climate and culture. The design abandons the simple imitation of "form" with the appliance of seemingly local ingredients; it manages to construct real pile houses.

The façades highlight white walls; glass window and black steel railing reflect modern flavour. In addition, part of the surface is furnished with local stone and wood to reduce cost as well as to show the features and spirit of Yangshuo County.

小面宽，大进深的"竹筒屋"的平面形式，以及"下店上住"的竖向划分是阳朔传统的城市居住与商业空间的单元模式。本方案继承和发展了这种传统空间单元，结合现代城市开敞空间的设计手法，重新组合，生成独具特色的空间形态，发扬整体阳朔居住文化，使建筑具有民族特色、阳朔风格。

"轻巧、通透"是桂林杆阑式民居建筑的主要特征，这是一种适应当地气候和文化特征的地方建筑形式。本设计抛弃了各种装饰符号拼贴的"形似"，而是力图追求非简单复古的"神似"。

建筑立面以白粉墙和体现时代气息的清玻和黑钢为主基调，局部以取自当地的阳朔石和木材做饰面，一方面节省运费降低造价，一方面以建筑作为当地文化精华和人文精神的载体，充分体现当地的风情。

Riverside Restaurant
天门山"山之港"临江餐厅

Location: Guilin, Guangxi Zhuang Autonomous Region
Architect: Liu Chongxiao
Photographer: Deng Xixun, Liu Chongxiao, He Rong, Song Yan
Completion Date: 2011

地点：广西壮族自治区，桂林
建筑师：刘崇霄
摄影：邓熙勋、刘崇霄、何蓉、宋彦
建成时间：2011 年

Guangxi, China
中国，广西

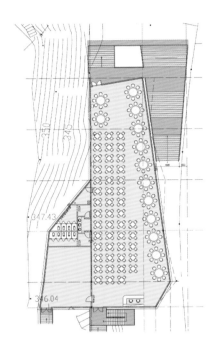

Riverside Restaurant is fantastic for its small scale and pure function which make it fused into the surroundings in a semi-transparent way as well. The solidness of structure and material and the illusory feeling from visual and tactual aspect are combined together in a surprising way. Sightseeing is not to be observed nor the architecture a shelter; they permeate into each other subtlety and the boundary between them becomes obscure.

Rather than submerged into the surroundings completely, the design aims to redefine the features of the surroundings by bringing architecture in and at the same time to dug out the potential of the surrounding background. What is more important, there is no clear difference between exterior and interior design. The starting point is the interior.

From the outside, one can feel the fusion of the architectural exterior and the surrounding bamboo forest, mountains and interior materials; once inside, one can enjoy the feeling of being enveloped by architectural materials and sightseeing under certain light. Such effect is reproduced by the reinterpretation of exterior scenic views through architectural design.

周边环境提供了条件。设计出乎意料地把结构与材料的真实性和视觉与触觉的虚幻感结合在一起。景色，已经不再是被观察的对象。建筑，已经不再是遮蔽所。它们以某种微妙的方式互相渗透着，建筑和环境之间的边界变得模糊起来。

与完全消隐不同，餐厅的设计期待通过建筑的嵌入对周遭环境的特征加以重新定义，揭示出原有环境中潜藏的某些活力。此外，非常重要的一点是，整栋建筑没有建筑设计与室内设计的分别，它是从内部开始并由内而外开始思考的。

人们在外部的某些角度观察建筑，通过建筑不同界面虚实关系的组合，会感受到某种建筑外表面与竹林、山体、室内材料质感的混合，而进入室内则将会体会到某种光线下被建筑材质和风景组合包围的感受，那是借助建筑界面对外部景色的重新编辑产生的。

Niyang River Visitor Centre
尼洋河游客中心

Tibet, China
中国，西藏

Location: Linchi, Tibet
Area: 430m²
Architect: Zhang Ke, Zhang Hong, Hou Zhenghua, Zhao Yang, Chen Ling / standardarchitecture
Photographer: Chen Su
Completion Date: 2009

地点：西藏，林芝
建筑面积：430 平方米
建筑师：张轲、张弘、侯正华、赵扬、陈玲 / 标准营造
摄影：陈苏
建成时间：2009 年

The construction of this building adopted and developed the techniques of the Tibetan vernacular. On top of the concrete foundation a 600mm thick load-bearing wall is erected. Most openings have deep recessions. The 400mm thick walls at both sides of the openings work as buttresses, increasing the overall structural stability and reducing the interior span as well. Beams for bigger spans are made from several small logs bonded together. A 150mm thick layer of Aga clay covers the waterproof membrane. Aga clay is a vernacular waterproofing material. It stiffens when tampered with water and works as another layer of waterproofing and heat insulation. Its plasticity allows gutters to be shaped. Roof drainage is well organised with these gutters and channel steel scuppers.

Colour is a crucial element of Tibetan visual culture. Architects introduce a colour installation into the building's inner public space. The local mineral pigments are directly painted on the stone surfaces. The transitions of colours highlight the geometric transitions of space. From morning to dusk, the sunshine changes its direction and altitude angle, penetrating through the different openings. When passing through the building, people perceive ever-changing colour combination from different perspective and at different time. There is no cultural symbolism in this colour concept. These colours are abstract. They multiply the spatial experience and also work as a performance of colours independent from the concept of architecture.

整个工程采用并发展了西藏民居的传统建造技术。混凝土基础以上便是 600 毫米厚的毛石承重墙体。大部分门窗洞口都深深地凹入墙面，洞口两侧的墙体作为扶壁墙在结构上增加了建筑的整体刚度，同时也减小了室内空间跨度。屋面采用简支梁和檩条体系的木结构，局部跨度较大的木梁用 200 毫米 X 300 毫米的木材拼合而成。卷材防水以上覆盖了 150 毫米厚的阿嘎土。阿嘎土是西藏建筑中常见的防水材料，疏松的粘土在加水反复拍打后板结，形成可靠的屋面防水层和保温层。建筑师还利用阿嘎土的塑性在檐口内侧拍打出檐沟，并用槽钢加工的雨水口形成有组织排水。

颜色是西藏视觉文化传统的重要元素。设计师把一个颜色的"装置"引入这个建筑内部的公共空间。西藏的矿物颜料被直接涂刷在毛石墙上，颜色的转换强化了空间几何的转变。从日出到日落，不同方向和高度角的阳光射入各个洞口。从建筑中穿过时，人们可以在不同的角度和时刻体验到不断变化的色彩效果。这些颜色没有西藏传统文化中的象征性，它们是单纯而抽象的，强化了建筑的空间体验，同时也作为独立于建筑的"装置"演绎颜色本身的魅力。

Namchabawa Visitor Centre
西藏林芝南迦巴瓦接待站

Location: Linchi, Tibet
Area: 1,500m²
Architect: Zhang Ke, Zhang Hong, Hou Zhenghua, Claudia Taborda / standardarchitecture
Photographer: Chen Su
Completion Date: 2008

地点：西藏，林芝
建筑面积：1,500 平方米
建筑师：张轲、张弘、侯正华、克劳迪娅·塔博达 / 标准营造
摄影：陈苏
建成时间：2008 年

Tibet, China
中国，西藏

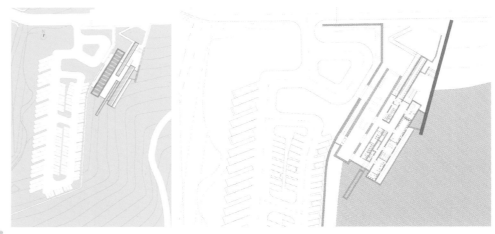

Like a group of rock slices extending out of the mountain the Namchabarwa Visitor Centre consists of a series of stone walls set into the slope. The sharp geometry of the stone volumes is accentuated by an absence of windows in the wall facing the incoming road in the west, giving the building the appearance of a scale-less abstract element in the natural landscape. Looked from afar the building neither hides itself, nor stands out from its background as a piece of Tibetan Architecture. Approached from a distance on the road, people can't be sure if the volumes form a building or a set of retaining walls or even a "Mani" wall at the foot of the mountain.

The building is a mixed structure of traditional masonry stone walls and concrete. The walls are all made of local stone and vary in width of 60cm, 80cm and 1m. The constructional columns, ring beams and lintels inside the self-supporting stone walls reinforce the anti-seismic function of the walls, while supporting the concrete roof above them. The Tibetan craftsmen who built the stone walls are mainly from Shigatse. Their special customs and methods of building stone walls are not what can be designed on paper.

There are no popular Tibetan decorations either on the windows and doors of the building or in the interior space. It is obvious that Namchabarwa Visitor Centre is contemporary architecture. The unique local characters embodied in this building should come from its local building materials and its unornamented construction process rather than disguises.

设计师对本地建筑形式的模仿和拼凑不感兴趣，也显然对移植一个时尚的外来建筑不感兴趣，接待站是一个形式上很不显眼的建筑，远远看去，当地人甚至看不出它有什么新奇，它由几个高低不一、厚薄不同的石头墙体从山坡不同高度随意地生长出来的体量构成，墙体内部是相应的功能空间，这样，多数功能空间实际上是半掩藏在山坡下的。

建筑的结构体系是传统砌筑石墙和混凝土混合结构。墙体从60厘米、80厘米到1米厚不等，全部由当地石材砌筑而成，石头墙是自承重的，墙内附设构造柱、圈梁和门窗过梁，起到增强石墙整体性的抗震作用，同时用来支撑混凝土现浇的屋顶。砌筑石墙的藏族工匠主要是来自日喀则，他们在石墙的砌筑上有很特别的习惯和方法，这一点是在图纸上设计不出来的。

建筑的门窗和室内没有使用任何常见的"西藏形式"的门窗装饰，设计师毫不回避南迦巴瓦接待站是一个当代建筑，如果人们可以感受到特殊的本地气质，设计师希望这种气质是通过本地的材料通过真实而朴素的建造过程自然形成的，而不是"乔装改扮"的结果。

The Apple Elementary School in Ali, Tibet

西藏阿里苹果小学

Location: Ali, Tibet
Area: 1,850m²
Architect: Wang Hui, Dai Changjing
Completion Date: 2005

地点：西藏，阿里
建筑面积：1,850 平方米
建筑师：王晖、戴长靖
建成时间：2005 年

Tibet, China
中国，西藏

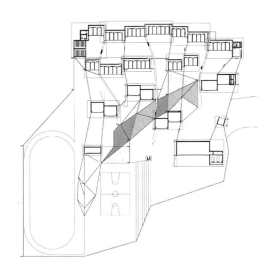

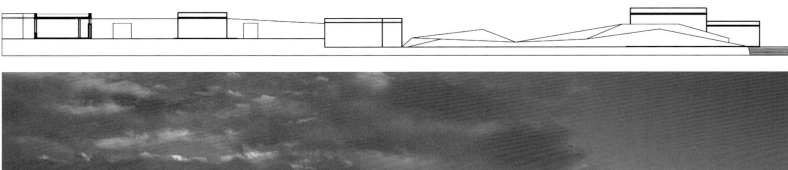

468 / SOUTHWEST CHINA

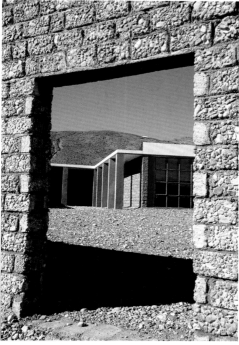

The Ali Apple Elementary School is located at the foot of the 4,800-metre-high Kangrinpoche Mountain, a Buddhism shrine. It is the first completed project funded by Apple Foundation donated by Antaeus Group.

Stones are prevalent in Tibet. All the "buildings" are constructed by piling up many biggish cobbles without mortar. A lot of self-made cobble-concrete blocks are applied to the Apple Elementary School. As the new building volume is homogeneous with the base in respect of material, they are integrated with each other compactly.

The cobble wall along the slope, together with the buildings scattered in groups, partition the campus into many courtyards. The walls, placed in north-south direction, undulate just like the natural shape of the mountain. Due to the average 149 windy days in a year, one of the important functions of the walls is to keep off the wind. With reference to the local buildings in Tibet, the walls are placed in irregular shapes and at irregular intervals.

The buildings are arranged in groups, on the three bases in different heights. The group arrangement is learned from the architecture groups in several villages within 100 kilometres around the school. In addition, by such arrangement, the construction can be carried out in different grounds at the same time, which can shorten the construction period effectively. The buildings and walls in groups form diversified relationships between courtyards. Over 20 courtyards, similar but not identical, enrich children's enjoyments in dwelling and school life.

阿里苹果小学位于佛教圣地——海拔4800米的冈仁波其峰脚下。这是由今典集团捐资的苹果教育基金所启动的第一个新建项目。石头在西藏随处可见，当地所有的"建筑"都是用较大的卵石干垒而成，没有任何砌筑的痕迹。阿里苹果小学大量地采用了自制鹅卵石混凝土砌块的这种材料，建筑的新建体量和原有的基地由于材料相同的关系，使建筑与基地紧密地结合在了一起。鹅卵石的墙体顺着坡地与群落式散布的建筑一起将整个学校划分成一个个院落，纵向布置的墙体起起伏伏，有着山体的自然形态。由于当地年均大风天气为149天，因此，墙体承担重要的作用——挡风。借鉴西藏当地的建筑模式，墙体呈不规则形态和间距自由布置。

建筑的布置随着大致三层高度不同的基地，采取了一种群落的方式。群落的布置方式源自对周边的100千米左右的几个村镇的建筑群体空间研究。此外，群落的布置方式，可以在施工操作阶段采取分片同时施工的方法，能够有效地缩短工期。建筑群落和成组出现的墙体形成更为丰富的院落关系，这些二十多个形态相近、但又各不相同的院子为孩子们的居住和学习生活增添乐趣。由于台地之间的高差形成建筑之间的南北向间距变化，所以建筑也分为三层标高布置。

St. Regis Lhasa Resort
拉萨瑞吉度假酒店

Location: Lhasa, Tibet
Area: 33,333m²
Architect: Jean-Michel Gathy / Denniston International Architects & Planners Ltd.
Photographer: St. Regis Lhasa Resort
Completion Date: 2010

地点：西藏，拉萨
面积：33,333 平方米
建筑师：让－米歇尔·加蒂 / 丹尼斯顿国际建筑规划有限公司
摄影：拉萨瑞吉度假酒店
建成时间：2010 年

Tibet, China
中国，西藏

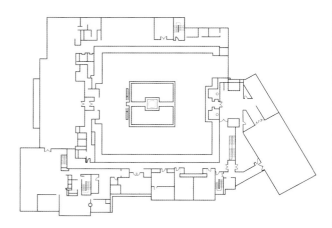

Located on Jiangsu Road and situated in the famous ancient Barkhor area, The St. Regis Lhasa Resort is just minutes away from the holy Potala Palace and 75km away from Gongga Airport.
With the opening of Tibet's first international five-star resort, plus the country's new four-lane expressway, Lhasa moves from a remote backpacker's getaway to an emerging luxury destination. The St. Regis Lhasa (elevation: 12,000 feet) took its style inspiration from a fifteenth-century monastery. The result is a pagoda-roofed, 162-room resort with solar panels and an underground water recycling system at the edge of a serene lake. The lobby's red lacquer pillars and glass walls frame views of the Himalayas and grand Potala Palace. It may be an inside secret, but the cloud-patterned carpet in the lobby sets an unspoken promise – the Chinese name of the motif translates to "all the best".

拉萨瑞吉度假酒店位于西藏著名的八角街内，距布达拉宫仅有几分钟的路程，距贡嘎机场 75 千米。
拉萨不仅拥有西藏第一家国际化的五星级度假村，并新建了一条四车道高速路，不断地发展进步。酒店设计风格灵感源于 15 世纪的寺院，塔状屋顶、太阳能板以及地下水循环系统构成了主要特色。大堂内红漆柱子以及玻璃墙将喜马拉雅山及布达拉宫的景致"定格"在内。内部装饰给人神秘的感觉，云朵图案的地毯让人不禁联想到一句成语：一帆风顺。

Banyan Tree Lijiang
丽江悦榕庄

Yunnan, China
中国，云南

Location: Lijiang, Yunnan Province
Area: 25,000m²
Architect: Architrave Design and Planning Pte Ltd.
Photographer: Banyan Tree
Completion Date: 2006

地点：云南，丽江
建筑面积：25,000 平方米
建筑师：悦榕酒店集团设计部
摄影：悦榕酒店集团
建成时间：2006 年

Location: Lijiang, Yunnan Province
Area: 25,000m²
Architect: Architrave Design and Planning Pte Ltd.

The pulse radiating from Lijiang, Yunnan resonates a symphony full of soul, peace and ethnic harmony. To the north of this beautiful city rises the majestic Jade Dragon Snow Mountain, a sacred sanctuary which has protected the ethnic minorities and their unique cultures since ancient times. Another splendor is the Old Town of Lijiang, regarded as the Venice of the Orient because of the many bridges and canals that lace through the maze of cobbled streets. This historic town was declared a UNESCO World Heritage Site in 1997 for its ethnic charm, historical milieu and architectural landscape.

The luxurious resort villas in Banyan Tree Lijiang, Yunnan reflect the rich fabric of this locale through their design and furnishings. For souvenirs, visit the inimitable Banyan Tree Lijiang Gallery and bring home exclusive handicrafts – each traditional memento has a story to tell about the misty hamlet of Lijiang, Yunnan. Romance beckons in the form of an intimate suite dressed in an oriental palette of red, gold and black. The suites follow traditional sweeping roof designs, adapted from the local Naxi culture.

丽江散发着一种和谐的美，独特的民族风情与平和的精神风貌完美融合。位于北部的玉龙雪山堪称圣地，从远古时期便担负着保护民族风情与文化的重任。而丽江古城被称作"东方"威尼斯，幽静的小桥与河道遍布在街头小巷，民族特色、悠久的历史以及独特的建筑使其于1997年被联合国教科文组织列入"世界文化遗产"。

悦榕度假酒店坐落于这里，通过设计以及装饰充分展示着浓郁的地域特色。酒店内部设有纪念馆，供游客选购特有的纪念品，其中每一件都讲述着传奇的故事。套房采用红色、金色及黑色装饰，让人不禁联想到浪漫。此外，套房采用传统的弯曲屋顶结构，灵感源于当地的"纳西文化"。

Water House
淼庐

Location: Lijiang, Yunnan Province
Area: 1,200m²
Architect: Li Xiaodong Atelier
Photographer: Li Xiaodong Atelier
Completion Date: 2009

地点：云南，丽江
建筑面积：1,200 平方米
建筑师：李晓东工作室
摄影：李晓东工作室
建成时间：2009 年

Yunnan, China
中国，云南

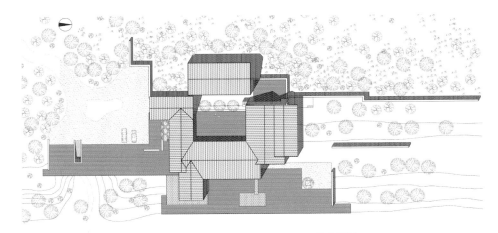

The Water House is a private house, located at the foot of snow mountain Yulong in Lijiang. Sitting on a wide open sloping site, which has a panoramic view towards Lijing ancient town and the surrounding landscape, the house has an open yet closed courtyard space. Closed in the sense, it is "secured" and "separated" physically from the "outside world" by design elements such as stone wall, reflective pool and levelling; yet it is open visually towards the outside environment.

The low profile architectural language by using local material and simple "non-experimental" tectonic skill is meant to blend the "artificial" into nature and local culture. Here, focus is given to space and atmosphere. Emphasis is on local-texture and non-decorative detailing, and the idea is to make the house a background, from which, nature could be appreciated to its extreme.

Besides the steel structure which was contracted to a professional team, the rest of the construction was done by local villagers, not simply to reduce the cost, but as a clear message of regional engagement with architecture practice in the local context.

淼庐是一个私人会所，坐落在云南丽江郊外，雪山脚下，房子在山坡之上，坐山拥水而建。建筑的坡屋顶与山势相和，中有水院，外有水池环绕，三块水面像盘子一样将建筑托起，和四面的山景一齐，从内到外为建筑创造了全角度的视野。

玉龙雪山脚下是一片开阔的平野，四周有山环绕，前有玉湖水库。斑驳的火成岩和绿树清水相映，为这一地区提供了丰富的颜色和质感。附近的村落建筑受传统影响，大多数是传统院落式布局，一般是封闭内向的。回顾中国的传统院落，无论是北方的四合院还是南方的天井式住宅，建筑边界几乎皆是墙壁，景色藏于院中。这种内敛、含蓄的建筑法则专注于营造建筑内部的空间关系，却与外界的山水天际保持对立。在此地，建筑则必须回答环境提出的特殊挑战：如何在平野之处能够安然独立，更与周围景色相和并纳之以为己用。

Gaoligong Museum of Handcraft Paper
高黎贡山手工造纸博物馆

Location: Tengchong, Yunnan Province
Area: 361m²
Architect: Hua Li / TAO (Trace Architecture Office)
Photographer: Hua Li
Completion Date: 2010

Yunnan, China
中国，云南

地点：云南，腾冲
建筑面积：361 平方米
建筑师：华黎 / TAO 迹·建筑事务所
摄影：华黎
建成时间：2010 年

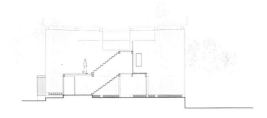
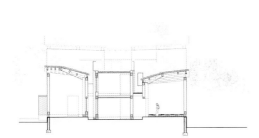

The museum is situated in a beautiful landscape next to Xinzhuang Village under Gaoligong Mountain of Yunnan, a world ecological conservation area in southwest of China. The village has a long tradition on handcraft paper making. To exhibit the history and culture of paper making, this museum will include gallery, bookshop, work space and guestrooms. The museum is conceived as a micro-village, a cluster of several small buildings. The spatial concept is to create a visiting experience alternating between interior of galleries and landscape outside when visitor walks through the museum, so as to provoke an awareness of the inseparable relationship between paper making and environment.

The design is aimed at making a climate-responsive and environmentally friendly building. Local materials such as wood, bamboo, handcraft paper and volcano stone are used for exterior finish, roof, interior finish and floor respectively. With time passed, these materials will worn and fade into a more harmonious colour with the landscape. The construction is to maximise the usage of local materials, technique and craftsmanship. The building combines traditional timber structural system featuring nail-less tenon (SunMao) connection and modern detailing. It was built completely by local builders.

博物馆建造在云南腾冲高黎贡山下新庄村边的田野中，建筑的目的是为了向来访者展示新庄古老的手工造纸工艺，及相关手工纸的文化产品，建筑内部也没有办公空间、茶室和客房等。设计将建筑做成由几个小体量组成的一个建筑聚落，如同一个微缩的村庄。而整个村庄连同博物馆又形成一个更大的博物馆——每一户人家都可以向来访者展示造纸的工艺。访问者对建筑的游览将是在内部展览和外部优美的田园景观之间不断转换的一种体验，以此来提示建筑、造纸和环境的不可分。展览部分由六个形状各异的展厅围绕中心庭院组成一条连续的参观路线，中间则是一个可向庭院完全开敞的茶室。二层是办公空间，通过一个室外楼梯联系到三层客房和一个可以观山的屋顶平台。

设计采用当地的杉木、竹子、手工纸等低能耗、可降解的自然材料来减少对环境的影响。在建构形式上真实反映材料、结构等元素的内在逻辑，以及建造过程的痕迹与特征。建筑适应当地气候，充分利用当地材料、技术和工艺，结合了传统木结构体系和现代构造做法，全部由当地工匠完成建造，使项目建设本身成为地域传统资源保护和发展的一部分。

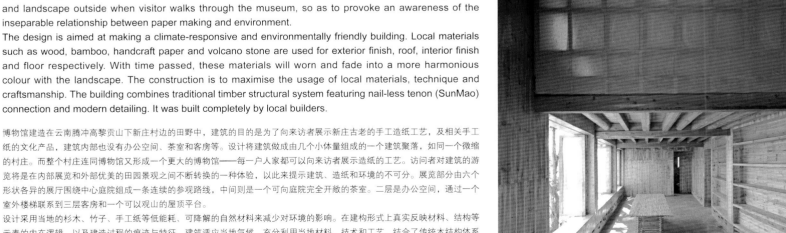

Yun Tianhua Group Headquarters Office Building
云天化集团总部办公楼

Location: Kunming, Yunnan Province
Area: 14,733m²
Architect: Meng Jianmin, Chen Hui / Shenzhen General Institute of Architectural Design and Research Co., Ltd.
Photographer: Fu Xing
Completion Date: 2003

地点：云南，昆明
建筑面积：14,733 平方米
建筑师：孟建民、陈晖 / 深圳市建筑设计研究总院
摄影：傅兴
建成时间：2003 年

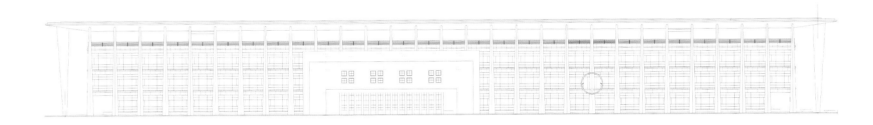

The site is divided into three zones, the front being the office and research area, the central hotel reception area, and the rear living area. The design of the entire office and research area focuses on the geometric composition of points, lines, surfaces, the contrasts between square and rectangular, square and round, straight line, last but not least the simple yet varied spatial levels formed by smooth and clean layout patterns. The sequence of colonnades along the buildings must first meet the requirements for supporting the roof cornice framework. Columns with deformation along the cantilevered direction not only reduce the relative cantilevered distance of architecture but provide shades against the blazing sunshine in Kunming. The irregularly curved cylinders caused by elliptical cross section changing along the vertical direction are in stark contrast to the regular pattern of the main building, highlighting its rich rhythm and strong visual impact.

A progressive sequence of space is formed in this case by the multi-layered transition of the water space, the entrance space, the grey colonnade space, the lobby space, the interior space, achieving the effect of great tension. As an important spatial element, water surface is flexibly applied in a variety of spatial separation, transition and integration. When people enter the building through the water channel, it is not only a space transition completed, but also a psychological transition in which people fully engage with the water surface, and merge into the architectural surroundings.

The change in space brings people and buildings closer to each other. The tranquil waters, stretching lines, simple shapes and the mathematical beauty of divided space, complement one another in the change of light and shades.

总用地分为三个区域，前部为办公科研区域，中部为宾馆接待区域，后部为生活居住区域。整个办公科研区域的设计着眼于点、线、面的几何构成，正方与矩形、方形与圆形、直线与曲线的对比，干净利落的布局形态形成简洁大气且变化丰富的空间层次。建筑周边的柱廊序列首先满足对于屋面构架挑檐的支承要求。截面沿悬挑方向变化的变形柱，既减少了构架的相对悬挑距离，又成为建筑抵挡昆明高强度日照的遮阳体系。沿垂直方向变化的椭圆截面构成不规则的曲面柱形与建筑主体的横平竖直形成鲜明对比，得以凸显其丰富的韵律和强烈的视觉冲击力。

本案利用水面空间、入口空间、柱廊空间、门厅空间、建筑内部空间等若干层次的空间过渡，形成递进的空间序列，实现极具张力的空间效果。水面作为一种空间要素，被灵活运用于各种空间的分隔、过渡与整合。当人们经过水面通道步入建筑时，不但完成空间过渡，更完成一次心理过渡，使人充分亲近水面，融入建筑氛围，变化的空间感受更拉近人与建筑的距离。静谧的水面、舒展的线条、简洁的形体以及空间划分的数学美，在光与影的变化之中相辅相成。

Yinchuan Cultural and Art Centre
银川文化艺术中心

Ningxia, China
中国，宁夏

Location: Yinchuan, Ningxia Hui Autonomous Region
Area: 28,000m²
Architect: Zhou Xuhong, Yuan Xiaoqing, Wang Jibin / Zhonglian Cheng Taining Architecture Design and Research Institute, Ltd.
Photographer: Zhou Xuhong
Completion Date: 2008

地点：宁夏，银川
建筑面积：28,000 平方米
建筑师：周旭宏、袁晓庆、王纪斌 / 中联程泰宁建筑设计研究院有限公司
摄影：周旭宏
建成时间：2008 年

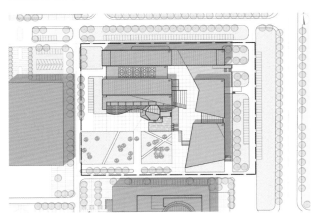
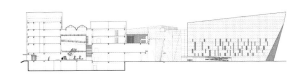

According to the urban planning of Yinchuan Square, the cultural centre is situated at the land use group on the east side of the square, including five planned building monomers of library, museum, art and cultural centre, the Grand Theatre, City Planning Exhibition Hall. Yinchuan Art and Culture Centre is located at the northeast corner of the group. The structure has an 'L' shaped layout, with neat profile at the east and north side, and vibrant curves and slashes of great intensity on the west and south.

The Art and Culture Centre together with the Library and the Grand Theatre form an inner court opening to the south, which has become an important part of the eastern square. In order to integrate the Art and Culture Centre with the organic group and enliven the courtyard space, its main body and the entrance are designed to face the inner court. Dynamic elements as well as the sculptural objects are placed on the one side of the inner court, while the indoor architectural style extends to the outdoor plaza.

As for the architectural details, the vertical separation of glass curtain wall employed on the external walls of regularly striped north side of the building volume is the natural facade manifestation of the small indoor office space, in stark contrast to the mass of huge close stones at the south and east side of the large space; the fine strip mullions at the museum's main entrance in the very south add to the human scale of the entrance, enhancing the image of a welcoming site; windows and holes in the exterior walls are basically consistent with the granite pattern; abstract metal decoration are installed at the building eaves to reflect the Hui influence.

根据银川广场的城市设计，广场东侧组团用地为文化中心，共规划5个建筑单体，包括图书馆、博物馆、文化艺术中心、大剧院、城市规划展览馆。银川文化艺术中心位于组团的东北角，平面呈"L"形布置，建筑东、北侧界面严整，而西、南面则采用了活跃的曲线与富有力度感的斜线，文化艺术中心与图书馆、大剧院3个单体围合成一个向南开口的三合内院，成为广场东区一重要节点。为使文化艺术中心能与整个组群有机融为一体，使院落空间更具活力，其主要部分及入口面向内院布置，设计将活跃元素以及富有雕塑感的形体置于内院一侧，并把室内的空间氛围延伸到室外广场。

建筑细部上，北侧的规整条形建筑体量外墙采用的竖向分隔、间隙玻璃幕墙的手法，是室内办公小空间在室外立面的自然体现，和南侧与东侧的大空间所形成的巨大封闭石材体量产生明确的对比；美术馆最南侧主入口处设计的条状细密竖挺使入口空间更贴近人的尺度，加强了整个建筑的亲切感；建筑外墙的开窗和洞口基本上与花岗岩石材分缝相一致；在建筑檐口处饰以抽象的金属装饰来呼应回民建筑的细部处理。

The Silk Road Site Museum of West Market in Chang'an, Sui and Tang Dynasties

隋唐长安城西市及丝绸之路博物馆

Shaanxi, China
中国，陕西

Location: Xi'an, Shaanxi Province
Area: 32,000m²
Architect: Liu Kecheng, Xiao Li / Shaanxi Monuments Sites Protection Engineering Technology Research Centre, Xi'an University of Architecture & Technology Architectural Design and Research Institute LiuKeChengStudio
Completion Date: 2009
Award: 2010 Chinese Building Media Prize for the Best Building Award, nominee works; 2011 6th China Architecture Academy of Architecture Creation Honourable Mention, etc.

地点：陕西，西安
建筑面积：32,000 平方米
建筑师：刘克成、肖莉 / 陕西省古迹遗址保护工程技术研究中心、西安建筑科技大学建筑设计研究院刘克成工作室
建成时间：2009 年
获奖：2010 年中国建筑传媒奖最佳建筑奖入围作品，2011 年第六届中国建筑学会建筑创作佳作奖等

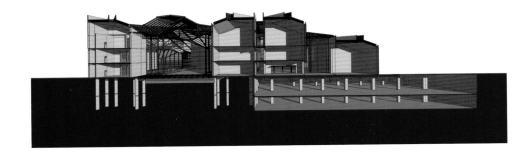

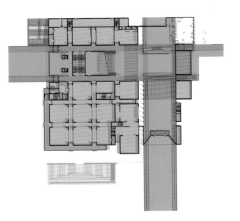
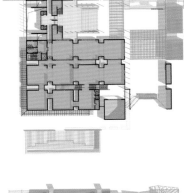

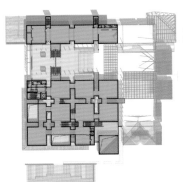
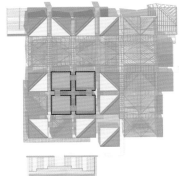

482 / NORTHWEST CHINA

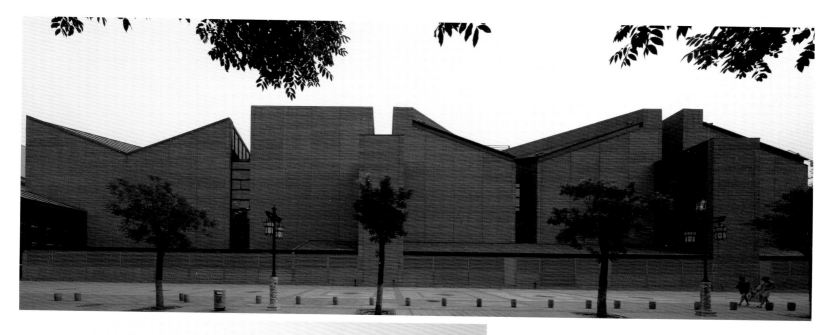

The museum is located over the Chang'an West City site of Sui and Tang Dynasties, as part of the commercial development project of the West City area. The project planning is led by Zhang Jinqiu, academician, covering an area about one ninth the size of the original West City. According to the original design, the West City is divided into nine squares with the museum in the fourth square. It is planning such to conserve the cross road heritage in the West City site, and to display the unearthed objects from the West City site and the Silk Road.

The architectural design abides by the advanced concepts of international cultural heritage conservation. With elaborate design and scientific arrangement, it manages to conserve and display the cross road heritage and the size, scale and scene of the original road system while finding harmony with the surrounding commercial buildings of renovated Tang style. Features of the residence area and the square road network are employed in the museum design in the form of 12m×12m exhibition units, carrying forward the heritage of Chang'an City at a profound level. At the same time, a series of exploration is conducted in terms of the mass, scale, material and colour of construction, creating a well organised multi-levelled space with a varied and colourful effect, restoring the magnificent historic streets in West City in all dimensions.

Specialised decorative concrete siding is added to enhance the historic value of the site and coordinate with the surrounding buildings. It is a main feature of the museum design as its texture resembles the earth wall of the original West City architectures and implies the expressiveness of modernity as well.

隋唐长安西市及丝绸之路博物馆位于原隋唐长安城西市遗址之上，属于大唐西市商业开发项目中的一项内容。整个西市商业项目由张锦秋院士规划设计，占地规模约为原隋唐西市的九分之一，按隋唐西市格局划分为九宫格，博物馆位于其中的第四格，主要用于保护隋唐西市十字街遗址，以及展示隋唐西市及丝绸之路出土文物。

建筑设计遵循国际文化遗产保护及展示的先进理念，通过精心设计，合理布局，创造性地保护和展示了隋唐西市十字街遗址，以及隋唐西市十字街原有道路格局、尺度、规模及氛围，并以现代方式与周边新唐风商业建筑形成了良好的和谐。建筑师通过采用尺寸为12米×12米的展览单元，将隋唐长安城里坊布局、棋盘路网的特点，纳入到博物馆的空间设计中，从深层结构继承了伟大隋唐长安城的传统。同时，对建筑的体量、尺度、材料和色彩等方面进行了一系列新探索，创造出高低错落、丰富有序的空间层次和效果，立体地还原了唐西市历史街道的恢弘气势与繁华景象。

为了进一步突出遗址的价值，体现隋唐长安城西市的繁荣景象，协调周围建筑，建筑师特别为博物馆设计了专门的装饰混凝土外墙挂板。挂板从肌理、质感等方面既呼应了隋唐西市建筑的夯土墙体，又展现出富于表现力的现代魅力，成为博物馆的一项突出特点。

Fuping Museum of Ceramic Art
富乐国际陶艺博物馆群主馆

Shaanxi, China
中国，陕西

Location: Fuping, Shaanxi Province
Area: 1,600m²
Architect: Liu Kecheng, Fu Qiang / Xi'an University of Architecture and Technology
Completion Date: 2004

地点：陕西，富平
面积：1,600 平方米
建筑师：刘克成、傅强 / 西安建筑科技大学
建成时间：2004 年

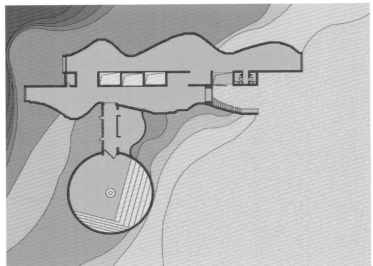

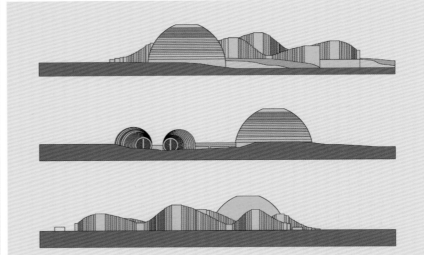

Architectural design of the Fuping Museum of Ceramic Art is based on two points: firstly, the architecture must be integrated into the site, and secondly, the architecture itself should be a piece of contemporary ceramic art.

Brick dome structure is adopted for the main building, accidentally yet predestined. Brick comes naturally as the main material because ceramic factories produce bricks and they are really cheap. The brick domes are constructed with 49cm thick bricks, with a span ranging from 10m to 3.6m. Finally we see the two 72m long giant domes. What's interesting is the construction process, in which mistakes happened frequently due to the constructors' inadequacy of experience. Sometimes the baselines were set inaccurately, or the brickwork was not trimmed well. Fortunately, the project didn't require much accuracy. The varied diameters of the long dome produce a dynamic rhythm, which happened to offset the disadvantages of the raw material, the rough brickwork and the mistakes in construction. The rhythm endows the building with a strong artistic sense. Fuping County has hot summers and cold winters. Therefore, the architects designed a building resembling the structure of brickkiln and acting as an air channel. The doors and windows can help adjust the ventilation. With thick cob walls, the architecture is proved to have a good thermal performance and energy efficiency. Knowing that the main building was liable to constructive errors, the architects tried to take advantage from the brickwork aesthetics. As for the French Pavilion, the façade was clad in reversed recycled tiles, creating random patterns, crude yet delicate, just like contemporary ceramic art.

陶艺博物馆设计之初有两点基本设定：一、建筑必须属于这片土地；二、建筑本身就应当是一个现代陶艺作品。

主馆采用砖拱既出自偶然，也有必然。用砖是一件最自然的选择，陶艺厂自己生产砖。砖拱厚49厘米，每49厘米一道，一端对齐，另一端沿长轴变径。砖拱的最大跨度10米，最小3.6米。这样两条长72米的变形"长虫"就出现了。有趣的是施工过程，工人们由于施工经验不足，经常出错，或放线不准，或砌筑有出入，但这个方案均予以包容。变径砖拱所形成的强烈的韵律，使材料的粗糙、工艺的简陋、施工的错误得以合法化，并丰富了建筑的整体艺术感染力。富平地区夏季闷热，冬季寒冷，建筑的整个建筑形体设计的像窑炉一样，成为一个良好的风道，门窗风量可以调节，建筑墙体采用半掩土厚墙。从实际使用情况检测，热工性能良好，节能效果显著。主馆将建筑在施工方面可能出现的错误，有意识纳入建筑的美学表现；法国馆外墙选取窑变面砖废品，正砖反贴，不同组合，形成一种粗旷而又细腻的趣味，与现代陶艺的美学追求同出一辙。

Jia Pingwa Literary Art Museum
贾平凹文学艺术馆

Location: Xi'an, Shaanxi Province
Area: 1,200m²
Architect: Liu Kecheng, Xiao Li / Xi'an University of Architecture and Technology
Photographer: Liu Kecheng
Completion Date: 2007

地点：陕西，西安
建筑面积：1,200 平方米
建筑师：刘克成、肖莉 / 西安建筑科技大学
摄影：刘克成
建成时间：2007 年

Shaanxi, China
中国，陕西

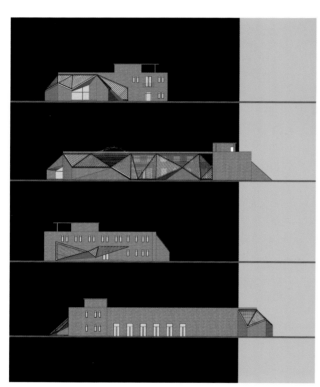

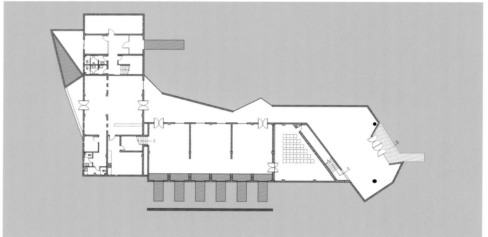

The museum is located at one side of the central axis on the campus of Xi'an University of Architecture and Technology, which belongs to the historical site. This project is rehabilitated from a printing factory built in 1970s. The structure of the printing factory used to be a simple two-storey brick and concrete building. The surface of the building is plain brick wall with grey paint.

Jia Pingwa Literary Art Museum positions itself that the building should retain the look and features of 1970s after renovation. Here, the history and reality should be dialogue relations, but not submissive relations.

The original inspiration of the design comes from the sunlight. Through several investigations on the spot, the designer finds that the lingering look of the shadow is the most appealing feature of the plain and simple building. The designer then takes photos of the building from six o'clock in the morning to seven o'clock in the afternoon. The shadow area changes in a very interesting way if you play the photos continually.

The renovation retains the original plain brick wall and the deep colour painting. Glass, steel frame and concrete are introduced as the new elements of the old building. The structure of the old architecture remains almost as it used to be, and the new elements are positioned together with the old building in a dialogue way. Steel frame, glass and concrete are unified into identical form logic according to the changes of the shadow, which breaks the mediocrity and stiffness of the old building. The steel frame includes three layers, namely, the main frame, the second frame and ornamental frame. Harmonious dialogue between the old and new elements is set up from different angles and densities. The reinforced concrete wall uses the common building waste materials – bamboo splint as the template to form a rough but fine-textured surface, which is in harmony with the plain brick wall in density. At the same time, the design also achieves an effect of "dry brick building" wall commonly used in the rural areas of Shaanxi province from the aspect of culture.

馆址位于西安建筑科技大学校园中轴线的一侧，属于学校的历史地段。选择学校历史地段中建于20世纪70年代的印刷厂进行改造。印刷厂建筑是一栋朴素的二层小楼（局部三层），砖混结构，局部框架。表面清水砖墙，外刷灰色涂料。

贾平凹文学艺术馆基于这样一种定位，在改造完成以后，这栋建筑还应保持20世纪70年代的面貌和特征。在这里，历史和现实应当是一种对话关系，而不是服从关系。

设计的最初启示来源于阳光。经过多次现场体验，设计者发现：对这个平淡、朴实无华的建筑，光影的流盼是其最具魅力的特征。从早晨6:00到下午19:00，每隔一小时拍一张照片，然后进行连续播放，可以看到由阴影构成的一个体量发生着有趣的变化。

建筑保留原印刷厂老建筑清水砖墙、外刷深色涂料的基底，选择玻璃、钢架和混凝土三种原建筑没有的词汇作为新因子介入。老建筑基本维持不变，新构件以对话的方式与老建筑并置。钢架、玻璃与混凝土依据光影变化，统一到同一形式逻辑，打破原建筑的平庸和呆板。钢架分主框架、次框架和装饰性框架三层，以不同角度和密度，形成新老元素的和谐对话。钢筋混凝土墙采用俯首可得的建筑废料——竹条作为模板浇注，形成粗糙而又富于肌理的表面，在密度上与清水砖墙和谐，在文化上造成一种与陕西农村普遍使用的"干打垒"墙体类似的效果。

Xi'an Horticultural Exposition Garden Reception Centre
西安园艺博览会精品酒店

Location: Xi'an, Shaanxi Province
Area: 26,990m²
Architect: MADA s.p.a.m.
Photographer: MADA s.p.a.m.
Completion Date: 2011

地点：陕西，西安
建筑面积：26,990 平方米
建筑师：马达思班建筑设计事务所
摄影：马达思班建筑设计事务所
建成时间：2011 年

Shaanxi, China
中国，陕西

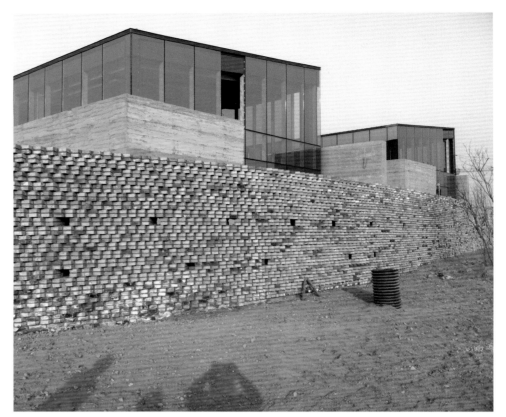

Inspired by the historic walled city of Xi'an, the Xi'an Horticultural Exposition Garden Reception Centre prominently features a stone wall that surrounds the hotel and spa complex that is intended to emulate the form of a floating lily on a pond. The stone wall forms five "petals" or zones for the Centre's hotel and spa facilities and the landscape design expresses the Wu Xing, or the Five Elements – Earth, Metal, Water, Wood and Fire in each of the five petals. Each lobe of the flower petal presents a distinct landscape character, while the central plaza forms a large public gathering space for cultural events and the display of public art.

The Xi'an Horticultural Exposition Garden Reception Centre landscape features spa pools and outdoor therapy rooms, garden meeting areas, plazas, walking paths, and venues for events and public art. The Centre sculptural paths, plazas and walls are inspired by traditional and regional methods of craftsmanship and natural landforms, yet present comptemporary interpretations in the details of stone, wood, and metals. Throughout the Centre, the garden experience includes a broad range of plant species, which are composed to create a variety of artful landscape experiences throughout the year. Each zone of the Centre has a distinct planting character, which reflects colours and textures of the Five Elements in the garden patterns and tree bisques. During the site development, many large specimen trees were preserved and salvaged as part of an initiative to promote the reuse of plant species within the Horticultural Exposition.

During the design process, important aspects of environmental stewardship were considered including an emphasis upon improving water quality of the lake surrounding the Centre through the use of bioswales, reducing water consumption through a high-efficiency irrigation system, and improved energy efficiency through the use of LED lighting fixtures and lighting control systems.

西安园艺博览会精品酒店的设计灵感来自对水流动包容的观察思考。浐灞新区的独特区位和周边河流、岛屿、山脉的自然景观和西北大地独特的气候植在设计过程中，被引发出无限可能。整个设计随小岛地形流动的闭合曲线通过青砖墙勾勒出建筑的边界，内侧则以连续变化的木条边墙进行围护。两道围墙形成了私密又富于趣味变化的廊道空间，墙体不仅仅作围合之用，还是交通游憩的休息之所。

建筑面向水面的方向延伸出呈花瓣状五个区：公共接待区、高端会所区及三个客房区。五个区设计了五种不同的建筑形态，并由此产生了无论是内部使用体验，还是外部视觉观察看来都各有特色的系列空间。酒店被设计成低能耗，被动式节能的，大量采用天然材料。设计结合了地源热泵、智能灯控温控等一系列技术，并在建筑、景观、室内采用 LEED 认证标准控制。

建筑材料选择尽可能考虑当地传统材料，对各种材料精心设计，细致的呈现出它们自身独特的质感，其中包括大面积粗砾黄土堆积而成的夯土墙；精致细长木片织缀而成的木条边墙；以及无数三角形玻璃折面构成的内庭接待长廊。

Shangri-La Hotel, Xi'an
西安香格里拉酒店

Shaanxi, China
中国，陕西

Location: Xi'an, Shaanxi Province
Area: 20,346m²
Architect: Solari Design Limited
Photographer: Shangri-La Hotels and Resorts
Completion Date: 2007

地点：陕西，西安
建筑面积：20,346 平方米
建筑师：昇辉设计有限公司
摄影：香格里拉度假酒店集团
建成时间：2007 年

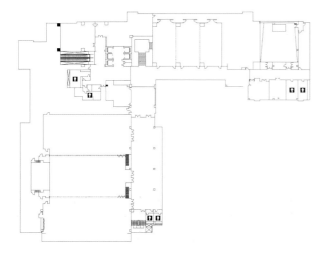

Situated amidst beautifully landscaped gardens, Shangri-La Hotel Xian is located in the Hi-Tech Zone within a city rich in historical and cultural significance. The hotel is convenient to the regions of many historical sites and museums – including the Terracotta Warriors Museum and breathtaking city walls – and is within walking distance of Century Ginwa and Golden Eagle shopping malls. Among the hotel's deluxe facilities are a sparkling 25-metre indoor pool (a luxurious day spa), a fully-equipped health club with Jacuzzi (steam room) and sauna. The hotel's three world-class restaurants offer a delectable selection of exquisite cuisine, while the Lobby Lounge provides an elegant backdrop for afternoon tea or evening cocktails. Shangri-La Xi'an features 386 beautifully-appointed guestrooms and suites – each with generous amenities. Spectacular floor-to-ceiling windows afford picturesque city views.

香格里拉酒店位于历史文化氛围浓郁的西安古城高新技术开发区内，并选址在风景秀丽的景观公园之间，毗邻秦始皇兵马俑博物馆及著名的古城墙。酒店内最为奢华的即为25米的室内泳池、水疗馆以及设备齐全的健身俱乐部。三个世界级别的餐厅提供各式菜肴，大堂休息室更是饮茶及举办酒会的最佳场所。此外，365间客房全部采用落地窗，尽享城市的美丽景致。

Flowing Gardens
流动花园

Shaanxi, China
中国，陕西

Location: Xi'an, Shaanxi Province
Area: 12,000m²
Architect: Plasma Studio / Groundlab Eva Castro, Holger Kehne, Alfredo Ramirez, Eduardo Rico, Liu Dongyun
Photographer: Plasma Studio
Completion Date: 2011
Award: Zhulong "10 Years of Architecture" Experts Attention Award

地点：陕西，西安
建筑面积：12,000 平方米
建筑师：Plasma 事务所 / 格兰德拉伯·伊娃·卡斯特罗、霍尔格·肯尼、阿尔弗雷多·拉米雷斯、爱德华多·里克、刘东运
摄影：Plasma 事务所
建成时间：2011 年
获奖：筑龙网"建筑十年"——专家关注奖

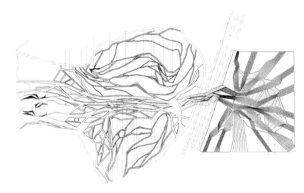

The project proposes a hybrid of both natural and artificial systems, brought together as a synergy of waterscapes. With consideration to the amount of water required for irrigation, the project seeks to introduce various technologies and designs found in nature, but it is enhanced to meet the specific needs of the new population. Rainwater is collected and channelled into the wetland areas, where natural plants and reed beds clean and store the water, which is then dispersed and used for irrigation.

Guangyun Gate operates as infrastructure and fulfils the role of bridging the main road that dissects the site. It channels visitors from the plaza at the entrance where they congregate and orient themselves, plotting their direction. Their path over the bridge rises 7m and offers vantage points to gain an overview of the different zones of the Expo displayed ahead.

The Creativity Pavilion is located on the edge of the lake as the endpoint to the central axis that starts with the Gate Building, and is the starting point for the water crossing by boat. It ties in with a series of piers that follow the landscape jutting out into the water. The built volume is interwoven with the articulating ground, producing continuities on many levels integrating the landscape and building together.

The Greenhouse is formed as a precious crystal, semi-submerged in splendid isolation reached by boat across the lake followed by a short walk from the shore. The greenhouse blends into the hillside even more so than the other two structures.

该项目提出了一种混合自然和人工系统的想法，并引入水景的协同作用。考虑到用于灌溉所需的水量，该项目旨在引入在自然界中发现的各种技术和设计，并加强用于满足新人口的具体需求。雨水被收集导入到湿地区域，在那里，天然植物和芦苇河床会清洁和储存水，之后这些水则被疏散用于灌溉渠道。这些综合性的湿地和池塘也可成为游客个人休憩的安宁绿洲。

广运门扮演着基础设施的角色和衔接这条剖切整个场地主要道路的作用。它把游客从入口广场引导进入聚集并自我寻找和计划行进方向。在桥上，道路被抬升了7米，以获得预览世园会不同区域的有利位置。

创意馆位于湖的边缘，是从门户建筑开始的中轴线端点，又是乘船过湖的起点。它位于一系列从景观里突出入水的码头之上。建筑体量与地面交织在一起，产生一种连续性，在各种层面将景观与建筑结合为整体。

温室的造型就像一颗精致的水晶，半淹没在乘船过河后几步之遥的与世隔绝的"桃花源"。这个温室比起其他两个结构体，更加融合于山坡之上。

Xi'an Television Broadcasting Centre
西安广播电视中心

Location: Xi'an, Shaanxi Province
Architect: MADA s.p.a.m.
Photographer: Jin Zhan, Chen Zhanhui
Completion Date: 2009

地点：陕西，西安
建筑师：马达思班
摄影：金霁、陈展辉
建成时间：2009 年

Shaanxi, China
中国，陕西

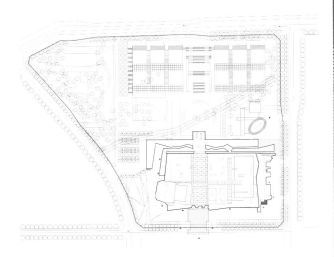

This project inherited the extraordinary and vigorous Han and Tang culture in Xi'an, as the buildings stand out and exaggerate the scale concept within the fixed construction area. Different function areas of the building are separated by symbolic 'walls' with solid boundary. This wall image is connected with the historical Xi'an city wall, and more importantly focuses on the modern approach in materials and construction methods. The architectural facade comprises two layers: the inner layer works as the functioning part of the building while the outer layer integrates the horizontal lines of different density. Clay shutters outside the windows with gaps of different width not only embodies the sense of weight gained through the classical way of masonry, but also expressed the lightness in modern architecture. However, despite unity of the sculptural wall of art and media, some of the functions are designed as outside, such as the media theatre at the west side poking out of the wall as a major landmark of the west public space, echoing to the urban spaces.

The concept of grand scale and the techniques of folding timberwork of classical Chinese halls are used in the lobby roof of the building, in the form of geometric triangle in structure and rationality. These triangular frameworks provide a satisfying support system through certain arrangement. They not only reduce the number of pillars needed, but also maximise the breathtaking effect within the space between the roof and ground. With scientific treatment of light blocking and filtering, construction of the 'hall of sun and media' is completed.

本项目承袭了西安汉唐文化中存有的舒展大度、简洁雄浑的特质，建筑群突出并夸张了给定的建筑面积条件下的尺度概念。整体建筑中诸多功能由一道有象征意义的"墙"围合起来，墙体的边界以实体感为主，这个意念上的城墙虽透着西安城墙的历史气质，但强调的却是用材及构造方式的现代化。建筑立面分内外两层，内皮为建筑物内功能的真实体现，外立面则被统一在不同疏密的横向线条中，开窗处设置不同宽度间隙的陶土百叶窗，既体现了古典砌筑方式的分量感，又表达了现代建筑的轻盈属性。但在这座具有雕塑感艺术媒体墙的统一维护中，也存有一些特殊功能被设计成"出墙"状态，譬如西侧的"媒体剧场"，从墙中跳出来，形成西侧公共空间的核心标志物，并与城市空间遥相呼应。

建筑中厅的大堂屋盖在概念上引用中国古典殿堂的宏大尺度及木架构的折面逻辑，选择了具有结构和理性的几何三角形。这些三角构架面，有上有下、有平有倾，形成良好的空间支撑体系，不但减少了空间中所需要的柱子，又最大化体现了屋顶与地面形成的空间界面的艺术震撼力，这一手法在经过科学地遮光、滤光处理后，便完成了"阳光媒体殿堂"这一创意。

Well Hall, Lantian

蓝田井宇

Location: Lantian, Shaanxi Province
Area: 192.8m²
Architect: Ma Qingyun, Sun Daha, Wang Shan / MADA s.p.a.m.
Completion Date: 2005

地点：陕西，西安
建筑面积：192.8 平方米
建筑师：马清运、孙大海、王山 / 马达思班
建成时间：2005 年

Shaanxi, China

中国，陕西

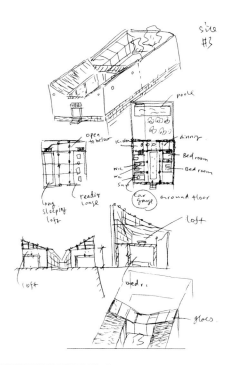

496 / NORTHWEST CHINA

The Jade Valley Winery is located in the county of Lantian, to the southwest of Xi'an City, between the Qingling Mountain Range and the Guanzhong Loess Plateau. As the accessorial guestroom building for Jade Valley Winery, the "Well Hall" presents a time-space splicing between indigenous style and exotic life, which overthrows the inheritance system for history and traditional culture.

Situated on the top of a plateau, the "Well Hall" imitates the civilian house in Guanzhong area, which features with small-bay living rooms, wing-rooms with single slope roof, and few narrow and long courtyards. Based on these features, the architect makes some changes in the "Well Hall" with regard to extension of the scale, layers of the surface and transition of the functions. By these changes, the architecture can reflect the surrounding landscape context with exotic flavour and adjust itself to the exotic way of life. The "Well Hall", with the length extended and the wall much heightened, reflect the landscape of surrounding mountains and rivers in scale. Consequently, its two courtyards look much longer and narrower. The lower half of the north wing-room is hidden in the shadow, which is in sharp contrast with another half that is exposed to the shining sunlight. Under sunlight and shadow, homogeneous materials show different textures and tones.

"Fabrication" is applied to the surface of the "Well Hall". The two skin layers, inner bricks in red and outer bricks in grey, are "weaved" together by interlaced header bricks. The grain of the rough wall surface and the resulting blocky shadow moderate the highness of the wall, which makes the architecture friendlier. The fabrication is applied to the exterior wall only, while the two inner courtyards are partitioned just by a common brick wall. Therefore, it seems that the architecture is "wrapped" by the surface. As result, the inner courtyards become a space of "internalised exteriority".

玉川酒坊是一个以葡萄酒坊为主体的文化艺术栖居之所，位于西安城西南的蓝田县玉川镇，地处秦岭与关中平原的交界地带。"井宇"作为玉川酒坊的附属客房，其建筑呈现出本土风格与异乡生活的时空拼接，颠覆了传统的文化历史传承机制，在设计与手工的结合之中构筑出新生活方式的滥觞。

"井宇"坐落在一座土塬之巅，据高俯远。"井宇"以关中乡间民居为类型蓝本。关中俗语把当地民居的特点概括为"房子半边盖"，厢房一律采用单坡，厅房开间小，庭院窄长，院落数目少。针对这个窄院高墙、青砖灰瓦的传统，"井宇"从尺度的延展、表皮的层次与功能的变通切入，在这几个层面加以侵蚀与扰动，使建筑摆脱单纯的文化标示，呼应周边带有异乡色彩的景观环境，借以容纳外来的生活方式。"井宇"的纵向被拉长，墙体增高了许多，在尺度上与周围映山带河的风景互相呼应。"井宇"内的两进院子因此显得更加狭长，北侧厢房的下半部被掩在阴影中，与烈日照射下的另一半形成强烈的对比。光影之下同质的材料平添了不同的质感与色调，蓝田独特的丽日艳阳随之介入而成为一种建构元素。

"井宇"的表皮作了"织造"处理，形成以红砖为内侧砖，灰砖为外侧砖的两道皮层，再被交错的灰红丁砖缝合在一起，形成一种"编织"的效果，赋予表皮质感与纵深。墙面的凹凸产生细碎的纹理与斑驳的光影，消解了高墙超人的尺度，使得建筑物更有亲和感。"井宇"内院两进院落之间的隔墙则是由普通的灰砖墙隔467，织造手法只应用于外墙处理上，使编织的表皮呈现包裹状态。而"井宇"内部的庭院也顺势成为一种户内的室外空间，由此暗示着马达思班对传统庭院的解读角度。

Mount Hua Forum and Ecological Square
华山论坛及生态广场

Location: Xi'an, Shaanxi Province
Area: 8,667m²
Architect: Zhuang Weimin / Architectural Design and Research Institute of Tsinghua University Co., Ltd.
Photographer: Zhang Guangyuan, Wang Chenggang
Completion Date: 2011

地点：陕西，西安
建筑面积：8,667 平方米
建筑师：庄惟敏 / 清华大学建筑设计研究院有限公司
摄影：张广源、王成刚
建成时间：2011 年

Shaanxi, China
中国，陕西

The site for the Mount Hua Forum and Ecological Square project is located 5km to the south of Huayin City, Shaanxi Province, in the south of State Road 310, and at the north feet of the famous national resort of Mount Hua.

The main structures are set on both sides facing the mountain peak, nestling the exhibition space under the rising natural steps along the terrain. The ticket office and service hall feature deformed slope roof pointing to the ground, lying at the feet of the Mount Hua like greenery coming out of the earth. The building and terraced platform bring the mountain peak into the overall composition, and with landscape design sweeping against the mountain profile, natural beauty and artificial construction merge as a whole. In this design, the environment and the site are fully respected, while the building bridges the natural landscape and local culture, the traditional culture and modern civilisation.

The main building consists of two parts. The west side is smaller and serves as the entrance to the mountain, with a range of facilities including ticket office, information desk, and tour guide services. The east side is used for dining, shopping, management, other supporting facilities, and exiting the mountain. A smooth, gradually rising platform in the middle connects the two buildings, allowing the tourists a panoramic view on the splendid mountain peak scenery from any place on the site.

华山论坛及生态广场项目建设用地位于陕西省华阴市城南5000米处，310国道以南，著名的国家风景名胜区华山的北麓。主体建筑布局面向华山主峰分两边设置，将论坛展示空间隐于依山势地形逐级升高的大台阶之下，两侧的游客票务大厅和游客集散服务大厅造型采用直接坡向地面的异型坡屋顶，建筑匍匐在华山脚下，仿佛由自然地势生长出来一般。建筑和台阶平台将华山主峰纳入整体构图，人工环境设计轻触大地山势，将自然景色与人工建筑融为一个整体。充分体现了对环境和场所的尊重，使得建筑真正成为联系自然景观与人文特色、传统文化与现代文明的一个载体。

主体建筑分为两个部分。其中，西侧部分体量较小，为游客进山通道，包括购票、咨询、导游服务等功能，东侧为餐饮、购物、管理以及其他配套用房，同时兼顾了出山通道。东西侧两个单体建筑的中间则用一个平缓的、逐渐升起的平台作为连接，使得游客无论站在用地的入口或者场地内的任何一个位置，均能够一览无余地看到华山主峰的秀美风光。

Aler Museum
三五九旅屯垦纪念馆

Location: Aler, Xinjiang Uyghur Autonomous Region
Area: 10,000m²
Architect: Pierre Chambron, Zhou Wenyi, Luis Sanchez, Yan Meng / Arte Charpentier (Shanghai)
Completion Date: 2009

地点：新疆维吾尔自治区，阿拉尔市
建筑面积：10,000 平方米
建筑师：向博荣、周雯怡、路易斯·桑切斯、严萌 / 夏邦杰建筑设计咨询（上海）有限公司
建成时间：2009 年

Xinjiang, China
中国，新疆

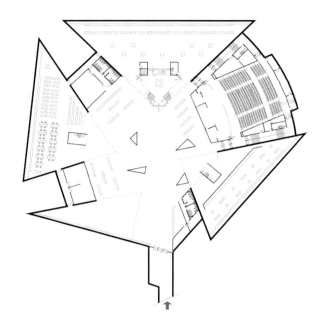

500 / NORTHWEST CHINA

The Aler Museum boasts its "rising slope" and is merged into the surrounding landscape. The main structure resembles pyramid and is scattered and organised in fine order. It is the silhouette of Tianshan Mountain whose water raises all the creatures on the land and at the same time it is built to symbolise the tenacious life. The translucent glass installed on the inner side is intentionally employed to reflect the sparkling and clear features of the icy water while the material used on the outer side is to highlight the solid and tough characteristics of mountainous stone. The façades are tidy and well refined with sides of the different "pyramid" structures reflecting different scenes of the sky, light and shadow in the same day, rendering the whole building into an ever-changing sculpture.
The museum has three floors in total with one underground, compassing eight exhibition halls, a 400-square-metre multi-functional hall and a café.

纪念馆建筑采用了升起斜面的造型手法，使之成为景观的一部分。错落有致的金字塔体量，是对天山山脉的一种抽象，冰山之水孕育着这片土地，它是顽强生命的象征。三角锥形内侧通透的玻璃表现了冰水的晶莹透彻，而外侧实面则体现了岩石的坚硬刚强。立面处理简洁而精密，不同三角锥体的不同斜面在一天中反射着不同的天空和光影，使整组建筑拥有变幻无穷的雕塑效果。
纪念馆共有地下一层和地上二层，设8个大型展厅，一个400平方米多功能厅及一个咖啡厅。

INDEX 索引

010 Herzog & de Meuron, Li Xinggang, China Architeture Design & Reserch Group

012 Ptw Architects + CsCEC+ CCdi design + Arup — John Bilmon, Mark Butler, Chris Bosse, Zhao Xiaojun, Zheng Fang, wang Min, Shang Hong, Tristram Carfrae, Peter Macdonald, Kenneth Ma, Haico Schepers

014 Paul Andreu Architect Paris

016 KsP Jürgen Engel Architekten

018 Zhu Pei, Wu Tong, Wang Hui / Studio Pei-Zhu, Urbanus Architecture & Design

020 RMJM

022 Kang Kai, Jin Dayong, Li Yu

024 Yang Ke / sparch architecture

026 Kohn Pedersen Fox Associates PC (Design Principal: James von Klemperer)

028 Steven Holl Architects

030 Zhu Pei

032 gmp Architekten – von Gerkan, Marg and Partners

034 Goettsch Partners

036 Simone Giostra & Partners

038 Crossboundaries Architects+BiAd international studio

040 Ricardo Bofill Architecture

042 SAKO Architecture

044 Cui Kai, Ye Zheng, Li Xiaomei, Tao Jingyang, Peng Bo

046 Zhu Xiaodi, Mi Ning, Zhong Fei, Yang Bo

048 gmp Architekten-von Gerkan, Marg and Partners

050 Henn Architekten

052 RTKL

054 Institute of Architectural Design, Chinese Academy of Sciences

056 MAD

058 Atelier FCJZ

060 Ma Yansong, Yosuke Hayano

062 Sunlay Architecture Design co., ltd.

064 Liu Yuyang, Zhao Gang / Atelier Liu Yuyang Architects

066 Zhu Xiaodi, Gao Bo

068 Atelier 100s 1

070 Xu Lei, Yu Haiwei, Li Lei, Meng Haigang / Atelier 11

072 Xu Tiantian

074 Xu Tiantian, Chen Yingnan, Zhu Junjie / dnA _design and Architecture

076 Atelier 100s 1 / Peng Lele, Huang Yi, Wan Lu

078 Kang Kai, Wang Jiang, Zhang Liping

080 China Architecture Design & Reserch Group/ Cui Shihong, Yu Guanfei, Deng Ye, Lian Li

082 SAKO Architects

084 Kengo Kuma & Associates

086 SAKO Architects

088 Wang Yun / Atelier Fronti

090 WSP Architects

092 Mario Cucinella Architects

094 Hu Yue

096 TSDI in Collaboration with TFP Farrells Limited

098 Zhang Hua, Fan Li / Sunlay Design Group Co., Ltd.

100 KSP Jürgen Engel Architekten

102 Liu Shunxiao, Zhou Xiangjin / ZPLUS

104 ABSTRAKT Studio Inc.

106 Meinhard von Gerkan and Stephan Schütz with Stephan Rewolle

108 Beijing Mastubara and Architects (BMA) / Hironri Mastubara, Daijiro Nakayama, Zhang Tingting

110 Liu Shunxiao, Zhou Xiangjin / ZPLUS

112 Liu Shunxiao / ZPLUS

114 ZPLUS

116 Liu Shunxiao / ZPLUS

118 HHd_FUN

120 Zhou Kai

122 Dong Gong, Xu Qianhe / Vector Architects + Lü Qiang / CCDI

124 Zhou Kai

126 Urbanus Architecture and design

128 Yu Kongjian, Xiang Jun, Zhangyuan / Turen Landscape Planning Co., Ltd, the Graduate School of Landscape Architecture, Peking University

130 Wang Hui, Cheng Zhi, Du Aihong, Chen Chun, Hao Gang, Wei Yan, Zhang Yongjian, Chen Lan, Zheng Na, Yang Qing, Wu Wenyi, Liu Yinyan, Liu Nini

132 amphibianArc

134 Zhang Yonghe, Xu Yixing, Liu Xianghui, Zhang Lufeng, Wang Hui, Yu Lu, Dai Changjing / Atelier FCJZ

136 Atelier 11, China Architecture Design & Research Group / Cui Kai, Xu Lei, Yang Jinpeng, Wang Yuhang

138 C T design + Associates

140 ARTE-Charpentier / Zhou Wenyi, Pierre Chambron

142 Cheng Dapeng

144 Zhang Yonghe / Atelier FCJZ

146 Li Xinggang, Qiu Jianbing, Sun Peng, Yi Lingjie, Zhang Yuting, Zhao Xiaoyu

148 Xu Tiantian, Guillaume Aubry, Chen Yingnan / DnA _Design and Architecture

150 MAD Architects

152 Beijing New Era Architectural ltd.

154 China Architecture Design Institute

156 Lv Panfeng / New World Architecture Design Co., Ltd., Shenyang

158 Ji Peng / New World Architecture Design Co., Ltd., Shenyang

160 Auer+Weber+Assoziierte, Munich

162 Charles Debbas / Debbas Architecture

164 gmp Architekten – von Gerkan, Marg and Partners

166 URBANUS/ Wang Hui, Tao Lei, Zhao Hongyan, Du Aihong, Hao Gang, Chen Chun, Liu Shuang, Zhang Yongjian, Zhang Yongqing

168 gmp Architekten – von Gerkan, Marg and Partners

170 Kohn Pedersen Fox Associates PC, East China Architectural Design & Research Institute

172 B+H Architects

174 gmp Architekten – von Gerkan, Marg and Partners

176 Arquitectonica

178 TFP Farrells

180 gmp Architekten – von Gerkan, Marg and Partners

182 Arquitectonica

184 As Architecture-studio

186 ABSTRAKT studio inc. / Voytek Gorczynski Architect, OAA / Zhong Ke Design institute

188 HPP international Planungsgesellschaft mbH

190 SPARCH

192 Meinhard von Gerkan / gmp Architekten – von Gerkan, Marg and Partners

194 gmp Architekten – von Gerkan, Marg and Partners

196 logon | urban.architecture.design

198 Zhu Xiaofeng / Scenic Architecture Office

200 Wang Yan / Architects Ring

202 Atelier Liu Yuyang Architects

204 Tong Ming, Huang Xiaoying

206 KH Architect

208 Zhu Xiaofeng, Ding Penghua / Scenic Architecture

210 Zhang Bin, Zhou wei / Atelier Z+

212 Zhang Yonghe / Atelier FCJZ

214 HPP Architects

216 gmp Architekten – von Gerkan, Marg and Partners

218 Hu Yu studio

220 Atelier Deshaus

222 Atelier Deshaus

224 Auer+Weber+Assoziierte

226 Zhou Wei, Zhang Bin / Atelier Z+

228 Kris Yao / Artech Architects & Designers (shanghai) Limited

230 Woods Bagot

232 Perkins+Will inc., Architectural Design & Research Institute of Southeast University

234 RMJM

236 Wang Degang, Bu Yuanyuan / W2 Architects

238 Zhang Lei, Jeffrey Cheng, Wang Wang, Wang Yi

240 Zhang Lei, Meng Fanhao, Cai Menglei, Lu Yuan, Tang Xiaoxin

242 Zhang Lei, Shen Kaikang, Zhang Ang

244 Zhang Lei, Shen Kaikang, Yang Hefeng

246 Zhu Xiaofeng, Cai Jiangsi, Xu Lei, Xu Ye, Ding Xufen/Scenic Architecture Office

248 Miao Pu, Jiang Ninqing / Shanghai Yuangui Structural Design Inc.

250 Miao Pu / Shanghai Landscape Architecture Design Institute

252 Miao Pu / Shanghai Landscape Architecture Design Institute

254 Zhang Lei, Shen Kaikang

256 I. M. Pei Architect with Pei Partnership Architects (New York, NY), Suzhou Institute of Architectural Design Co., Ltd.

258 Cheng Taining, Cheng Yuewen, Wu Nina, Yang Tao, Li shutian, Wu Wenzhu

260 KSP Jürgen Engel Architekten

262 Steven Holl Architects

264 9-town Architects, College of Architecture Design and Urban Planning at Tongji University, Li Li

266 He Jingtang / Institute of Architectural Design at South China University of Technology

268 WSP Architects

270 tvsdesign

272 Tong Ming

274 Preston Scott Cohen, inc. / Zhang Lei

276 Zhang Lei, You Shaoping, Yuan Zhongwei

278 Kokai Studio

280 MADA s.p.a.m

282 Pysall Ruge Architekten

284 Xu Lei / Atelier 11, China Architecture Design & Research Group

286 David Chipperfield Architects

288 He Jingtang / Institute of Architectural Design at South China University of Technology

290 AZL Architects

292 David Chipperfield Architects

294 Atelier 11, China Architecture Design & Research Group

296 Architrave Design and Planning Pte Ltd.

298 Bu Bing / Ningbo civil architectural design institute

300 Zhang Lei, Qi Wei, Zhong Guanqiu, Zhang Guangwei, Guo Donghai

302 Wang Weiren

304 Wang Shu & Lu Wenyu, The Amateur Architecture Studio, Contemporary Architecture Creation Study Centre, China Academy of Art

306 HASSELL

308 Tong Ming

310 Chen Jiajun, Sunqun / Vector Architects

312 Meng Jianmin

314 MengJianmin, Xing Lihua, Li Jinpeng, Yi Yu, Huang Chaojie / Shenzhen Architectural Design Research Institute Ltd.

316 He Jingtang, Liu Yubo, Zhang Zhenhui, Liang Weijian/ Institute of Architectural Design at South China University of Technology

318 CCDI

320 gmp Architekten – von Gerkan, Marg and Partners

322 Hong Zhongxuan

324 Cui Kai, Wu Bin

326 Wei Yeqi

328 MulvannyG2 Architecture

330 Li Xiaodong, Chen Jiansheng, Li Ye, Wang Chuan, Liang Qiong, Liu Mengjia

332 Kris Yao, Artech Architects

334 J.J. Pan & Partners, Architects & Planners

336 P&T Architects

338 Shu-Chang Kung, Josh Wu, Lisa Chen / AURA Architects & Associates

340 UNStudio

342 Kris Yao, Artech Architects

344 TAO ARCHITECTS and PLANNERS

346 J. J. Pan and Partners

348 G.A. Design International Ltd., London

350 Zhang Lei, Zhou Suning, Wang Liang

352 Yu Kongjian

354 In+of Architecture, Studio Wang Lu, Tsinghua University

356 Zhang Yonghe / Atelier FCJZ

358 Zhang Ke, Zhang Hong, Hao Zengrui, Han Xiaowei, Yang Xinrong, Liu Xinjie, Li Linna, Jing Jie, Lin Lei, Han Liping

360 Shan Jun, Lu Xiangdong, Tie Lei, Wang Xin, Sun Xian, Luo Jing, Liu Si

362 J.M. Duthilleul, E. Tricaud / AREP Architects

364 Shan Jun, Lu Xiangdong, Wang Xin, Tie Lei, Sun Penghui

366 Zhang Lei, Qi wei, Guo Donghai, Cai Zhenhua

368 Li Li / Tongji University Architectural Design and Research Institute

370 Wu Gang

372 Nadel Architects

374 gmp Architekten – von Gerkan, Marg and Partners

376 gmp Architekten – von Gerkan, Marg and Partners

378 gmp Architekten – von Gerkan, Marg and Partners

380 America Teamzero Design & Planning Group

382 Wei Yeqi

384 Steven Holl Architects

386 WSP Architects

388 HBA

390 Patel Architecture, inc

392 Terry Farrell and Partners

394 Urbanus Architecture Design

396 Urbanus Architecture Design

398 Rocco Design Architects Ltd

400 Studio Pei-Zhu

402 RMJM

404 Atelier Deshaus

406 Xiao Yiqiang, Liu Huixie, Shi Liang, Qi Baihui

408 Zaha Hadid Architects

410 Mark Hemel, Barbara Kuit / Information Based Architecture

412 Gavin Erasmus, Stefan Krummeck

414 Liu Yuyang, Larry Tsoi, Charles Lam

416 WOHA

418 SDL

420 Ad+RG Architecture Design and Research Group Ltd.

422 The CAAU Studio

424 WMKY

426 Grace Cheng

428 Philip Liao, Partners Limited / WMKY

430 Wong & Ouyang (HK) Limited

432 Arquitectonica

434 Wong Tung & Partners Ltd

436 Arquitectonica

438 Perkins Eastman

440 Tanghua Studio / Tang Hua, etc.

442 Architecture-Studio

444 gmp Architekten – von Gerkan, Marg and Partners

446 Hong Zhongxuan / HHD East Holiday International Design Institute

448 Li Xinggang, Zhang Yinxuan, Fu Bangbao, Liu Aihua

450 Kengo Kuma & Associates

452 Architectural Design and Research Institute of Tsinghua University Co., Ltd / Zhang Weimin, Ge Jiaqi, etc.

454 PATTERNS

456 Zhuang Weimin, Ren Fei, Cai Jun, Wang Xiaoxia / Architectural Design and Research Institute of Tsinghua University Co., Ltd

458 Meinhard von Gerkan and Nikolaus Goetze

460 Guangzhou Tianzuo

462 Liu Chongxiao

464 Zhang Ke, Zhang Hong, Hou Zhenghua, Zhao Yang, Chen Ling / standardarchitecture

466 Zhang Ke, Zhang Hong, Hou Zhenghua, Claudia taborda / standardarchitecture

468 Wang Hui, Dai Changjing

470 Jean-Michel Gathy / Denniston International Architects & Planners Ltd.

472 Architrave Design and Planning Pte Ltd.

474 Li Xiaodong Atelier

476 Hua Li / TAO (Trace Architecture Office)

478 Meng Jianmin, Chen Hui / Shenzhen General Institute of Design and Research Co., Ltd.

480 Zhou Xuhong, Yuan Xiaoqing, Wang Jibin / Zhonglian Cheng Taining Architecture Design and Research Institute, Ltd.

482 Liu Kecheng, Xiao Li / Shaanxi Monuments Sites Protection Engineering Technology Research Center, Xi'an University of Architecture & Technology Architectural Design and Research Institute LiuKeChengStudio

484 Liu Kecheng, Fu Qiang / Xi'an University of Architecture and Technology

486 Liu Kecheng, Xiaoli / Xi'an University of Architecture and Technology

488 MADA s.p.a.m.

490 Solari Design Limited

492 Plasma Studio / Groundlab Eva Castro, Holger Kehne, Alfredo Ramirez, Eduardo Rico, Liu Dongyun

494 MADA s.p.a.m.

496 Ma Qingyun, Sun daha, Wang shan / MADA s.p.a.m.

498 Zhuang Weimin / Architectural Design and Research Institute of Tsinghua University Co., Ltd.

500 Pierre Chambron, Zhou Wenyi, Luis Sanchez, Yan Meng / Arte Charpentier (Shanghai)

图书在版编目（CIP）数据

中国当代建筑地图 / 支文军，戴春编；常文心，贺丽，张晨译. -- 沈阳：辽宁科学技术出版社，2014.8
ISBN 978-7-5381-6587-6

Ⅰ. ①中… Ⅱ. ①支… ②戴… ③常… ④贺… ⑤张… Ⅲ. ①建筑设计－中国－现代－图集 Ⅳ. ①TU206

中国版本图书馆CIP数据核字(2013)第184593号

出版发行：辽宁科学技术出版社
　　　　（地址：沈阳市和平区十一纬路29号　邮编：110003）
印　刷　者：利丰雅高印刷（深圳）有限公司
经　销　者：各地新华书店
幅面尺寸：240mm×330mm
印　　张：63
插　　页：4
字　　数：200千字
印　　数：1～1200
出版时间：2014年 8 月第 1 版
印刷时间：2014年 8 月第 1 次印刷
责任编辑：陈慈良　高　巍　杜丙旭　殷　倩
封面设计：周　洁
版式设计：周　洁
责任校对：周　文
书　　号：ISBN 978-7-5381-6587-6
定　　价：528.00元

联系电话：024-23284360
邮购热线：024-23284502
E-mail: lnkjc@126.com
http://www.lnkj.com.cn
本书网址：www.lnkj.cn/uri.sh/6587